Lecture Notes in Mechanical Engineering

Lecture Notes in Mechanical Engineering (LNME) publishes the latest developments in Mechanical Engineering—quickly, informally and with high quality. Original research reported in proceedings and post-proceedings represents the core of LNME. Volumes published in LNME embrace all aspects, subfields and new challenges of mechanical engineering. Topics in the series include:

- Engineering Design
- Machinery and Machine Elements
- Mechanical Structures and Stress Analysis
- Automotive Engineering
- Engine Technology
- Aerospace Technology and Astronautics
- Nanotechnology and Microengineering
- Control, Robotics, Mechatronics
- MEMS
- Theoretical and Applied Mechanics
- Dynamical Systems, Control
- Fluid Mechanics
- Engineering Thermodynamics, Heat and Mass Transfer
- Manufacturing
- Precision Engineering, Instrumentation, Measurement
- Materials Engineering
- Tribology and Surface Technology

To submit a proposal or request further information, please contact the Springer Editor of your location:

China: Ms. Ella Zhang at ella.zhang@springer.com
India: Priya Vyas at priya.vyas@springer.com
Rest of Asia, Australia, New Zealand: Swati Meherishi at swati.meherishi@springer.com
All other countries: Dr. Leontina Di Cecco at Leontina.dicecco@springer.com

To submit a proposal for a monograph, please check our Springer Tracts in Mechanical Engineering at http://www.springer.com/series/11693 or contact Leontina.dicecco@springer.com

Indexed by SCOPUS. All books published in the series are submitted for consideration in Web of Science.

More information about this series at http://www.springer.com/series/11236

C. K. Mukhopadhyay · Ravibabu Mulaveesala
Editors

Advances in Non-destructive Evaluation

Proceedings of NDE 2019

Indian Society for Non-Destructive Testing (ISNT)

Springer

Editors
C. K. Mukhopadhyay
Metallurgy and Materials Group (MMG)
Indira Gandhi Centre for Atomic Research
Kalpakkam, Tamil Nadu, India

Ravibabu Mulaveesala
Department of Electrical Engineering
Indian Institute of Technology Ropar
Rupnagar, Punjab, India

ISSN 2195-4356 ISSN 2195-4364 (electronic)
Lecture Notes in Mechanical Engineering
ISBN 978-981-16-0185-9 ISBN 978-981-16-0186-6 (eBook)
https://doi.org/10.1007/978-981-16-0186-6

This Springer imprint is published by the registered company Springer Nature Singapore Pte Ltd.
The registered company address is: 152 Beach Road, #21-01/04 Gateway East, Singapore 189721, Singapore

Preface

We take this opportunity to welcome you all to the Proceedings of NDE 2019; Conference & Exhibition on "Non-Destructive Evaluation"; NDE 2019 organized by Indian Society for Non-destructive Testing (ISNT) during December 5–7, 2019, at Bengaluru, India. The theme of the conference was "eNDEavours from Detection to Prediction".

Throughout the world, nations have started recognizing that NDE is now acting as a catalyst in speeding up the testing and evaluation process of various industrial, defense, and nuclear products/materials for inspection of products for their functionality and its in-service inspection capabilities for empowerment and in improving the quality. Recent developments in NDE have touched almost every conceivable area of industrial, defense and nuclear sectors. In light of this, the theme, "eNDEavours from Detection to Prediction," is very much relevant. The objective was to bring the eminent academicians, scientists, researchers, industrialists, technocrats, government representatives, and experts from all strata of society, under one roof, to explore the new horizons of innovative technologies to identify opportunities and define the path forward. This new path should eliminate isolation, discourage redundant efforts, and promote scientific progress aimed to accelerate country's overall growth to prominence on the international front and contribute effectively to realize and achieve the goals. The conference featured regular paper presentation sessions, invited talks, keynote addresses, panel discussions, and poster presentations. It attracted researchers and practitioners from academia, industry, and government agencies in order to exchange ideas and share their valuable experiences. Several research, review, and case study papers written by authors, from all over of the world, were received, till the last date, thereby making the job of the technical program committee extremely difficult. After a series of tough review exercises by a team of experts, only high-quality manuscripts from different parts of the country and abroad were recommended for presentation in NDE 2019 during the three days. Out of these papers, a chosen set of papers were further recommended by the editorial team after undertaking further blind review process for publication as this conference proceeding brought out by Springer Nature. These papers represent wide variety of research topics in all the emerging areas besides traditional and core NDE areas. We are sure, these contributions will definitely enrich our knowledge and motivate many us to take up these challenging application areas and contribute effectively for the NDE challenges. At this point of time, on behalf of the organizing committee, we thank to esteemed authors for having shown confidence in us and considered NDE 2019, a platform to showcase and share their original research work. We wish to express our gratitude to the focused and dedicated team of chairs, co-chairs, members of the advisory committee, technical program committee, local organizing committee, web administration and content development committee, members of the executive committee of ISNT Bangalore chapter, members of NGC and ISNT. We express our deep sense of gratitude to ISNT for giving us the opportunity to edit the first ever conference proceeding of ISNT being brought out in association with Springer Nature. We will fail in our duty if we do not thank the session chairs of all the sessions, who undertook the challenge and worked so hard along with their team members. My heartfelt appreciation to each of the participants, industry partners for coming forward so actively, taking ownership and together making this conference a great success. With these great associations and extended ownership, we are confident that in the years to come future editions of NDE will achieve a much greater and unparallel heights. Finally, we are thankful to one and all, who have contributed directly or indirectly in making this conference successful. Tremendous number of efforts put in to compile this huge and voluminous proceeding will be successful only if these papers could motivate some of us in taking up the major NDE challenges in the coming years. Last but not least, we take this opportunity to give the credit of successfully bringing out this proceedings to my team, one and all, and personally own the responsibility of all the errors, deficiencies, and shortcomings.

About NDE 2019

Indian Society for Non-destructive Testing (ISNT), the Official Technical Society for NDT/NDE in India, organized the annual flagship event of ISNT, Conference & Exhibition on NDE (NDE 2019), during December 5–7, 2019, at Bengaluru, India. The theme of NDE conference in 2019 was "eNDEavours from Detection to Prediction," which was carefully chosen to reflect the changing philosophies in the world of industrial inspection.

In NDE 2019 conference, more than 30 invited talks delivered by experts both from India and abroad and about 193 presentations were made by contributing authors. The conference was attended by more than 800 delegates from India and abroad. In the annual conference NDE-2019, the talks covered various topics of conventional, advanced, and emerging NDE technologies. Among the full papers received in NDE 2019, only the selected papers are included in the proceedings "Advances in Non-Destructive Evaluation-Proceedings of NDE 2019" being published by Springer Nature, Singapore. The papers are selected after following the blind review process by two reviewers along with the editorial review process.

More than 35 technical sessions were conducted covering the wide spectrum of topics such as conventional and advanced NDE, imaging in NDE, QA and

QC, modelling and simulation, SHM and condition monitoring, automation and robotics, codes, standards and training, safety, NDE systems and instrumentation, NDE for composites, in-service inspection covering the applicability of different industries including aerospace, oil and gas, nuclear, power, defense, petrochemical, and fabrication industries.

About ISNT

Indian Society for Non-destructive Testing (ISNT) is a non-profit organization and the Official Technical Society for NDT/NDE in India and is registered under the Tamil Nadu Societies, Registration Act, 1975 (Tamil Nadu Act 27 of 1975) Regd. No. 49 of 1981. ISNT is the society for NDT professionals and practitioners which offers invaluable resources, information, and linkages for industrial quality development and professional development to its members. ISNT has 20 chapters spread all over the country with headquarters at Chennai.

The objective of the society is to promote the awareness of NDT Science and Technology through education, research, and exchange of technical information within the country and internationally to its members and other NDT professionals. The family of ISNT has more than 6000 strong members. It is a diverse and dynamic family of professionals representing NDT technicians, scientists, engineers, researchers, manufacturers, and academicians—all dedicated to improve product safety and reliability. These specialists represent virtually every industry and discipline that may benefit from NDT technology.

ISNT holds periodic seminars and workshops on topics relating to NDT methods and applications, as well as exhibitions displaying cutting-edge NDT products and services. National Certification Board of ISNT works for certification of NDT professionals in India and has been periodically conducting Level-I and Level-II courses through ISNT chapters. The Quality through Non-Destructive Evaluation Science and Technology (QUNEST) trust of ISNT identifies NDE issues and thrust areas; fosters NDE science and technology nationally with international inputs; and enhances international standing and works for making ISNT a global player.

Kalpakkam, India C. K. Mukhopadhyay
Rupnagar, India Dr. Ravibabu Mulaveesala

Contents

Study of Effect of Surface Roughness in Single Lap Joint (SLJ) of Carbon-Fiber-Reinforced-Polymer (CFRP) Composites Using Radiography and Thermography 1
Laxmikant Sarjerao Mane and M. R. Bhat

Non-invasive Detection of Extent of Corrosion in Steel Reinforcing Bars by Magnetic Force Measurement 11
Durgesh Tamhane, Sauvik Banerjee, and Siddharth Tallur

Phase-Sensitive Detection of Extent of Corrosion Using Anisotropic Magnetoresistive (AMR) Sensor in Steel Reinforcing Bars (Rebars) 19
Indrani Mukherjee, Jinit Patil, Sauvik Banerjee, and Siddharth Tallur

A Thermographic System for Quantitative and Automated Subsurface Detail Visualization 27
A. Vijaya Lakshmi, V. S. Ghali, B. Suresh, G. T. Vesala, and R. B. Naik

Inspection of Friction Stir Welded Joint Using EMAT Generated Fundamental Shear Horizontal Guided Wave Mode (SH0) 41
Nived Suresh, Sreedhar Puliyakote, and Krishnan Balasubramanian

Proper Orthogonal Decomposition-Based Coating Thickness Estimation in Quadratic Frequency Modulated Thermal Wave Imaging 51
G. T. Vesala, G. V. P. Chandra Sekhar Yadav, V. S. Ghali, B. Suresh, and R. B. Naik

High Temperature Eddy Current Sensor for Real-Time Structural Health Monitoring of Critical Engineering Components 63
Chandan Dutta, Tarun K. Das, Alok Kumar, Jayendra Kumar, and S. Palit Sagar

Damage Assessment for a Sandwich-Like Panel Using Experimental and Numerical Analysis of Guided Waves 73
Kaleeswaran Balasubramaniam, Piotr Fiborek, Shirsendu Sikdar, and Pawel H. Malinowski

Development of Sectored External Pipeline Inspection Gauge for Health Monitoring of Industrial Carbon Steel Pipelines 85
Shilpi Saha, Debmalya Mukherjee, Y. Chandra, S. C. Ramrane, S. K. Lahiri, and P. P. Marathe

Ultrasonic Waveguide Technique for Temperature Measurement Using $T(0,1)$ Wave Mode 97
Suresh Periyannan and Krishnan Balasubramaniam

Rapid Non-destructive Evaluation of Fireproof Coating Degradation Using Terahertz Line Scanner 107
A. Mercy Latha, A. S. Nirmala Devi, Harkrishnan Kaimal, and Bala Pesala

A Non-destructive Approach of Dynamic Characterization of CFRP Composite Subjected to Hygrothermal Ageing 117
Nilesh Bagale, M. R. Bhat, and S. B. Kandagal

Pore Evaluation and Distribution in Cement Mortar Using Digital Image Processing 125
Sreelekha Gargepuram, Smrithy Subash, and Sumedha Moharana

Thickness Estimation of Marine Structures Using an ROV-Based Pulsed Eddy Current Technique 133
Ashish Antony Jacob, Santhosh Ravichandran, Vineet Upadhyay, Prabhu Rajagopal, and Krishnan Balasubramaniam

Prediction of Failure Process in Geo-Material Using Finite Element Based Digital Image Correlation (DIC) Method 145
Chamanth Sai Reddy Vemulapati and Debasis Deb

Detection of Cracks in Stainless Steel Structures Using Transient Eddy Current Oscillations Method 153
Sudhakar J. Ongole, Lokesh Andavarapu, and Chandra S. Angani

Damage Assessment of Concrete Bridge Structures Using Ultrasonic Tomography Technique 163
Thirumalaiselvi A. and Saptarshi Sasmal

An Analytical and Experimental Study of Nonlinear Intermodulation in Delaminated Composite Plate 175
N. S. V. N. Hanuman, Subhankar Roy, and Tanmoy Bose

Ultrasonic Inspection of Weld Defects Using Total Focusing Method 187
Lalith Sai Srinivas Pillarisetti and Subramanian Anandamurugan

Modelling the Propagation of Partial Discharge Signals Inside Gas Insulated Transmission Line Sections ... 203
Yugandhara Rao Yadam, Sarathi Ramanujam, and Kavitha Arunachalam

Assessment of Sensitivity of Radioisotope-Based Radiometry Data During Inspection of Large Sized Solid Rocket Motors ... 211
S. Khan Lubna, Amali Umesh, N. K. Ghosh, S. C. Bhattacharyya, and S. Srinivasan

Planar Printed E-Field Sensor Array for Microwave NDE of Composites ... 219
Jayaram Kizhekke Pakkathillam, Nitheesh M. Nair, Parasuraman Swaminathan, and Kavitha Arunachalam

Studies on Using FBG Sensor as a Tool for Identifying Corrosion in Pipelines ... 229
B. Arun Sundaram, S. Parivallal, and K. Kesavan

Numerical Modelling of Remote Field Eddy Current Testing of Steam Generator Tubes ... 239
Ranjani Jayaraman, J. Selva Solomon, N. Sridhar, Chitti Venkata Krishnamurthy, and Kavitha Arunachalam

Asymmetrical Analysis of Breast Thermal Images for Detection of Breast Cancer ... 249
N. Kavya, N. Sriraam, N. Usha, D. Sharath, Prabha Ravi, Bharathi Hiremath, B. Venkatraman, and M. Menaka

Numerical Analysis of Partial Discharge Source Localization Using Time of Arrival Measurements and Nonlinear Least Squares Search ... 261
Krishna C. Ghanakota, Sarathi Ramanujam, and Kavitha Arunachalam

Characteristics of Acoustic Emissions Generated During Steel Rebar Corrosion in Reinforced Concrete ... 271
P. Anjali and R. Vidya Sagar

Implementation of Acoustic Emission Testing to Study the Type of Cracking in Reinforced Concrete Beams ... 283
Swarnangshu Ghosh, Rithik Agrawal, and R. Vidya Sagar

Vacuum and Helium Leak Testing Techniques Used for Very Large Size Vacuum Chambers ... 297
Venkat N. Ramani, M. Satheesh Kumar, and S. Sachuthan

Coded Excitation for Low-Power Operation in Guided Ultrasonic Wave Non-destructive Evaluation ... 307
Shashvat Jayakrishnan, Dileep Koodalil, Nived Suresh, and Krishnan Balasubramaniam

Mixing of Lamb Waves at Delamination Defect in a Unidirectional GFRP Laminate 317
Yamnesh Agrawal, Akhilendra S. Gangwar, and Dhanashri M. Joglekar

Parametric Study on Resolution Achieved Using FMC-TFM-Based Phased Array Ultrasonic Imaging 329
Sumana and Anish Kumar

EMAT-Phased Array Inspection of Thick Austenitic Stainless Steel and Dissimilar Metal Welds 341
R. Dhayalan, Anish Kumar, and C. K. Mukhopadhyay

Computed Tomography (CT) Is an Asset to Ensure the Quality and Reliability of Parts in Aerospace Applications 353
Manu Joseph, M. Arumugam, Regi Varghese, and G. Narayanan

Improvements in Quality of Neutron Radiography Images of Pyro Components Used in Aerospace Applications Using Image Processing Tools 371
Girish N. Namboodiri, V. Shaheer Ali, M. C. Santhosh Kumar, K. K. Moideenkutty, M. Nallaperumal, S. Umasankar, and G. Levin

The Promise of Metamaterials for Ultrasonic Nondestructive Evaluation 381
Mohamed Subair Syed Akbar Ali and Prabhu Rajagopal

About the Editors

Dr. C. K. Mukhopadhyay is currently Head, Non Destructive Evaluation Division of Metallurgy and Materials Group, Indira Gandhi Centre for Atomic Research (IGCAR). He is also Professor at Homi Bhabha National Institute, India. He pursued his B.E. from the University of Calcutta (1982) and Ph.D. from Indian Institute of Technology Kharagpur (2003) in Metallurgical Engineering. His research interests include non-destructive evaluation techniques, materials and defect characterization, deformation and crack growth, structural integrity monitoring and failure analysis. He has over 150 publications in peer-reviewed journals and conferences. He has guided nearly 20 M.Tech. and B.Tech. theses and is currently guiding 6 Ph.D. scholars.

Dr. Ravibabu Mulaveesala is currently Associate Professor with the Department of Electrical Engineering, Indian Institute of Technology (IIT) Ropar, India. He received the B.Sc. in Electronics from Maha Raja College in 1996, M.Sc. in Electronics from Nagarjuna University in 1998, M.Tech. from National Institute of Technology (NIT), Tiruchirappalli, in 2000, and Ph.D. from Indian Institute of Technology (IIT) Delhi, India, in 2006. His research interests include development of thermal, acoustical and optical non-invasive/non-destructive testing technologies for bio-medical and industrial imaging applications. He serves on the editorial/advisory boards of several refereed journals of national and international repute.

Study of Effect of Surface Roughness in Single Lap Joint (SLJ) of Carbon-Fiber-Reinforced-Polymer (CFRP) Composites Using Radiography and Thermography

Laxmikant Sarjerao Mane and M. R. Bhat

1 Introduction

Composites have high specific strength and high specific modulus. Composite materials find wide applications for the components of aerospace and automobile vehicles. Components are often required to be assembled for the final product. Joining of the composites with rivets and mechanical fasteners requires holes to be drilled. Drilling holes in the composite laminates could lead to delamination, or they lead to stress concentration. Adhesive joint offers advantages like uniform distribution of stress, large stress-bearing area, good strength-to-weight ratio, dampen vibration and absorb shock, sealing properties, fewer components in the assembly, resist fatigue, and cyclic loads [1]. Though adhesive joint presents advantages over other joining techniques, it has the following demerits: visual examinations are required and need defect control at manufacturing stages. Fixture, presses, and autoclaves are not usually required for other techniques, but adhesive joining demands it. The adhesive joint becomes inseparable after curing, artifacts need to be detected before the joint gets into service, and environmental conditions affect the joint during service; and hence, NDE plays an important role in the evaluation of status and the degradation of joint.

Adhesive joint strength is mainly affected by different factors. They are; (1) type of joint: single lap joint, double lap joint, scarf joint, bevel joint, step joint, butt joint, tubular joint, (2) geometric parameters: overlap length, bondline thickness, adherend thickness for single lap joint, (3) environmental conditions: temperature, moisture, pressure, (4) material properties modulus of adherend and adhesive, ductility of adhesive, and (5) surface preparation. In this study, surface preparation is varied, and the other factors are kept the same. The single lap joint for the investigation is selected as other types of joints can be deduced to single lap joint in analyses except

L. S. Mane (✉) · M. R. Bhat
Indian Institute of Science, Bangalore 560012, India
e-mail: manel@iisc.ac.in

C. K. Mukhopadhyay and R. Mulaveesala (eds.), *Advances in Non-destructive Evaluation*, Lecture Notes in Mechanical Engineering,
https://doi.org/10.1007/978-981-16-0186-6_1

for butt joint, and SLJ is the most common joint in assemblies and used for strength determination of adhesive joint [2].

Performance enhancement of the adhesive joint is carried out with pre-treatment of adherend prior to bonding. Surface preparation modifies or removes the contamination, oxidized layers, increases the area of the adherend, and increases the interlocking. Jin Kook Kim et al. studied the sandblasting, sandpaper abrasion, and plasma treatment on the carbon-epoxy composites [3]. Optimal conditions for the sandblasting and sand abrasion were studied by them. Peel ply treatment on carbon-epoxy and glass–epoxy was studied with the help of contact angle measurements, laser profilometry, SEM, and ToF–SIMS analysis by Bernard et al. [4]. Wingfield studied different surface treatments for composites. Treatments used were mechanical, chemical, and energetic methods [5]. This study is carried with sand abrasion with different grades of emery cloth. Roughness provides the site for bonding, and boding happens with adsorption, diffusion, and mechanical interlocking [6].

Defect detection, defect size, type, and defect area in composite structures are studied in literature with NDE tools. The strength of the specimens is frequently related to these parameters of the defect. X-rays are commonly used for defects like debonds, delamination, voids, foreign material, and measurement of thickness [7]. Absorption of X-ray depends on the atomic number, density, and thickness of the material [8]. In this study, images obtained from X-ray are used for obtaining the intensity values. The potential of thermography is explored in literature with monitoring of adhesive joint during manufacturing [9–11], rate of cooling for different volume present [12], and depth of defect [13], and defect types like kissing bond [14], interlaminar dis-bonds [15], and dis-bond growth during repairs [16]. In this work, the thermographic investigation has been carried out for the temporal curve and analyzed for heat interaction with the adhesive bond.

2 Materials and Specimens

2.1 Laminate Fabrication

Hand-layup and vacuum bagging processes were used for manufacturing of ten-layer laminate (see Fig. 1a) of 0–90 ° woven fabric (200GSM and 3 k filament). Huntsman araldite LY5052 and Aradur 5052 were the epoxy-based matrix used. The volume fraction of fiber was 53.16%, and thickness of the cured laminate was 2.5 mm.

2.2 Adherend Preparation

Adherend was cut using water jet machining as per the dimensions given in ASTM D5868. Surface preparation was carried out using guidelines of the ASTM D2093 and

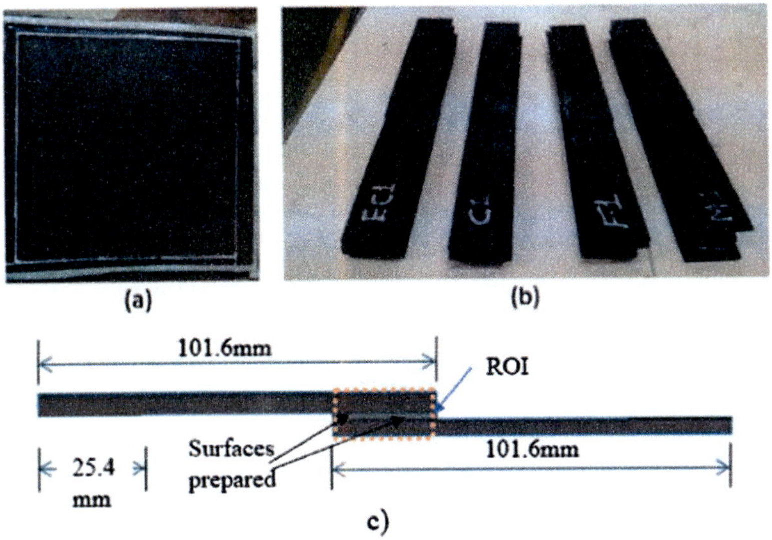

Fig. 1 **a** Fabricated laminate, **b** SLJ samples of different roughness grades, and **c** region of interest ROI (bonded region with hidden roughness)

emery cloth with different roughness grades which were used. Emery cloth of fine, medium, coarse, and extra coarse types were utilized to achieve different grades of the surface roughness. Markings of roughness profile on the surface of the adherend were generated in such a way that it had an angle of 45° to the longitudinal axis. The area which was abraded in this manner was 25.4 mm by 25.4 mm. F1–F10 were set of fine category samples, and similarly, M1–M10, C1–C10, and E1–E10 were medium, coarse samples, and extra-coarse sets of samples, respectively. The region of interest for this study was the bonded region (shown in Fig. 1c).

2.3 *Single Lap Joint Fabrication*

Adhesive AW106-HV953 IN was used for the bonding of adherend with the help of fixture to maintain constant bondline thickness. The bonding between adherends with similar grades of the roughness was carried out, and five samples were obtained for each roughness category. Area of interest for study is interfaces shown in Fig. 1c.

3 Experimental Methods

3.1 Roughness Measurement

Contact-based measurements were performed along lines perpendicular to the lays of roughness. Roughness measurements *Ra* (arithmetic average of ordinates of roughness profile) were carried out according to ISO 4288. According to ISO 4288, cutoff length, evaluation length, and traversing length for aperiodic profiles with $Ra > 2\ \mu m$ and $Rz > 10\ \mu m$ are 2.5 mm, 12.5 mm, and 17.5 mm, respectively. The instrument used was Mahr MarSurf PS1 with a PHT pickup stylus.

3.2 Digital X-ray Radiography

Vidisco foX Rayzor system was utilized to carry out radiography of the adhesively bonded joints. Setup as shown in Fig. 2 has the imager at one meter from the X-ray source. The system used was consisting of CDU (control and display unit), interface control unit (ICU), interface box, AC/DC converter, imager, and source. The source output parameters were 80 kV and 1 mA. Exposure timing of 6000 ms was arrived at after a few trial and error attempts. 14-bit images were obtained by the system with an imager of resolution 3.5 lp/mm. These images were post-processed by image-J software. Intensity values were obtained from the bonded region.

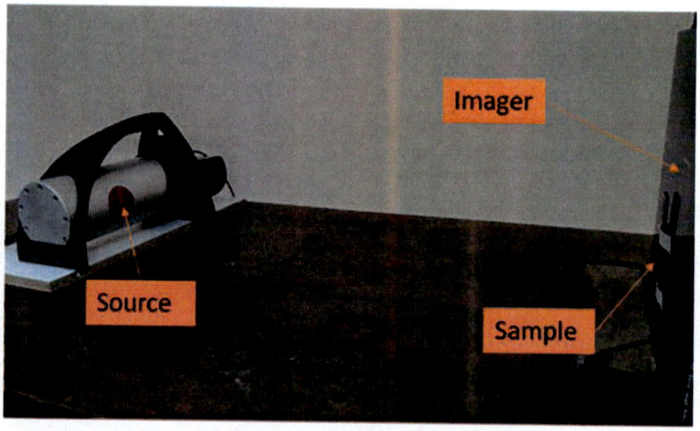

Fig. 2 X-ray setup with source and sample on the imager

3.3 Thermography

Research IR software with FLIR AC655Sc infrared camera was utilized for the thermography. The setup used is shown in Fig. 3. Through technique of heating was used in which blower was used to heat the sample from the back, and live monitoring of the bonded region was carried out from the front side of the sample with an IR camera to ensure uniform heating. Samples heated up to 60 °C then video recorded with a frame rate of 50/s. As Newton's cooling law says that the rate of cooling is directly proportional to the temperature difference of temperature $T(t)$ and ambient temperature T_{ambient} $(T(t) - T_{\text{amb}})$, hence any temperature value $T(t)$ will fall on an exponentially decaying cooling curve with a constant negative exponent of 'e'. So, the cooling curve for the sample is fixed, and starting temperature for the plotting graph is independent of exponents of e. From 60 °C to ambient temperature, the number of temperature values are available to plot the temporal curve. In this study, temporal curve from 46.6 °C was selected as the starting temperature, so that changes with surface roughness can be seen in decaying of the temperature of the specimen.

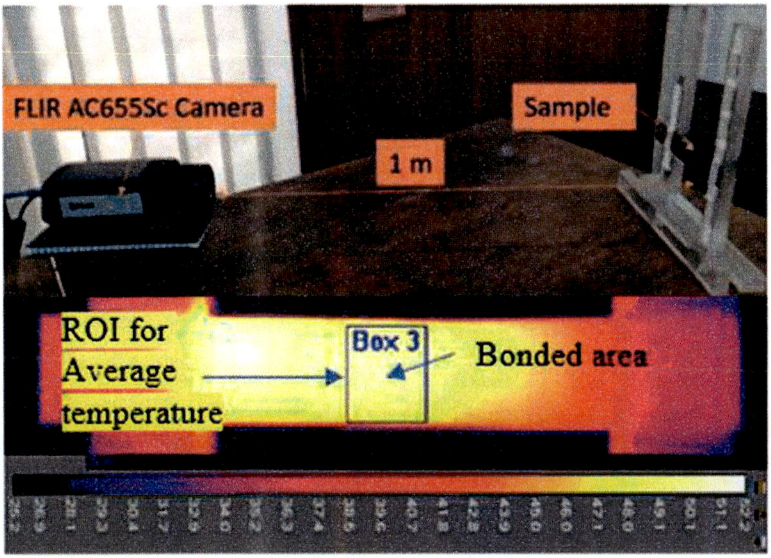

Fig. 3 Temporal curve acquisition with thermography setup

4 Results and Discussion

4.1 Roughness Result

Roughness profile was measured in terms of Ra values. With increasing coarseness of the emery cloth or decreasing size of grit, roughness Ra increases. The values of the roughness grades are given in Fig. 4. The increasing coarseness of grit deepens the valleys. Hence, the arithmetic average of the roughness profile increases. Figure 5

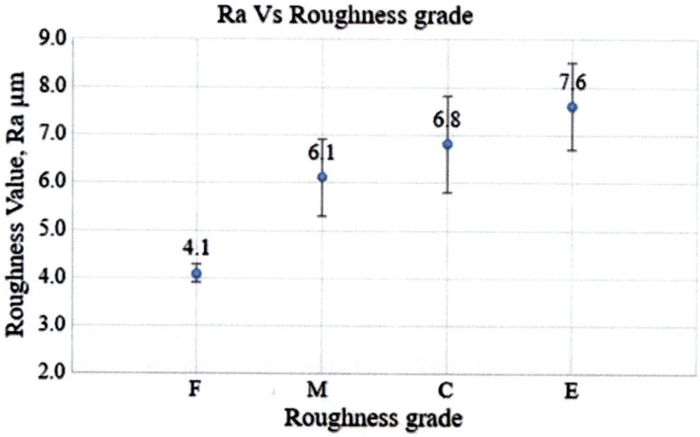

Fig. 4 Roughness Ra with emery grade of fine (*F*), medium (*M*), coarse (*C*), and extra coarse (*E*)

Fig. 5 Normal distribution of roughness Ra with emery cloth of different grades

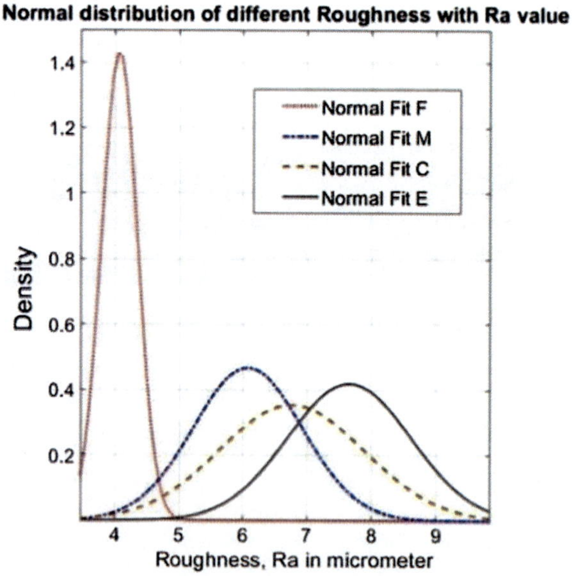

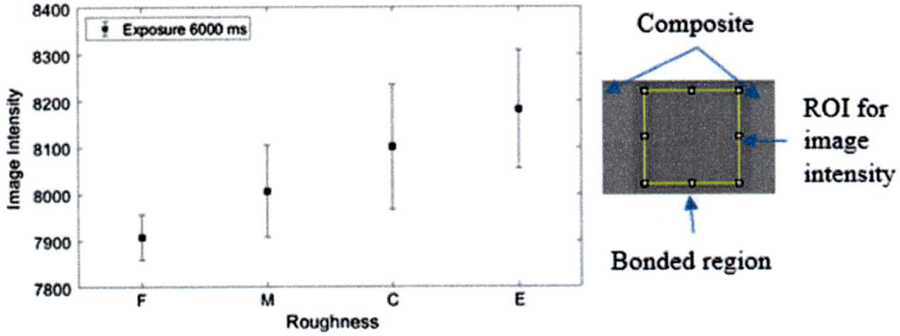

Fig. 6 Exposure variation and its effect on different roughness

shows the information of the normal distribution of the roughness Ra with emery cloth used.

4.2 *X-ray Radiography*

Intensities are measured over the bonded area for the exposure time of 6000 ms. With increasing roughness, reach of adhesive in peaks and valley is difficult, and hence, decreasingly dense bond at the interface is expected. Thus, the increase in roughness is associated with less absorption of X-ray (Fig. 6). For a sample from roughness group *F*, noise present in an image of a sample was ±102.55. For a given range of roughness values, intensity 7850–8300 has given the separation of bonded joints of different roughness grade.

4.3 *Temporal Curves Using Thermography*

Temperature decay with respect to time was extracted from the average temperature over the bonded region. As shown in Fig. 3, the average pixel temperature within ROI was considered for the temperature of the specimen. The sample will have a definite cooling curve as decaying exponents of 'e' are constant for a given sample. If we consider Newton's law of cooling for material, the exponential coefficient of material remains constant, independent of the starting temperature at which you start cooling. So even if each pixel is at a different temperature, decaying constant of temporal curve does not change. These experimental temporal curves of each roughness grade were fitted with a two-term exponential model, so that the parameters for cooling can be obtained. R-square values for roughness grade *F*, *M*, *C*, and *E* are 0.9998, 0.9999, 0.9997, and 0.9999, respectively. The temporal curve decays with two simultaneous components, i.e., slow-decaying and fast-decaying constant (given in Eq. (1)). The two-term model is the solution of Eq. (2).

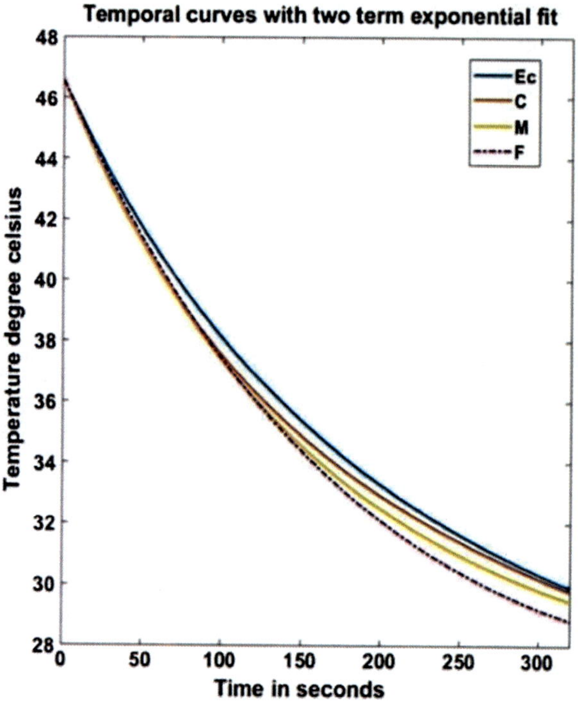

Fig. 7 Temporal curves for different roughness grade after two-term exponential fit

$$T = Ae^{-ct} + Be^{-dt} \tag{1}$$

$$\frac{d^2T}{dt^2} - (c + d)\frac{dT}{dt} + cd(T) = 0 \tag{2}$$

where T = temperature (°C), t = time in seconds, c = fast decaying coefficient, d = slow decaying coefficient, and A, B are constants.

With increasing roughness, samples are getting cooled slowly. Temporal curves provide evidence of less dense bond at the interface with increasing roughness. Hence, transmission of heat is slower in high roughness than in fine samples. As closer the particles, higher will be the transmission of the heat. Results of curve fittings for different grades of joints are shown in the Fig. 7.

5 Conclusion

X-ray radiography and thermography results show promising outcome in terms of NDE parameters for different grades of surface roughness present in the adhesively bonded joint. Hidden surface information about the roughness is exploited by NDE techniques. Temporal curves have differentiated surface preparation changes present in the adhesively bonded joints. X-ray has shown a range of energies to separate the different roughness samples. Thus, this study demonstrates the capability of NDE tools used to pick even roughness changes which are in microns. Further investigations are on to quantify the correlation between the NDE parameters and adhesive bond strength.

References

1. Ebnesajjad S (ed) (2008) Adhesives technology handbook. William Andrew Inc.
2. Mechanics of adhesive bonded lap-type joints: survey and review distribution statement A. Approved for Public Release Distribution Unlimited (1964)
3. Kim JK, Kim HS, Lee DG (2003) Investigation of optimal surface treatments for carbon/epoxy composite adhesive joints. J Adhes Sci Technol 17:329–352. https://doi.org/10.1163/156856103762864651
4. Bénard Q, Fois M, Grisel M (2005) Peel ply surface treatment for composite assemblies: chemistry and morphology effects. Compos Part A Appl Sci Manuf 36:1562–1568. https://doi.org/10.1016/j.compositesa.2005.02.012
5. Wingfield JRJ (1993) Treatment of composite surfaces for adhesive bonding. Int J Adhes Adhes 13:151–156. https://doi.org/10.1016/0143-7496(93)90036-9
6. Kinloch AJ (2012) Adhesion and adhesives: science and technology. Springer Science and Business Media
7. Buragohain MK (2017) Composite structures: design, mechanics, analysis, manufacturing, and testing. CRC Press
8. Hull JB (2015) Non-destructive testing. Macmillan International Higher Education
9. Schroeder JA, Ahmed T, Chaudhry B, Shepard S (2002) Non-destructive testing of structural composites and adhesively bonded composite joints: pulsed thermography. Compos Part A Appl Sci Manuf 33:1511–1517. https://doi.org/10.1016/S1359-835X(02)00139-2
10. Hung MYY (2007) Review and comparison of shearography and pulsed thermography for adhesive bond evaluation. Opt Eng 46:051007. https://doi.org/10.1117/1.2741277
11. Grosso M, Lopez JEC, Silva VMA, Soares SD, Rebello JMA, Pereira GR (2016) Pulsed thermography inspection of adhesive composite joints: computational simulation model and experimental validation. Compos Part B Eng 106:1–9. https://doi.org/10.1016/j.compositesb.2016.09.011
12. Waugh RC (2016) Development of infrared techniques for practical defect identification in bonded joints
13. Ibarra-castanedo C (2005) Clemente ibarra castanedo quantitative subsurface defect evaluation by pulsed phase thermography: résumé. Time
14. Tighe RC, Dulieu-Barton JM, Quinn S (2016) Identification of kissing defects in adhesive bonds using infrared thermography. Int J Adhes Adhes 64:168–178. https://doi.org/10.1016/j.ijadhadh.2015.10.018
15. Vavilov VP, Burleigh DD (2015) Review of pulsed thermal NDT: physical principles, theory and data processing. NDT E Int 73:28–52. https://doi.org/10.1016/j.ndteint.2015.03.003
16. Genest M, Martinez M, Mrad N, Renaud G, Fahr A (2009) Pulsed thermography for non-destructive evaluation and damage growth monitoring of bonded repairs. Compos Struct 88:112–120. https://doi.org/10.1016/j.compstruct.2008.02.010

Non-invasive Detection of Extent of Corrosion in Steel Reinforcing Bars by Magnetic Force Measurement

Durgesh Tamhane, Sauvik Banerjee, and Siddharth Tallur

1 Introduction

Corrosion of steel reinforcing bars in structures built with concrete leads to high repair costs and reduced serviceable lifetime of such infrastructure through the deterioration of the overall structural integrity. The extent of corrosion needs to be monitored, preferably by non-destructive inspection, to prevent costly repairs. Several non-destructive testing methods [1] based on techniques like ultrasonic testing [2], magnetic flux leakage [3], eddy currents [4] are employed today as a means of inspection of the structural integrity of civil infrastructure, with significant deployment toward the detection of internal corrosion [5].

One of the prominent issues today is the detection of corrosion of steel rebars inside reinforced cement concrete (RCC). Non-destructive techniques face several challenges, when applied to the detection of corrosion in RCC. These challenges include detection under an insulation layer and that of reliability of the detection process with varying ambient conditions, e.g., temperature variation. Accordingly, there is a necessity for a simple, practical and a non-destructive test that allows for easy and reliable assessment of corrosion of steel rods in RCC. Measurement of magnetic force to detect anomalies and defects in a ferromagnetic structure could fulfill this necessity.

Several techniques have been proposed for the detection of corrosion by exploiting the changes in ferromagnetic property of the sample due to corrosion. An inexpensive method [6] was proposed to extract information of corrosion in a non- or weakly ferromagnetic test objects by using a permanent magnet applied to the surface of

D. Tamhane · S. Tallur (✉)
Department of Electrical Engineering, Indian Institute of Technology Bombay, Mumbai, India
e-mail: stallur@ee.iitb.ac.in

S. Banerjee
Department of Civil Engineering, Indian Institute of Technology Bombay, Mumbai, India
e-mail: sauvik@civil.iitb.ac.in

C. K. Mukhopadhyay and R. Mulaveesala (eds.), *Advances in Non-destructive Evaluation*, Lecture Notes in Mechanical Engineering,
https://doi.org/10.1007/978-981-16-0186-6_2

the test object. By measuring the force required to detach the magnet from the test object (or by measuring the force of attraction with which the test object attracts the magnet), the carburization, nitriding, and chromium depletion in test objects was detected. A similar magnetic force based non-destructive testing apparatus [7] was proposed, where a sensor measures the interaction of an applied magnetic field and the sample to determine the extent of corrosion on the surface. A magnetic force sensor [8] based on a simple Fiber Bragg Grating (FBG), based on the principle of change of wavelength of light in response to the magnetic pull force was used to detect a natural corrosion pit formed in the bottom plate of a fuel storage tank. This sensor can detect cavity depths in the sub-millimeter range. Another magnetic force-based measurement [9] describes a ferromagnetic distance sensor based on the principle of fiber optic macro-bend loss. The motion of a magnet in response to the ferromagnetic sample couples to a pre-bent optical fiber and the change in the light intensity due to the bending of the optical fiber from its reference position was used to detect ferromagnetic metal loss due to corrosion. Another fiber optic sensor [10] was proposed that was applied as a strain gauge on a cantilever beam with a magnet at one end to detect internal metal loss in steel pipes and storage tanks where the authors have exploited the weak FBG nature of an optical fiber to respond to strain. The cantilever assembly similar to [7] on which the optical fiber resides functions as a strain gauge and measures the change in the thickness of a steel plate. The authors have also proposed another strain gauge for temperature compensation but have not included it with the measurement apparatus.

The constraints faced by non-destructive testing techniques mentioned above and elsewhere were to detect corrosion in ranges of depth in millimeter to centimeter range with or without an insulating coating over the test sample. In this work, we propose an apparatus and technique of detecting corrosion in the underlying steel rods of RCC structures. Our method, based on the detection of the change in the magnetic force exerted by a magnet on a cantilever, correlates to the varying degree of corrosion of the steel rods.

2 Method

In this work, a Wheatstone bridge circuit with a full-bridge strain gauge sensor configuration is employed to detect changes in magnetic force due to corrosion. The magnetic force exerted by a permanent magnet on the sample under test is proportional to the volume magnetic susceptibility, the area of the sample, the applied magnetic field and the distance of the magnet from the sample [11]. For a ferromagnetic material, like the steel bars used in RCC, this effect is more pronounced given the material has volume magnetic susceptibility of approximately 200, 000. Corrosion of these rods results in loss of ferromagnetic material, since the by-products of corrosion (oxides of iron) are not ferromagnetic. The material loss and the corresponding increase in the distance between the magnet and the steel bar due to the reduction in its diameter decreases the magnetic force with which the magnet attracts the steel bar. This magnetic force is measured using a temperature-compensated full-bridge strain gauge configuration mounted on a cantilever to give output as a voltage, which is proportional to the degree of corrosion in the steel rod. The voltage output across the full bridge configuration taken at various positions on the steel bar correspond to varying deformation of the cantilever due to magnetic force arising from varying degrees of corrosion along the length of the bar.

2.1 Experimental Setup

The fixture shown in Fig. 1 is designed to detect corrosion using strain gauges. This apparatus consists of a rectangular steel plate B fixed to an immovable base G by screws H. Four strain gauge sensors C, D, E, and F are fixed on the steel plate B. Strain gauge sensors E and F are fixed on top of the steel plate, while strain gauge sensors C and D are fixed on its bottom surface. A permanent magnet (here, a neodymium magnet) A is attached to the free end of the rectangular steel plate B. This setup forms an assembly equivalent to a cantilever beam with a point force applied to its free end. Figure 2 shows the steel rebar sample of diameter 20 mm used to test the detection apparatus. The sample has three distinct regions: non-corroded

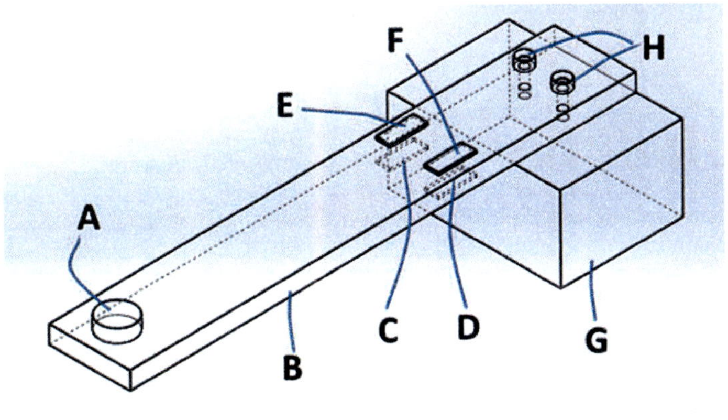

Fig. 1 The strain gauge fixture: *A* Neodymium magnet, *B*: Rectangular steel beam, *C, D, E, F*: Strain gauge sensors, *G*: Immovable steel base, *H*: Fixing screws

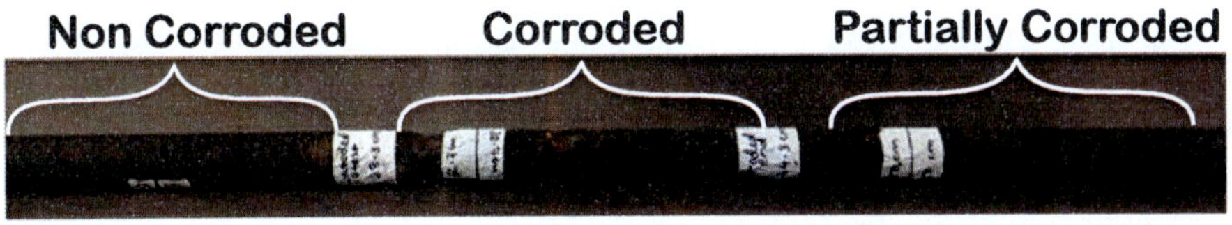

Fig. 2 Steel bar of 20 mm diameter with non corroded, corroded and partially corroded regions

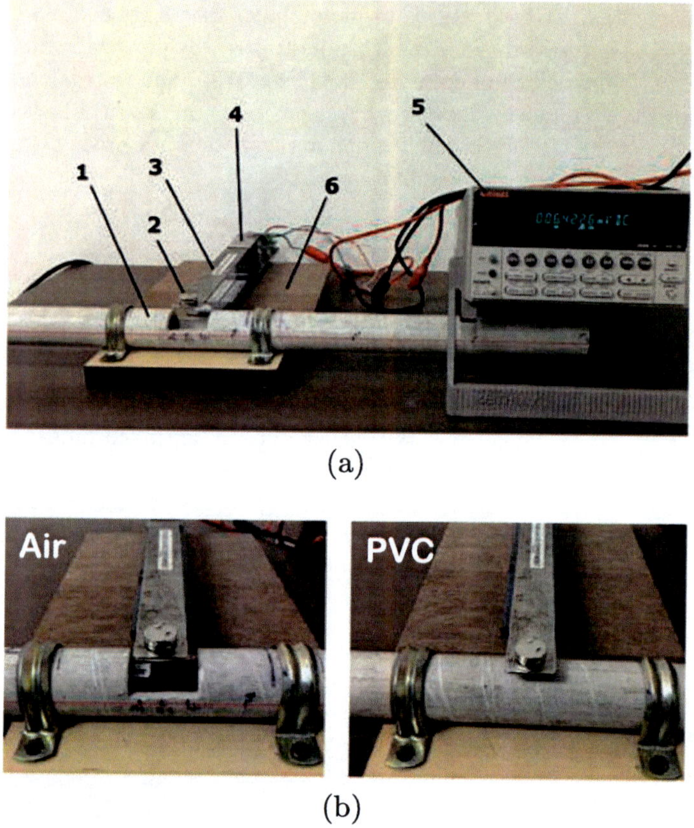

Fig. 3 **a** Experimental setup: **1**: PVC pipe with a window cut-out *3*: Steel cantilever with *4* Strain gauges and *2*: Magnet, *5*: Digital multimeter (DMM), *6*: Wooden stage, **b** Left: The magnet on the cantilever is directly over the sample with only air in between, Right: The sample is under a 5 mm thick PVC coating

(left end), corroded (center), and partially corroded (right end) labeled accordingly in the figure. The apparatus depicted in Fig. 1 is then integrated into the complete experimental setup shown in Fig. 3a. A PVC pipe clamped to the wooden stage has a window cut-out, to allow either PVC insulation or air insulation between the magnet and the sample, as shown in Fig. 3b. The digital multimeter (DMM) is connected to measure the voltage output of the Wheatstone bridge configuration of the strain gauges.

The strain gauges C, D, E and F are connected in a Wheatstone bridge circuit with a full-bridge configuration, as shown in Fig. 4. Strain gauges E and F are mounted on the steel plate, and strain gauges C and D are mounted under the steel plate. This type of placement of the strain gauges and connecting them in full-bridge configuration improves the sensitivity of this system and also compensates for the ambient temperature variation. Under bending, the strain gauges E and F undergo

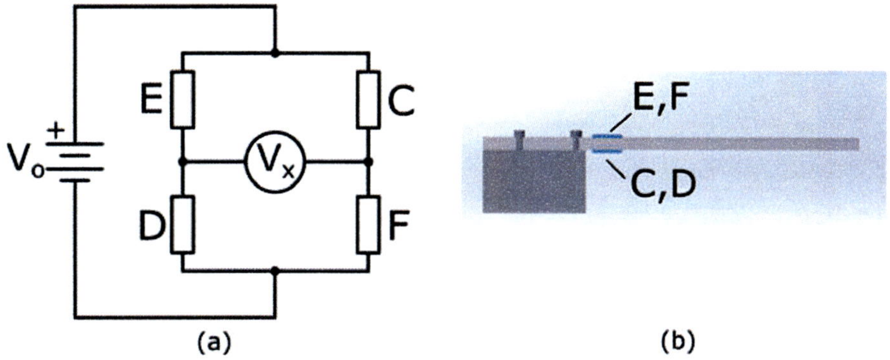

Fig. 4 **a** Block diagram of the system showing four strain gauges (*C*, *D*, *E* and *F*) in a Wheatstone bridge configuration. **b** The strain gauges *C* and *D* are mounted at the bottom, while E and F are mounted on top of the cantilever beam

tension, while strain gauges *C* and *D* undergo compression. The strain gauges are connected in the full-bridge configuration as shown in Fig. 4a. With V_x, V_0 being the measured and the bridge input voltage respectively, *k*, the gage factor and ε, the strain introduced due to the bending of the beam, the output of the bridge is given by [12]:

$$\frac{V_x}{V_0} = \frac{k}{4}[\varepsilon_{\mathrm{E}} - (-\varepsilon_{\mathrm{C}}) + \varepsilon_{\mathrm{F}} - (-\varepsilon_{\mathrm{D}})]$$

With $\varepsilon_{\mathrm{E}} = \varepsilon_{\mathrm{F}} = \varepsilon_{\mathrm{C}} = \varepsilon_{\mathrm{D}} = \varepsilon$, the measured voltage is

$$V_x = V_0 \cdot k \cdot \varepsilon$$

A +5 V DC voltage is applied across the full-bridge configuration, as shown in Fig. 4 using a DC power supply, and the differential voltage across the bridge is measured using a DMM with micro-Volt resolution. The strain gauges respond to the bending of the cantilever beam due to the force of attraction between the magnet and the steel bar. We record the voltage across the bridge without the sample and then with the sample inserted inside the PVC pipe, as shown in Fig. 2b. The first reading of voltage is recorded for the non-corroded region, followed by the voltage reading of the corroded and the partially corroded regions of the steel bar, and finally, a voltage reading is recorded with the rod removed. The measurements are performed by scanning the rebar along the pipe. Thirty such readings are repeated to acquire a statistically significant set of measurements. We repeat the above procedure for the steel rod with the 5-mm-thick PVC insulation coating in between the magnet and the steel rod sample.

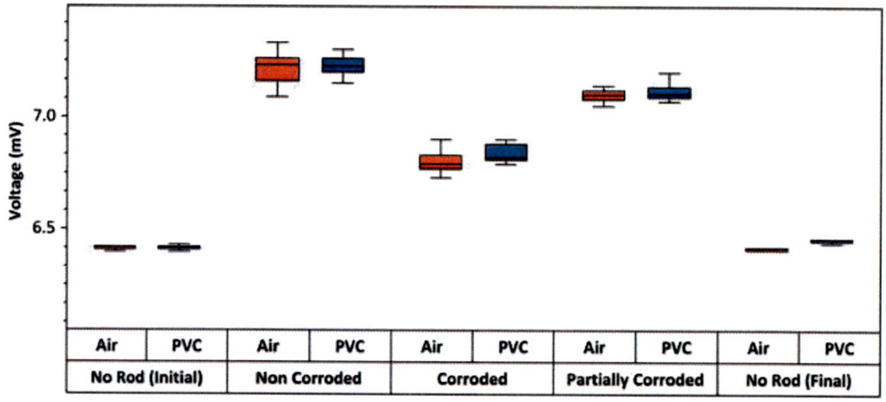

Fig. 5 Bridge voltage corresponding to the various corrosion states of the steel rod (with and without PVC insulation). The distinction in the sensor output for various regions is clearly observed

3 Results

Figure 5 shows the variation in the differential voltage across the bridge with respect to the degree of corrosion of the steel rod. The trend of decreasing bridge voltage for the corroded regions corresponds to reduced deflection of the cantilever due to the reduced magnetic force of attraction owing to loss of ferromagnetic material. The results do not show any significant effects of PVC insulation on the output voltage, and this technique is thus applicable to corrosion monitoring in insulated pipelines as well.

4 Conclusion

A concept of magnetic force of attraction of a magnet by ferromagnetic materials to detect the extent of corrosion in steel rebars is presented in this work. Detection of the degree of corrosion is achieved by measuring the output voltage across the full-bridge configuration of strain gauges mounted on a cantilever in a full-bridge, temperature-compensated manifestation. The measurement is insensitive to the presence or absence of PVC insulation over the sample and is thus applicable to corrosion monitoring in insulated pipeline use case as well. Our proof-of-concept demonstration suggests the ability of this setup to distinguish between different degrees of corrosion in several applications.

Acknowledgements The authors thank Mahesh Bhaganagare and Subrata Dutta at the Wadhwani Electronics Lab (WEL) at IIT Bombay for helping in designing the strain gauge apparatus, and Dr. Mangesh Joshi, CEO of Sanrachana Structural Strengthening Pvt. Ltd., for insightful discussions and support. The measurements were carried out partly at Wadhwani Electronics Lab (WEL) and Applied Integrated Microsystems (AIMS) lab at IIT Bombay.

References

1. Helal J, Sofi M, Mendis P (2015) Non-destructive testing of concrete: a review of methods. Electron J Struct Eng 14(1):97–105
2. Saint-Pierre F, Philibert A, Giroux B, Rivard P (2016) Concrete quality designation based on ultrasonic pulse velocity. Constr Buildi Mater 125:1022–1027
3. Shams S, Ghorbanpoor A, Lin S, Azari H (2018) Nondestructive testing of steel corrosion in prestressed concrete structures using the magnetic flux leakage system. Transp Res Rec 2672(41):132–44
4. Frankowski P (2011) Eddy current method for identification and analysis of reinforcement bars in concrete structures. in: Electrodynamic and mechatronic systems. Opole, pp 105–108
5. Sarja A, Vesikari E, Andrade C, Sellevold E, Sentler L, Sick L, Tomosawa F (2000) Durability design of concrete structures. Mater Struct 33:14–20
6. Ling A (2017) Methods for testing non or weakly ferromagnetic test objects. WO2017009309A1
7. Zamanzadeh, M, Kirkwood G, Scheinman S (2010) Non-destructive testing apparatus for the detection of corrosion. US7719266B1
8. Pacheco C, Bruno A (2013) A Noncontact force sensor based on a fiber Bragg grating and its application for corrosion measurement. Sensors 13(9)
9. Li W, Ho SCM, Luo M, Huynh Q, Song G (2017) Fiber optic macro-bend based sensor for detection of metal loss. Smart Mater Struct 26(4)
10. Almahmoud S, Shiryayev O, Vahdati N, Rostron P (2018) Detection of internal metal loss in steel pipes and storage tanks via magnetic-based fiber optic sensor. Sensors 18(3)
11. Brubacher LJ, Stafford FE (1962) Magnetic susceptibility: a physical chemistry laboratory experiment. J Chem Educ 39(11)
12. Hoffmann K (2001) Applying the Wheatstone bridge circuit. HBM Company

Phase-Sensitive Detection of Extent of Corrosion Using Anisotropic Magnetoresistive (AMR) Sensor in Steel Reinforcing Bars (Rebars)

Indrani Mukherjee, Jinit Patil, Sauvik Banerjee, and Siddharth Tallur

1 Introduction

Corrosion of steel reinforcement is a global problem that leads to deterioration of RC structures. Damage induced by steel corrosion usually requires proper repair, followed by maintenance. It is reported that the costs of repair and maintenance of corroded structures exceed billions of dollars per year [1]. For effective structural health monitoring (SHM), detection of incipient corrosion is of paramount importance. There are several techniques for non-destructive testing (NDT) [2] which are broadly captured in Table 1. Eddy current-based sensing is a good choice but for ferromagnetic substances with high permeability, however, small skin depth and large magnetic flux fluctuations coupled with presence of insulating sheaths in many applications lead to poor signal-to-noise ratio (SNR) reducing the detection range.

Conventional eddy current-based testing uses alternating current (AC) signal to create a harmonic magnetic field using a solenoid [3]. Due to mutual inductance of the solenoid and sample under test, the magnetic field produced by the eddy currents thus excited in the sample (typically metal) has a phase shift relative to the AC input to the solenoid, due to the inductive nature of the metal. The phase shift is proportional

I. Mukherjee · J. Patil · S. Tallur (✉)
Department of Electrical Engineering, Indian Institute of Technology (IIT) Bombay, Mumbai, India
e-mail: stallur@ee.iitb.ac.in

I. Mukherjee
e-mail: indrani@ee.iitb.ac.in

J. Patil
e-mail: jppatil@ee.iitb.ac.in

S. Banerjee
Department of Civil Engineering, Indian Institute of Technology (IIT) Bombay, Mumbai, India
e-mail: sauvik@civil.iitb.ac.in

C. K. Mukhopadhyay and R. Mulaveesala (eds.), *Advances in Non-destructive Evaluation*, Lecture Notes in Mechanical Engineering,
https://doi.org/10.1007/978-981-16-0186-6_3

Table 1 NDT techniques for SHM

NDT techniques	Description	Advantages	Disadvantages
Optical	Fibre optics, etc.	Non-contact, inexpensive, fast	Low noise immunity and penetration
Acoustic	Ultrasonic and acoustic methods	Large coverage, real time	Needs coupling and smooth surfaces
Radio-graphic	X-rays/gamma rays	Superior resolution	Access to both sides needed
Thermal (thermo-graphy)	Infrared, eddy current	Non-contact, imaging	Expensive, bulky, poor resolution
Electro-chemical	Half-cell potential, impedance spectroscopy	Fast,provides degradation rate	Requires expertise, not for in-situ detection
Electro-mechanical	Impedance spectroscopy	Light, low power	Low sensitivity
Electro-magnetic	Eddy current, magnetic flux leakage (MFL)	Economical, portable, high resolution, multilayer	Requires electrically conducting surfaces

to the conductivity of the sample [4] and can thus be used to distinguish non-corroded rebars (higher conductivity) from corroded rebars (lower surface conductivity due to presence of rust, i.e. iron oxide) (Fig. 1).

The following section specifies the details of the sensor assembly and subsequent sections cover the experimental setup, results and discussion.

2 Methodology

2.1 Sensor Configuration

A scannable and portable sensor assembly is envisaged to allow rapid translation from lab testing to field testing. A 3D printed housing is designed for holding two coils and the AMR (Honeywell HMC1001) sensor as shown in Fig. 2. The AMR sensor is fixed inside a slot in a cylindrical holder, on which a coil (denoted as 'cancellation coil') is wound. This cylinder is inserted inside another hollow cylinder that houses another coil (denoted as 'transmission coil' in Fig. 2). An external holder with a cylindrical cavity holds both coils in place. This design ensures that the sensitive axis of the AMR sensor is aligned to the axis of the coaxial cylindrical coils. The current direction and number of turns of the winding in both coils are varied such that the total magnetic field at the AMR sensor due to both coils cancels out. This allows the AMR sensor to selectively measure the magnetic fields due to the eddy currents with low output offset.

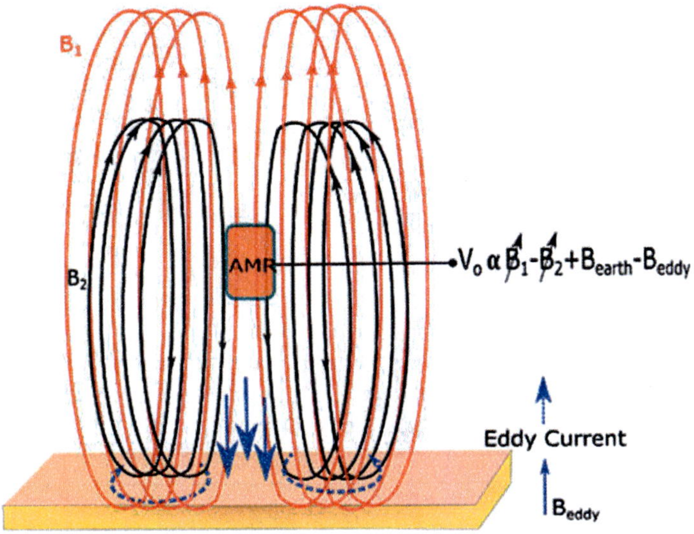

Fig. 1 Eddy current-based magnetic field sensing scheme using an anisotropic magnetoresistive (AMR) sensor (high sensitivity -typically order of few mV/V/Gauss used in a Wheatstone bridge configuration) to detect small changes in magnetic field strength [5] due to corrosion in rebars. The phase shift due to the inductive nature of the ferromagnetic rebars is accurately measured with a lock-in amplifier, and a well-resolved distinction between corroded and non-corroded areas of a rebar are observed in measurements. The currents in the coils are adjusted such that magnetic fields due to the primary and cancellation coils (B_1 and B_2, respectively) cancel out. Earth's magnetic field, B_{earth} is rejected by phase-sensitive detection

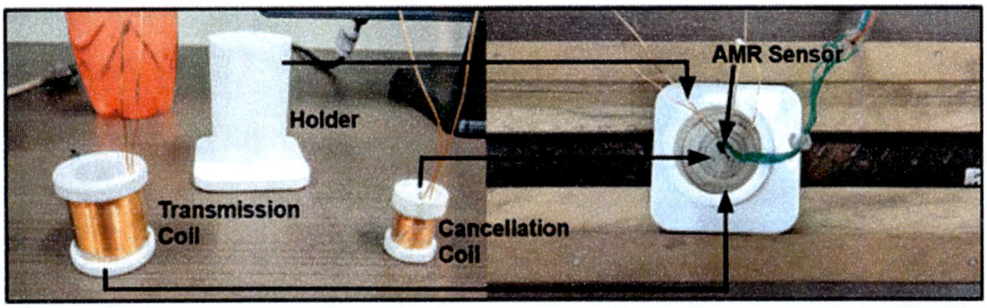

Fig. 2 Sensor assembly for corrosion monitoring in rebars. The left panel shows two coaxial coils and 3D printed holder that fit within each other as shown in the right panel. The net axial magnetic field in the centre of the coils is cancelled by adjusting the current in the inner (cancellation) coil. An AMR sensor is inserted in a slit in the inner coil. The magnetic field induced by eddy currents in the sample is measured by the AMR sensor. The phase shift of the AMR sensor output relative to the excitation phase for the AC currents in the coils depends on the conductivity of the sample and can be used to estimate the extent of corrosion

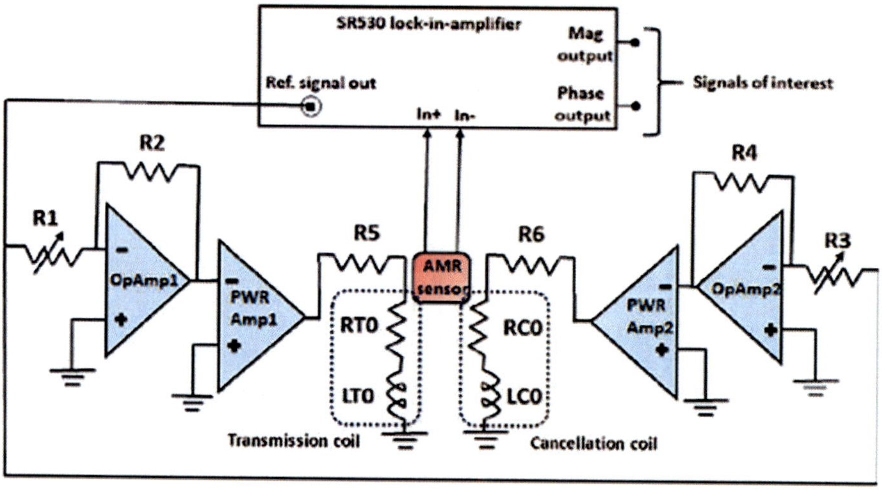

Fig. 3 Circuit diagram of the experiment. The gains of pre-amplifiers are controlled using potentiometers *R*1 and *R*3. LM380N audio (power) amplifiers (denoted as PWR Amp) are used to generate a large excitation current (~65 mA rms) in both coils. *R*5 and *R*6 are 100Ω, 5 W rated power resistors. The SR530 lock-in amplifier provides the excitation voltage to both coils and measures the magnitude (amplitude) and phase shift of the differential output of the AMR bridge sensor

The sensor interface circuit is shown in Fig. 3. For precision measurements, we use Stanford Research Systems SR530 lock-in amplifier that generates a reference AC signal for exciting harmonic magnetic fields in both coils. General purpose LM741 operational amplifiers (op-amps) are used to implement pre-amplifiers, and LM380N audio amplifiers are used to provide a large excitation current in both coils. The coils are represented by their equivalent circuit (series R-L circuit). The currents in the coils are adjusted by varying potentiometers *R*1 and *R*3. Current limiting power resistors R5 and R6 (100Ω, 5 W each) limit the current in both coils to ~65 mA rms. The AC excitation frequency of 1 kHz is provided using in-built function generator of SR530, and the AC signal excitation provided by lock-in amplifier is 1 V rms.

Both the transmission and cancellation coils are solenoidal by principle of operation. The parameters of the transmission coil are: resistance 0.4Ω, inductance 40 μH, outer diameter of cylinder 2.5 cm, length of solenoid 2.5 cm (L_1), diameter of wire 0.05 cm with 45 turns (N_1). The cancellation coil has resistance 0.2Ω, inductance 7.5 μH, outer diameter of cylinder 1.5 cm and length of solenoid 1.4 cm (L_2) with 25 turns (N_2). The turns-to-length ratio (N/L) of both coils has to be the same by design and is maintained to be approximately equal to 1800.

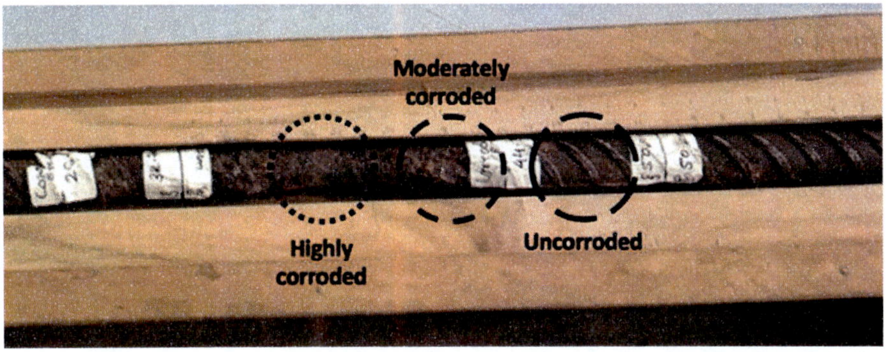

Fig. 4 The sample chosen for testing is a rebar of diameter 20 mm. Three regions are chosen for testing as shown in the figure, with average diameter in uncorroded, moderately corroded and highly corroded sections being 20.61 mm, 19.57 mm and 17.85 mm, respectively.

2.2 Experimental Setup

A steel rebar with regions of visibly varied levels of corrosion is used as the sample under test. Figure 4 shows a photograph of the specimen. Three regions on the sample are identified as uncorroded, moderately corroded and highly corroded with average diameters as 20.61 mm, 19.57 mm and 17.85 mm, respectively, as measured with a Vernier caliper. The sample is held in a wooden fixture that is designed with grooves to hold the sensor assembly such that the AMR sensor is directly atop the rebar, while allowing the sensor assembly to slide along the length of the rebar. This test setup design minimizes error due to sensor misalignment during scanning. The clearance (lift-off) between the AMR sensor and the top surface of the uncorroded area of the rebar is 5 mm. Figure 5 shows a photograph of the complete experimental setup. A dual channel DC power supply is used to provide supply voltage to the amplifier ICs and for providing bridge excitation voltage to the AMR sensor.

3 Experimental Results and Discussion

A bridge bias voltage of 5 V is applied to the AMR sensor, and an offset voltage of 7.6 μV is recorded. By tuning the drive currents in both coils using the potentiometers, the offset of the AMR sensor is minimized well within the dynamic range of AMR sensor. After inserting the AMR sensor in the 3D printed assembly, the drive currents in both coils are trimmed using the potentiometers $R1$ and $R3$ to minimize the sensor offset, in the absence of metal in the vicinity. Next, the sensor is scanned over the sample under test using the wooden fixture, and corresponding magnitude and phase outputs of the lock-in amplifier are recorded in all three regions. Ten measurements are conducted in each region.

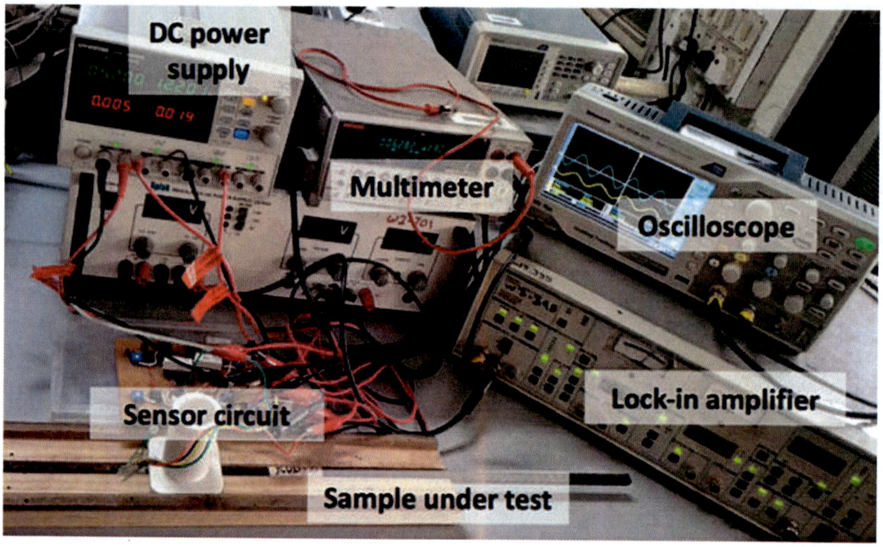

Fig. 5 Photograph of the experimental setup used for conducting measurements reported in this work. The wooden block is machined to dimensions required for scanning the sensor holder along the length of the rebar (sample under test). The 3D printed holder and the wooden block ensure that the AMR sensor is directly atop the rebar in all measurements, thus eliminating human error due to lateral misalignment

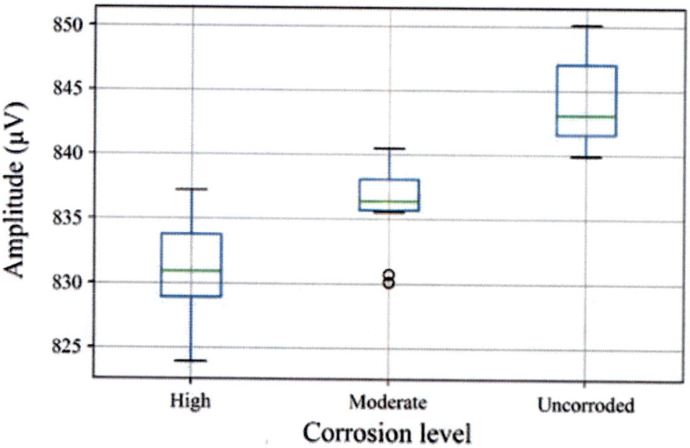

Fig. 6 Amplitude of the AMR sensor output recorded by the lock-in amplifier. The three regions with visibly different extent of corrosion on the rebar are clearly resolved based on the measured sensor voltage. Ten measurements are conducted in each region

Figures 6 and 7 show the amplitude and phase shift outputs of the lock-in amplifier, respectively, obtained for the three regions on the sample under test. While the amplitude shows a decreasing trend, the phase shift is observed to increase with the

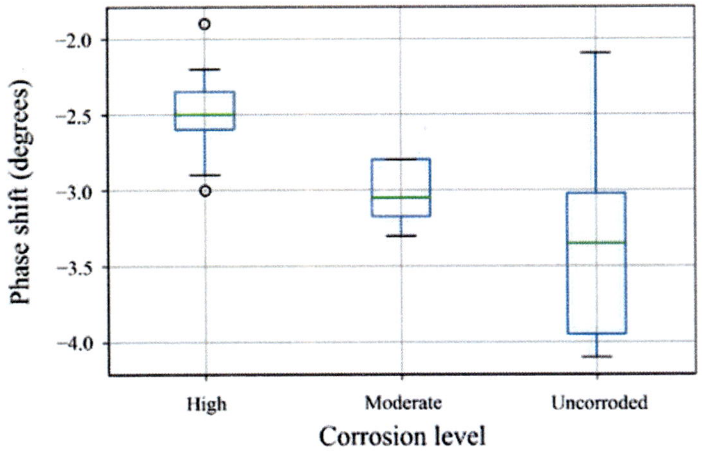

Fig. 7 Phase shift of the AMR sensor output recorded by the lock-in amplifier. The phase shift is calculated relative to phase of the sensor output offset in absence of any metal in the vicinity of the sensor. The highly corroded and uncorroded regions on the rebar are clearly resolved based on the measured sensor output phase. Ten measurements are conducted in each region

extent of corrosion. Since corrosion leads to loss of ferromagnetic material and its conversion into iron oxide, these trends are intuitively justified. Validation using a commercially available eddy current detector (EddyUSB32 manufactured by Technofour) was attempted, but yielded no inference due to unavailability of quantitative assessment in the instrument.

4 Conclusion and Future Work

The well-known principle of metal discrimination using eddy current has been exploited for detecting extent of corrosion in rebars. Since corrosion is a slow and gradual process, there are small changes in conductivity over long periods of time. The sensor should thus be highly sensitive, and hence, an AMR sensor was chosen as the prime sensing element. A custom sensor assembly is designed to emulate a scanning probe that can distinguish between different regions of a rebar with varying extent of corrosion based on the amplitude and phase output recorded using a lock-in amplifier. Proof-of-concept measurements demonstrate the potential of the system to adequately resolve the different regions both in amplitude as well as phase of the sensor output. Due to cylindrical structure of the rod, a small fraction of the magnetic flux is cut by the AMR sensor, hence creating the need of a high power, high resolution phase-sensitive detection system. Future work will focus on optimizing the sensor assembly, testing response at different AC excitation frequencies for lock-in amplifier, modifying the setup to introduce lift-off (separation between the sensor and the sample) as well as conducting measurements on rebars embedded in concrete. Efforts will also be directed towards miniaturizing the phase-sensitive detection scheme implemented by the lock-in amplifier to an on-board circuit to enable a truly portable and field-deployable sensor system.

Acknowledgements The authors thank TATA Centre for assistance in 3D printing, and Prof. K. L. Narasimhan at IIT Bombay for providing guidance to the student authors on usage and handling of SR530. The measurements were carried out partly at Wadhwani Electronics Lab (WEL) and Applied Integrated Microsystems (AIMS) lab at IIT Bombay. Authors also thank Mr. Sitaram Varak at IIT Bombay for building the wooden test fixture.

References

1. Sarja A (2002) Reliability based life cycle design and maintenance planning. JCSS workshop on reliability based code calibration
2. Schull PJ (2002) Nondestructive evaluation: theory, techniques, and applications, 1st edn. CRC Press
3. De Alcantara NP et al (2016) Corrosion assessment of steel bars used in reinforced concrete structures by means of eddy current testing. Sensors 16:E15
4. Yamazaki S, Nakane H, Tanaka A (2002) Basic analysis of a metal detector. IEEE Trans Instrum Meas 51(4):810–814
5. Tsukada K et al (2016) Detection of inner corrosion of steel construction using magnetic resistance sensor and magnetic spectroscopy analysis. IEEE Trans Magn 52(7):1–4

A Thermographic System for Quantitative and Automated Subsurface Detail Visualization

A. Vijaya Lakshmi, V. S. Ghali, B. Suresh, G. T. Vesala, and R. B. Naik

1 Introduction

Active infrared thermography is a feasible non-destructive testing method [1–3] used in various fields such as aerospace, civil, and mechanical fields to evaluate the test object. It makes use of the contrast in thermal response from the test object being energized by a controlled external stimulus either from recording end or from the opposite surface depending on the reflection or transmission modes [4]. Depending upon the external stimulus, active thermography is further classified into pulsed thermography (PT), lock in thermography (LT), pulsed phase thermography (PPT), and other modulated non-stationary thermal wave imaging methods like frequency modulated thermal wave imaging (FMTWI) and quadratic frequency modulated thermal wave imaging (QFMTWI).

Pulsed thermography (PT) [5, 6] employs a short duration high peak power pulse from a flash lamp as stimulus and corresponding thermal response in cooling phase is recorded and used for subsurface analysis, non-uniform emissivity, and non-uniform radiations are observed in thermograms limits the application of PT. Lock in thermography (LT) [7, 8] is a continuous modulated thermal wave imaging technique; low frequency low power sinusoidal stimulation is used to energize the object, and subsequent thermal response is processed with the phase-based analysis due to its deeper depth probing capability. Pulsed phase thermography (PPT) [9, 10], pulse type of input is used as stimulus and phase-based analysis considered for defect detection

A. Vijaya Lakshmi · V. S. Ghali (✉) · B. Suresh · G. T. Vesala
Infrared Imaging Center, Koneru Lakshmaiah Educational Foundation, Vaddeswram, Guntur, India
e-mail: gvs0raos@kluniversity.in

R. B. Naik
Naval Materials Research Laboratory, Ambernath (E), Thane, Maharashtra 421506, India

C. K. Mukhopadhyay and R. Mulaveesala (eds.), *Advances in Non-destructive Evaluation*, Lecture Notes in Mechanical Engineering,
https://doi.org/10.1007/978-981-16-0186-6_4

fascinated in a variety of industrial applications. Due to high peak power requirement in PT and repetitive experimentation in LT for the evaluation of realistic objects, frequency modulated thermal wave imaging (FMTWI) [11–13] has been emerging as a viable alternative by probing the test object with low power stimulation of a suitable band of frequencies with relatively low peak power sources in a single experimentation by providing the complete depth scanning of the object. Whereas a variant of FMTWI, quadratic frequency modulated thermal wave imaging (QFMTWI) [14, 15] imposes more energy on low frequency thermal waves and facilitate a deeper depth analysis along with all the advantages of FMTWI.

In this paper, neural network-based defect detection is proposed and further statistically estimated the corresponding characteristics of subsurface anomalies. This paper explores the use of a ANN-based subsurface analysis for defect detection and depth estimation of artificial flat bottom holes in mild steel sample (MS) and simulated carbon fiber rain forced polymers (CFRP) using quadratic frequency modulated thermal wave imaging.

In this paper, Sect. 2 gives theory for quadratic frequency modulated thermal wave imaging; the details of different post processing methods employed in this paper are described in Sect. 3, and to facilitate enhanced subsurface characterization, Sect. 4 provides the experimental results as obtained from neural net-based processing and further segmentation for automatic detection.

2 Theory of QFMTWI

In this section, a theoretical development for surface evaluation corresponding to quadratic frequency modulated thermal wave imaging as stimulation by solving one-dimensional heat equation. A quadratic chirped stimulus is provided to surface of test sample and is represented by [14]:

$$H(t) = H_0 \sin\left(at + bt^3\right) \tag{1}$$

where 'a' is initial frequency, 'b' as bandwidth of chirped excitation with peak value of stimulation 'H_0'. The stimulation creates a thermal perturbation on the surface of the sample and propagates deeper into test object as diffusive thermal waves depending on thermal properties of the surface temperature contributes over the surface of test sample. The surface temperature evolution of thermal wave can be obtained by solving one-dimensional heat equation to homogeneous, isotropic, and semi-infinite media in the absence of heat sink or source and is represented by:

$$\frac{\partial^2 T(x,t)}{\partial T^2} = \frac{1}{\alpha}\frac{\partial T(x,t)}{\partial t} \tag{2}$$

where 'α' is the thermal diffusivity of the material and boundary conditions to solve Eq. (2) for a specimen of thickness 'L' are given by:

$$-k\frac{\partial T}{\partial Z}\bigg|_{x=0} = Q_0 e^{j2\pi(a+bt^2)t}\ s \tag{3}$$

where 'k' is thermal conductivity of the material and 'Q_0' is amplitude of heat flux. By solving Eq. (2), one can obtain the diffusion length 'δ' as:

$$\delta \propto \sqrt{\frac{\alpha}{1.77\left(a + bt^2\right)}} \tag{4}$$

By Eq. (4), it is clear that quadratic frequency modulated stimulation provides depth resolution with time-varying frequency. And phase at any location is given by:

$$\phi_r = -\frac{3\pi}{4} - \sqrt{\frac{2\theta}{\alpha t}}d + \phi + \phi_1 - \frac{\theta\sqrt{\theta}}{\sqrt{6bt^3}} \tag{5}$$

where

$$\phi = \frac{\frac{\theta a^2}{\alpha t} + \frac{\theta z^2}{\alpha t} - \frac{2z\theta\left(\mu+\frac{1}{\sqrt{2}}\right)}{\left(\left(\mu+\frac{1}{\sqrt{2}}\right)^2+\frac{\theta}{\alpha t}\right)} + \frac{2\theta\left(\left(\mu+\frac{1}{\sqrt{2}}\right)^2-\frac{\theta}{\alpha t}\right)}{\alpha t\left(\left(\mu+\frac{1}{\sqrt{2}}\right)^2+\frac{\theta}{\alpha t}\right)^2}}{-1 + \frac{\theta\sqrt{\theta}}{\alpha t\sqrt{\alpha t}}\frac{1}{\left(\left(\mu+\frac{1}{\sqrt{2}}\right)^2+\frac{\theta}{\alpha t}\right)} + \frac{2\theta\sqrt{\theta}}{\alpha^2 t}\frac{1}{\left(\left(\mu+\frac{1}{\sqrt{2}}\right)^2+\frac{\theta}{\alpha t}\right)}};\quad and\quad \phi_1 = \frac{\sqrt{\frac{\theta}{\alpha t}}}{\left(\mu+\sqrt{\frac{2\theta}{\alpha t}}\right)}$$

It is clear that obtained phase value can be used to estimate the characteristics of the anomaly. Here, 'θ' are the polar coordinate system parameter. Thus, from Eq. (5), phase of the defect location is proportional to the depth (d) of the defect which has been used for depth estimation of the anomalies.

3 Methodology

During experimentation, response from the object is captured using IR camera and data corresponding to a pixel from all the captured thermograms can be collected as a thermal profile. Further, various post processing approaches are applied over this profile, and resultant response is preserved for further analysis. Same will be repeated with all other pixels in view as explained in sub sections below.

3.1 FFT Phase

To extract the subsurface details of the test specimen, various post processing approaches are employed over the captured temperature map. In this method, FFT is applied over the mean removed thermal profiles of each pixel delay between defective and non-defective profiles can be used for detection of anomalies from phasegrams.

3.2 Hilbert Phase

It is a multi-transform technique; where, cross-correlation takes place in between Hilbert transformed reference thermal profile and temporal thermal profiles of all pixels and corresponding phase contrast in time-domain is used to locate subsurface features.

3.3 Pulse Compression

Pulse compression is performed between preprocessed profiles of all the pixels with a reference profile, and corresponding normalized correlation coefficients of each pixel is placed in the corresponding location. The correlation coefficient contrast can be used for subsurface analysis.

3.4 Artificial Neural Networks

In the present work, an artificial neural networks (ANN)-based analysis has been proposed to provide qualitative and quantitative information about subsurface features in QFMTWI [15]. This work employs a multilayer perceptron neural network with 2500 input nodes with one hidden layer having 20 neurons. The hidden layers are used for training and analysis. Figure 1 represents the general architecture of artificial neural network input layer, output layer with one hidden layer.

Algorithm uses tan-sigmoid function-based activation along with back propagation algorithm (BPA) to discriminate also for quantification analysis. First record the temporal sequence of thermograms, then forms the training vector corresponding to data acquisition. The number of inputs is equal to the number of thermograms in a sequence. Then, from the training vectors, the back propagation adjusts its weights and provides pixels belongs defective locations and pixels belongs non-defective locations.

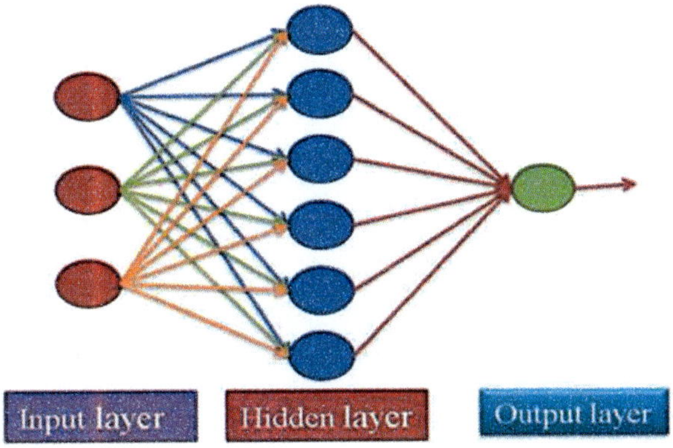

Fig. 1 Artificial neural network for defect detection

3.5 Particle Swam Optimization (PSO)

The particle swam optimization process was introduced to overcome the problem of slow convergence [16–18] and reduces misclassified patterns by adjusting the hidden layers as well as weights of hidden layers and is used for optimization task.

3.6 Segmentation

In present work, we adopted active contour-based image segmentation of Chan-Vese for automatic separation of subsurface anomalies. Along with chan-vese model, a level set-based modality has been applied and compared toward achieving the best sizing detections [19–21].

4 Results and Discussion

In this work, two samples one with mild steel sample having five flat bottom holes of same size and different depths and a carbon fiber reinforced polymer containing twenty five flat bottom holes with different sizes and different depths have been used for testing hypothesis. To evaluate the subsurface analysis, experimentation has been carried on mild steel (MS) sample and simulated CFRP test specimen. The corresponding layouts are shown in Fig. 2

During experimentation, in both the cases, the test specimen is kept opposite to infrared camera with 1 m distance and set of halogen lamps facilitating a uniform illumination with 1 kW each over the test sample with the energizing optical stimulus of 0.01–0.1 Hz in 100 s is shown in Fig. 3 with a frame rate of 25 Hz. Further, various post processing approaches like FFT phase (FFT), pulse compression (PC), Hilbert

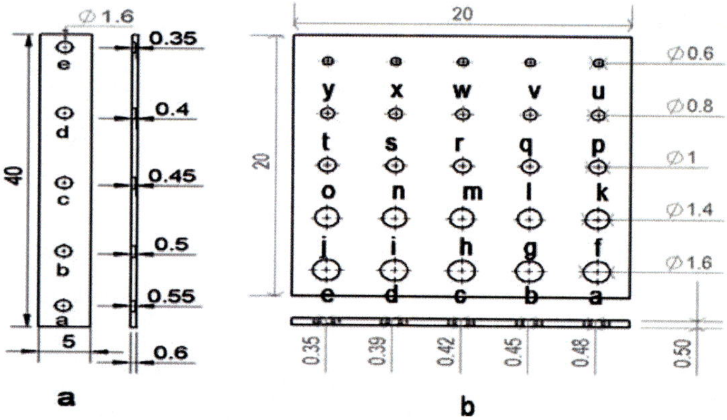

Fig. 2 **a** Layout of mild steel sample, **b** layout of CFRP sample

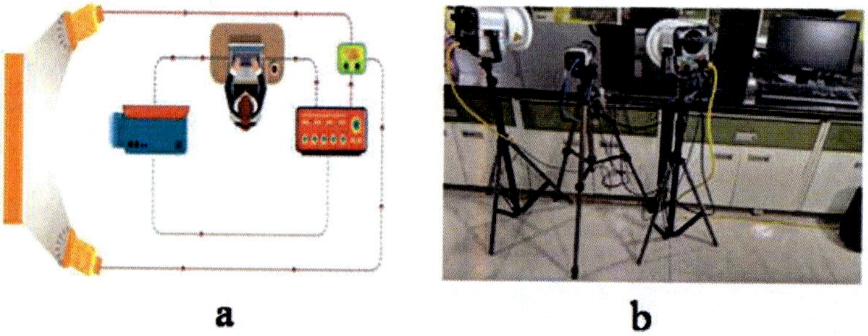

Fig. 3 **a** Schematic view, **b** experimental setup of active thermal wave imaging system

phase (HP), and artificial neural networks (ANN) are employed over the thermal response to extract fine subsurface details.

The post processing methods are applied over mild steel specimen for experimental data and corresponding results are shown in Fig. 4a–d, as well as simulated CFRP sample with flat bottom holes having different depths with different sizes are shown in Fig. 5a–d.

Among the various processing methods, artificial neural network approach gives better contrast. In order to measure the detectability of sample with different processing methods, signal-to-noise ratios for both experimental mild steel sample with five flat bottom holes having same size with different depths as well as simulated CFRP sample having different sizes with different depths have been calculated and plotted in Fig. 6. From these plots, it is clear that SNR with artificial neural network-based approach gives better SNR compared with other processing approaches. All the dimensions of defects are observed; artificial neural network approach gives better SNR and provides good detectability compared to other methods.

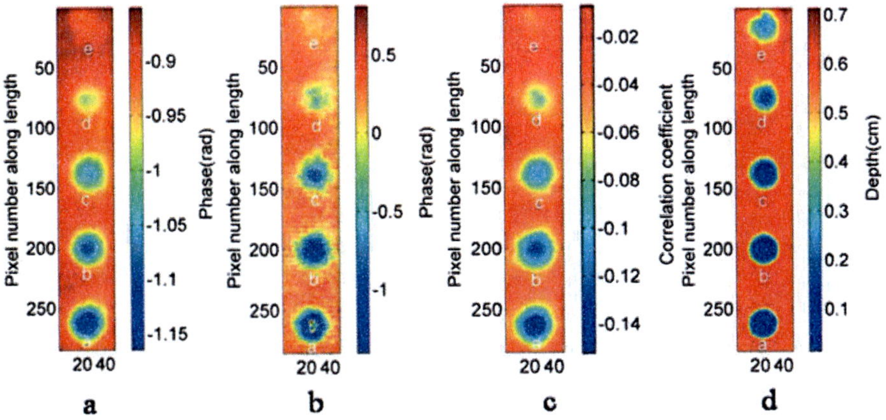

Fig. 4 **a** FFT phase, **b** Hilbert phase, **c** pulse compression, **d** artificial neural network

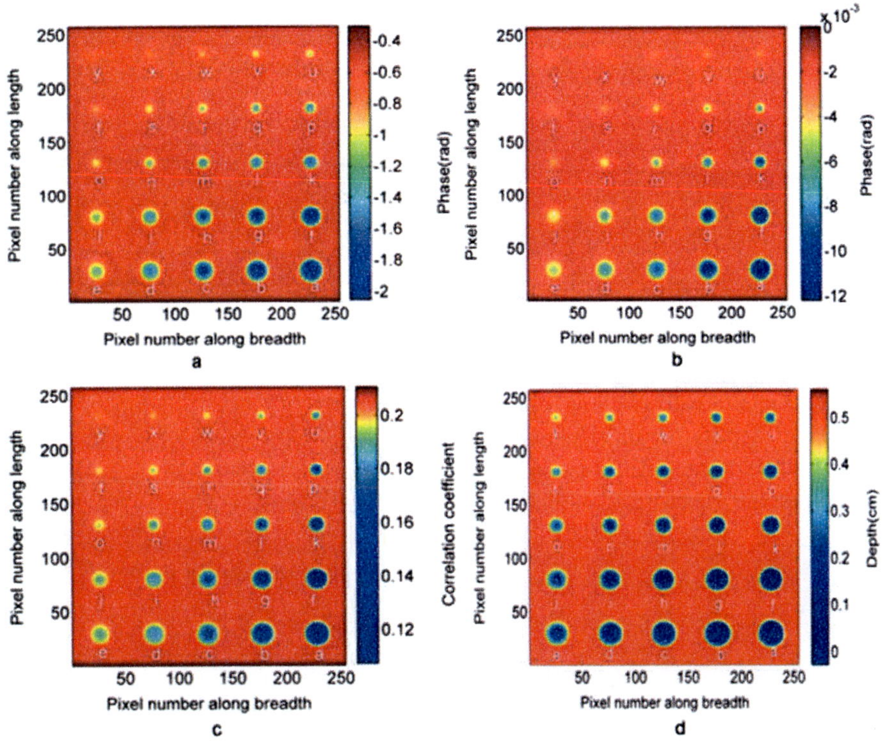

Fig. 5 **a** Hilbert phase, **b** FFT phase, **c** pulse compression, **d** artificial neural network

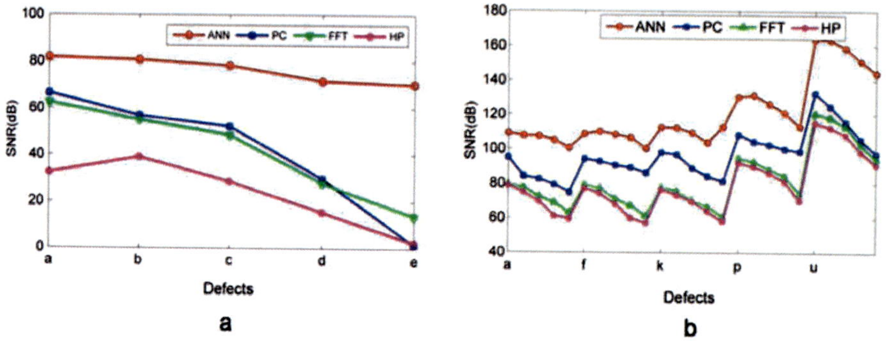

Fig. 6 **a** SNR of defects for mild steel sample, **b** SNR of defects for CFRP sample

In order to quantify the detectability of defects of materials, mean of defective and non-defective area as well as standard deviation of non-defective region are calculated, and further, signal-to-noise ratio of each pixel is calculated using following equation.

$$SNR(dB) = \frac{\text{mean of defective area} - \text{mean of non defective area}}{\text{standard deviation of non defective area}} \quad (6)$$

4.1 *Defect Sizing*

The detectability of defects also depends on the size and depth [22]. To estimate the size of defects, a full width at half maxima are used. The representation of sizing at different depths for mild steel and CFRP sample with flat bottom holes have been calculated using different processing methods are represented in Table 1 by considering the phase profile at the centers of defects for test sample.

4.2 *PSO Algorithm*

The parameters used in this PSO algorithm are number of generation cycles, inertia weight, and acceleration constants. The neural network was trained with PSO algorithm with acceleration constant of 2, inertia weight 0.9 for 1000 iterations. In the training process, the error rate should be decreased. The acceleration constant is used to pull the particle toward pbest and gbest positions. Figure 7 represents the validation performance plot of CFRP sample using particle swam optimization with 1000 iterations.

Table 1 Full width at half maxima of defects from center line passing through defects for mild steel and CFRP samples

Mild steel sample						CFRP sample					
Defect	Actual size (cm)	Estimated size (cm)				Defect	Actual size (cm)	Estimated size (cm)			
		ANN	PC	FFT	HP			ANN	PC	FFT	HP
a	1.6	1.65	1.96	1.68	1.82	a	1.6	1.61	1.45	1.56	1.5
b	1.6	1.64	1.84	1.57	1.68	f	1.4	1.46	1.32	1.36	1.24
c	1.6	1.66	1.75	1.82	1.40	k	1	1.00	0.88	1.01	0.75
d	1.6	1.46	1.16	1.42	1.72	p	0.8	0.84	0.69	0.85	0.61
e	1.6	1.45	0.42	0.63	0.81	u	0.6	0.65	0.53	0.64	0.65

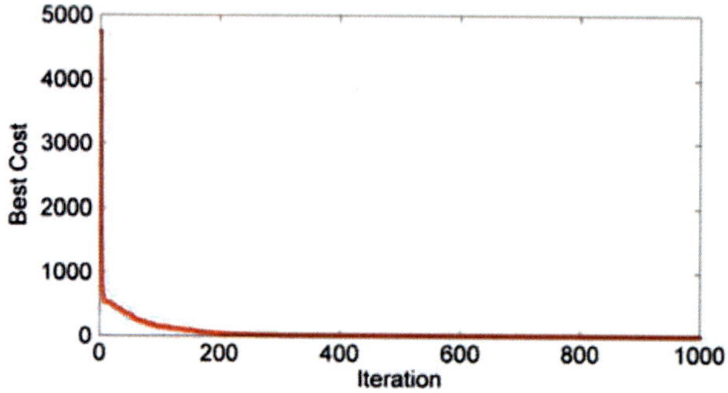

Fig. 7 Performance plot of ANN with PSO for CFRP sample

By using performance plot, we can conclude that the performance of neural network can be improved by using PSO algorithm. The results gives that the ANN trained with PSO gives better prediction and also reduce computational times. The time taken for training the neural network with back propagation algorithm as 1H 09 min and using PSO, the training time reduced to 42 s for CFRP sample. So, time consumption using PSO algorithm was very less compared to back propagation algorithm. The performance of particle swam optimization compared with back propagation neural network gives improved prediction as well as takes less time to train.

4.3 Depth Estimation

Figure 8 represents the quantitative depth analysis of simulated CFRP test sample having twenty five plot bottom holes. The experimentally obtained depth for CFRP sample is closed to the theoretical values obtained from the phase of defect location. The normalized ANN phase values and theoretical phase values are closer to each other, accurately with small difference at deeper depths because of diffusion effect.

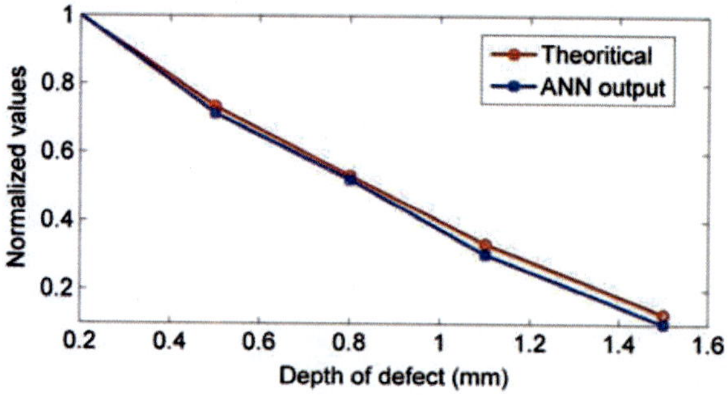

Fig. 8 Depth verses phase contrast profile of defects of simulated CFRP sample

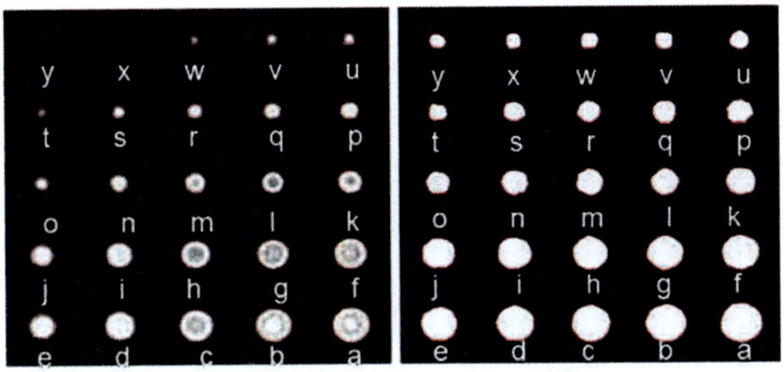

Fig. 9 **a** Level set, **b** active contour-based segmented images of CFRP sample with 500 iterations for ANN output

4.4 Segmentation Outputs

Region-based active contour image segmentation is performed based on the statistical modeling. The foreground and background-based mean intensity levels are compared for setting initial contour to zero level. After setting zero level expand the contour, and mean intensities of background and foreground have maximum difference. It makes the curve on global constraints along the boundary of foreground. In this paper, the level set and active contour image segmentation applied to results obtained from the artificial neural network-based image with 500 iterations and corresponding segmented results for CFRP sample are illustrated in Fig. 9. From this figure, segmentation using active contour method provides best detectability compared to level set method.

5 Conclusion

An artificial neural network-based analysis is employed on quadratic frequency modulated thermal wave imaging to detect and characterize subsurface defects on experimental mild steel and simulated CFRP samples with flat bottom holes with provided an enhanced defect detection using artificial neural network-based analysis as compared to pulse compression, Hilbert phase, and FFT phase analysis. The ability to defect identification using QFMTWI is verified experimentally, and it is observed that the proposed method provided best contrast. Defect detection is quantified with SNR, and sizing estimation can be cheeked for proposed method with full width at half maximum process. The active contour-based segmentation approach provided improved detection capability of artificial neural network-based output. Hence, an artificial neural network-based analysis has been experimentally produced to enhance depth resolution in non-stationary thermal wave imaging.

Acknowledgements This work was supported by Naval Research Board, India under grant no: NRB-423/MAT/18-19.

References

1. Maldague XPV (2001) Theory and practice of infrared technology for nondestructive testing. Wiley, New York, NY, USA. ISBN: 978-0-471-18190-3
2. Arora V, Siddiqui JA, Mulaveesala R, Muniyappa A (2015) Pulse compression approach to nonstationary infrared thermal wave imaging for nondestructive testing of carbon fiber reinforced polymers. IEEE Sens J 15(2). https://doi.org/10.1109/JSEN.2014.2361391
3. Subhani SK, Suresh B, Ghali VS (2016) Empirical mode decomposition approach for defect detection in non-stationary thermal wave imaging. NDT&E Int 81:9–45. https://doi.org/10.1016/j.ndteint.2016.03.004
4. Mulaveesala R, Ghali VS, Arora V (2013) Applications of non-stationary thermal wave imaging methods for characterisation of fibre-reinforced plastic materials. Electron Lett 49(2):1–2. https://doi.org/10.1049/el.2012.3844
5. Tang Q, Bu C, Liu Y, Qi L, Yu Z (2015) A new signal processing algorithm of pulsed infrared thermography. Infrared Phys Technol 68:173–178. https://doi.org/10.1016/j.infrared.2014.12.002
6. Guo X, Vavilov V (2015) Pulsed thermographic evaluation of dis-bonds in the insulation of solid rocket motors made of elastomers. Polym Test 45:31–40. https://doi.org/10.1016/j.polymertesting.2015.04.015
7. Liu J, Liu Y, Wang F, Wang Y (2015) Study on probability of detection (POD) determination using lock-in thermography for nondestructive inspection of CFRP composite materials. Infrared Phys Technol 71:448–456. https://doi.org/10.1016/j.infrared.2015.06.007
8. Song H, Lim HJ, Lee S, Sohn H, Yun W, Song E (2015) Automated detection and quantification of hidden voids in triplex bonding layers using active lock-in thermography. NDT& E Int 74:94–105. https://doi.org/10.1016/j.ndteint.2015.05.004
9. Ibarra-Castanedo C, Avdelidis NP, Maldague X (2005) Qualitative and quantitative assessment of steel plates using pulsed phase thermography. Mater Eval 63(11):1128–1133
10. Maldague X, Marinetti S (2005) X Pulsed phase thermography. J Appl Phys 79:2694–2698. https://doi.org/10.1063/1.362662
11. Mulaveesala R, Tuli S (2006) Theory of frequency modulated thermal wave imaging for non-destructive sub-surface defect detection. Appl Phys Lett 89:191913. https://doi.org/10.1063/1.2382738
12. Dua G, Mulaveesala R (2013) Applications of Barker coded infrared imaging method for characterization of glass fibre reinforced plastic materials. Electron Lett 49(17):1071–1073. https://doi.org/10.1049/el.2013.1661
13. Ghali VS, Mulaveesala R (2010) Frequency modulated thermal wave imaging techniques for non destructive testing. Insight 52(9). https://doi.org/10.1784/insi.2010.52.9.475
14. Ghali VS, Mulaveesala R (2012) Quadratic frequency modulated thermal Wave imaging for non-destructive testing. Prog Electromagn Res M 26:11–22. https://doi.org/10.2528/PIERM12062101
15. Vijaya Lakshmi A, Gopitilak V, Parvez MM, Subhani SK, Ghali VS (2019) Artificial neural networks based quantitative evaluation of subsurface anomalies in quadratic frequency modulated thermal wave imaging. Infrared Phys Techno 97:108–115. https://doi.org/10.1016/j.infrared.2018.12.013
16. Clerc M, Kennedy J (2002) The particle swarm-explosion stability and convergence in a multi dimensional complex space. IEEE Trans Evol Comput 6(1):58–73

17. Kennedy J (2011) Particle swarm optimization. In: Encyclopedia of machine learning. Springer, Berlin, pp 760–766. https://doi.org/10.1007/978-0-387-30164-8_630
18. Eberhart R, Kennedy J (1995) A new optimizer using particle swarm theory. In: Micro machine and human science, MHS'95, proceedings of the sixth international symposium on IEEE, pp 39–43. https://doi.org/10.1109/MHS.1995.494215
19. Gong J, Liu J, Qin L, Wang Y (2014) Investigation of carbon fibre-reinforced polymer (CFRP) sheet with subsurface defect inspection using thermal-wave radar imaging (TWRI) based on the multi-transform technique. NDT&E Int 62:130–136. https://doi.org/10.1016/j.ndteint.2013.12.006
20. Zhu W, Ha Kang S, Biros G (2013) A geodesic-active-contour-based variational model for short-axis Cardiac MR image segmentation. Int J Comput Math 90(1):124–139. https://doi.org/10.1080/00207160.2012.695355
21. Li C, Huang R, Ding Z, Gatenby JC, Metaxas DN, Gore JC (2011) A level set method for image segmentation in the presence of intensity inhomogeneities with application to MRI. IEEE Trans Image Process 20(7). https://doi.org/10.1109/TIP.2011.2146190
22. Suresh B, Subhani SK, Vijaya Lakshmi A, Vardhan VH, Ghali VS (2017) Chirp Z transform based enhanced frequency resolution for depth resolvable non stationary thermal wave imaging. Rev Sci Instrum 88:014901–014906. https://doi.org/10.1063/1.4973192

Inspection of Friction Stir Welded Joint Using EMAT Generated Fundamental Shear Horizontal Guided Wave Mode (SH0)

Nived Suresh, Sreedhar Puliyakote, and Krishnan Balasubramanian

1 Introduction

Friction stir welding (FSW) is a solid-state joining process. It uses a rotating pin tool to generate frictional heating on materials. The heat plasticises the material. It consolidates even before the temperature reaches the melting point of the parent material [1]. FSW joints have many advantages such as low distortion, good mechanical properties, and smaller heat-affected zone (HAZ). Researchers have investigated the formation of defects in FSW joints which deteriorates its mechanical properties. Improper process parameters, excess or insufficient heat input are some of the reasons for the defect formation [2–4]. FSW is currently used in aerospace and shipbuilding industries and is found in a wide range of material including aluminium, steel and titanium alloys.

Various approaches were implemented to evaluate the defects in FSW by different non-destructive evaluation (NDE) communities. Attempts were made using liquid penetrant testing [1], eddy current testing [1, 5–7] radiography testing [1, 4, 8] and ultrasonic testing [7, 9] to evaluate the quality of the joint.

This work describes a reliable, rapid inspection technique to evaluate onsite FSW joints. Electro-magnetic acoustic transducers (EMAT) generating shear horizontal (SH) modes are used for this inspection. Its non-contact, non-couplant nature helps to complete the inspection quickly and hence attractive to industrial applications. Here, the FSW of aluminium alloy sheets will be investigated further using EMAT-based shear horizontal (SH) guided ultrasonic wave mode.

N. Suresh (✉) · S. Puliyakote · K. Balasubramanian
Centre for Non-Destructive Evaluation, Indian Institute of Technology Madras, Chennai, India
e-mail: nivedkvr@gmail.com

C. K. Mukhopadhyay and R. Mulaveesala (eds.), *Advances in Non-destructive Evaluation*, Lecture Notes in Mechanical Engineering,
https://doi.org/10.1007/978-981-16-0186-6_5

2 SH0 Mode

Guided waves are widely used in modern non-destructive evaluation techniques [10]. SH guided waves are those guided waves in which the particle vibrations of these modes are in a horizontal plane parallel to the surface of the plate. This simple physical nature of SH modes makes the generation and mode identification process less complicated compared to other lamb wave modes [11].

Figure 1 shows the phase velocity dispersion curve for aluminium at shear velocity $C_T = 3080$ m/s. Other than SH0, all higher order modes are dispersive. Because of the non-dispersive behaviour, SH0 is preferred over other modes in many of the NDE inspections [12–14].

One of the key features of the displacement profiles of SH family of wave modes is that it does not change with frequency [15]. However, except for the fundamental SH0 mode, higher order SH modes are dispersive. For SH0, the displacement profile over the thickness is uniform as shown in Fig. 2. As a result of this, the sensitivity of SH0 to a defect at any position across the thickness remains the same. For other higher order SH modes, the sensitivity to defects changes with defect position, across the thickness of the sample. These peculiar specialities are the reasons why SH0 is preferred over other modes in this inspection.

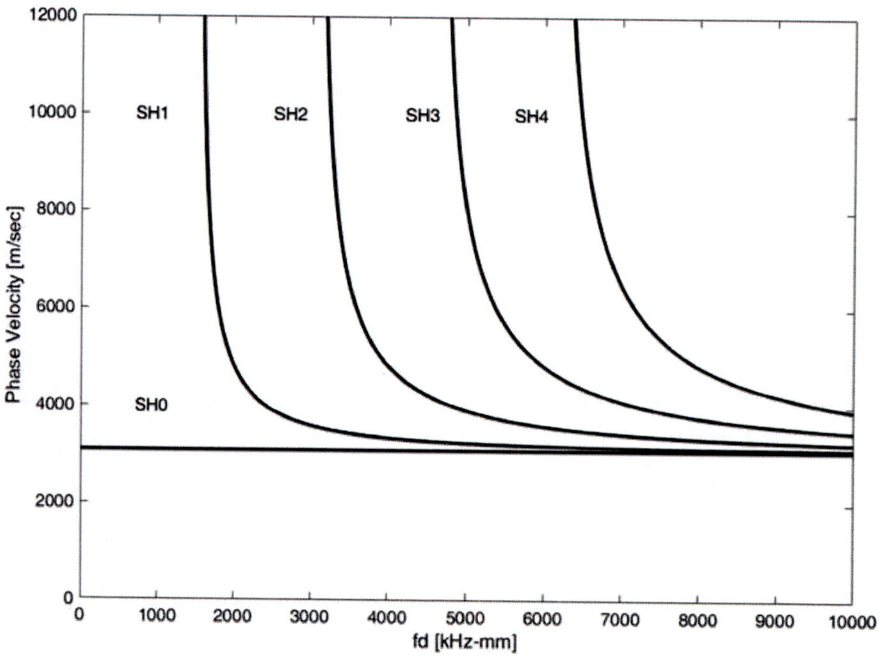

Fig. 1 SH phase velocity dispersion curve for CT = 3080 m/s

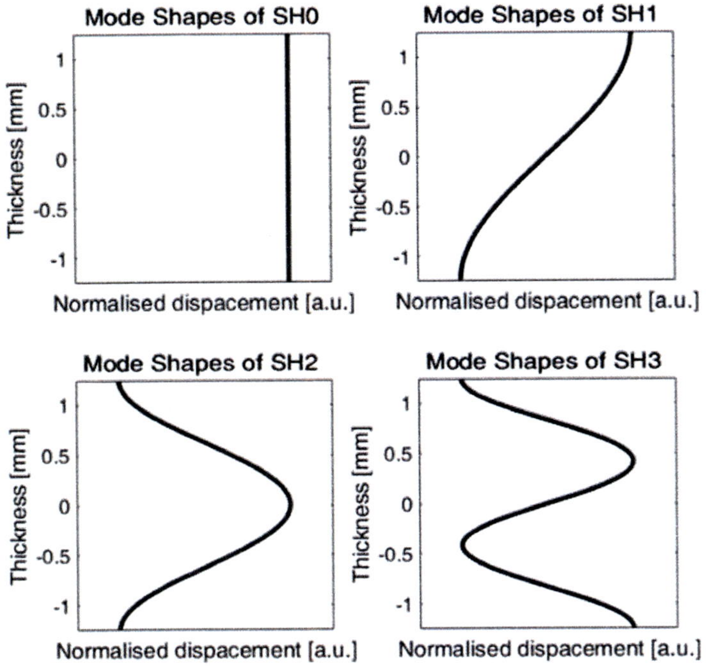

Fig. 2 Mode shapes of different SH modes. Particle displacements are in a horizontal plane parallel to the surface of the plate

3 EMAT Generation of SH0

Electromagnetic acoustic transducer (EMAT) working based on the principle of Lorentz force is used to generate SH0 in this inspection. Permanent magnets provide a magnetic field in the plate. A race track current carrying coil generates time-varying eddy current in the plate. This results in a time-varying Lorentz force inside the plate [16, 17]. Many researchers successfully modelled EMAT and studied the physics behind that [18–20].

In most of the guided wave inspection techniques, the interaction of a single mode with defects is preferred. Analysing more than one mode makes the process complicated. In the generation of the guided wave itself, additional care is given to generate a single mode alone. Comb transduction is one of the methods that is used in this inspection process to excite SH0 mode selectively [21]. In comb transduction, wavelength of generation of guided waves is constrained. The dashed line in Fig. 3 shows a constant wavelength line of 6.35 mm. As a result, whatever the excitation frequency is, guided wave generation takes place along the constrained wavelength line at that excitation frequency. This method enables to excite any preferred guided wave modes selectively [21].

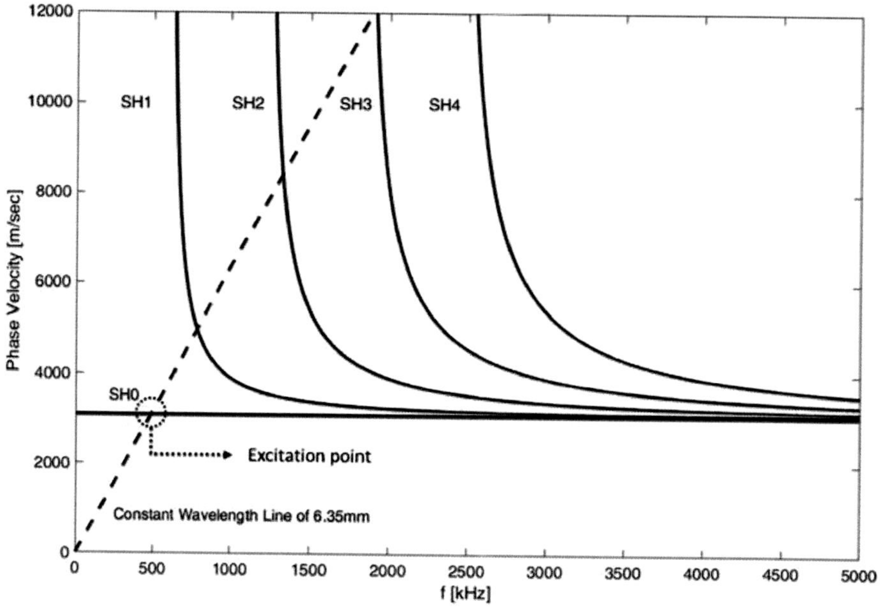

Fig. 3 Constant wavelength line of 6.35 mm in phase velocity dispersion curve of a 2.5 mm aluminium plate for $C_T = 3080$ m/s

In a Lorentz force-based EMAT, wavelength constraint is introduced by the alternate periodic spacing of the polarity of magnets. Here, the spatial periodicity represents the wavelength of the SH modes generated in the plate. EMAT with this kind of magnetic arrangement is known as periodic permanent magnet EMAT (PPM—EMAT). In this experiment, PPM—EMAT with a wavelength of 6.35 mm was chosen here for our experiments. In Fig. 3, the intersection point of SH0 line and constant wavelength line has a frequency of 485 kHz. This was chosen as the excitation frequency of the SH0 generation. Hence by this method, the generation of SH0 wave mode alone is expected. If other higher order modes are also present, they will have relatively low energy compared to the SH0 mode.

4 Experimental Setup

One FSW sample is inspected for this study. The sample is an aluminium plate of 2.5 mm thickness. The width of the weld is 10 mm. A portion of this FSW joint is inspected for defect detection.

FSW joint is evaluated by using the pitch-catch method. Transmitter and receiver are placed 60 mm away from the centre of the weld. Transducers are oriented in such a way that 45° specular reflection is collected by the receiver as shown in Fig. 4. Previous studies suggest that 45° angle is sufficient to provide significantly reflected

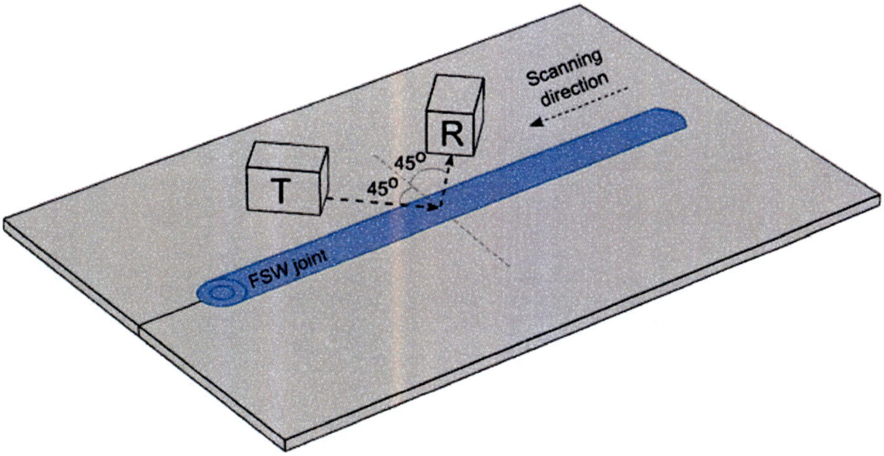

Fig. 4 Schematic diagram of the experimental setup

wave amplitudes [22]. Transducers are moved in a direction parallel to the weld. The relative position and orientation of the transducers kept the same with the help of a holder. Used a motion controller to move the holder and an encoder to record the position. The transducers are moved with a speed of 10 mm/s.

Innerspec POWERBOX H pulser-receiver was used for the experiment. PPM - EMAT of wavelength 6.35 mm was used as transmitter and receiver. Transmitter EMAT is excited with five cycles at 485 kHz pulse. Signal recorded at the receiver at each position is analysed further (Fig. 5).

5 Results and Discussion

The experiment is performed based on reflection of SH0 mode. More the intensity of SH0 reflection, more impedance miss match at the welded joint. Higher impedance miss match can occur as a result of voids, inclusions or insufficient fusion at the welded zone. Hence, the high intensity of SH0 indicates the presence of defects in the joint. Low intensity indicates healthy joints.

Post-processing is done on the signals obtained from the experiment to have a better view of defect distribution. Figure 6a shows the reflected SH0 from the welded joint. From the A-scan, SH0 can be well distinguished with significant amplitude. Taking a Hilbert transform of A-scans and arranging them will result in a B-scan as shown in Fig. 7a.

Throughout the experiment, the distance of wave travel from the transmitter to the receiver remains constant. As a result, all A-scans show SH0 within a fixed time of flight range. Here, the range is from 25 to 55 μs. Only this range is considered to analyse the frequency spectrum of SH0 (Fig. 6b). The frequency spectrum shows

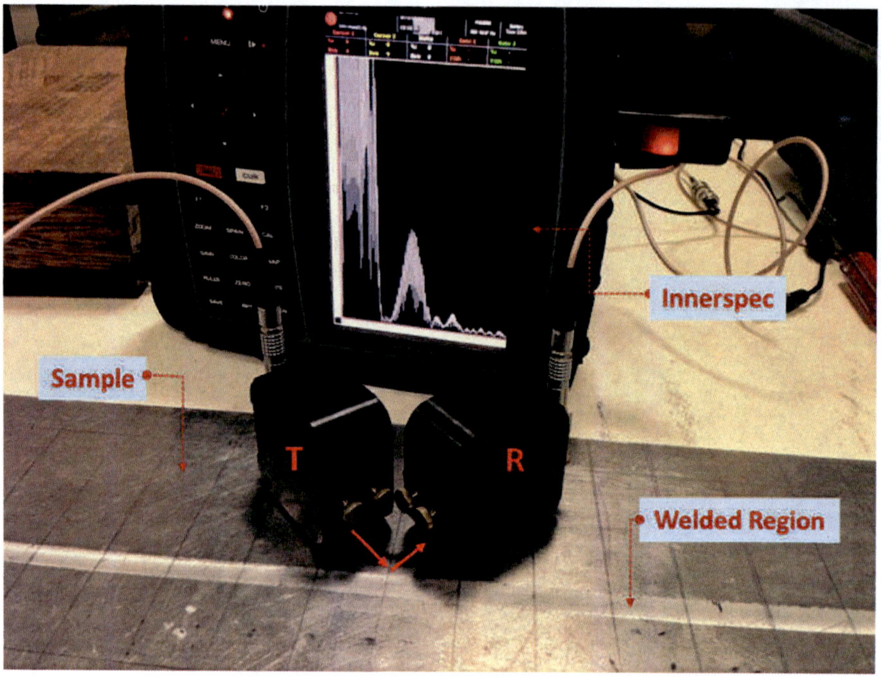

Fig. 5 Experimental setup using the PPM EMATs in pitch-catch mode on the FSW sample

a central frequency of 485 kHz which is exactly same as the excitation frequency (Fig. 6c). Post-processing of A-scans and finding the frequency spectrum results in a frequency domain B-scan shown in Fig. 7b. The positions of defects and perfectly welded regions can be distinguished clearly from the frequency domain B-scan image.

From Fig. 7b, frequency domain amplitudes at 485 kHz were selected and plotted against the position. To check the reliability of results obtained by this method, they are compared with X-ray images of the weld in Fig. 8b. It can be seen that Fig. 8a, b are in agreement with respect to defect position if the defects are gross defects and are continuously spread over a distance. If the defect is small and isolated, it is not well resolved using this wavelength. Based on the application of the FSW joint, non-defective amplitude threshold can be chosen from Fig. 8a.

6 Conclusions

Experiments performed on this study demonstrated the use of EMAT-based SH0 guided ultrasonic Lamb modes, in the reflection pitch-catch mode, is sensitive to, and capable of inspecting the tunnelling and porosity type defects in aluminium FSW joints. Since EMATs are non-contact, non-couplant transducers, this technique

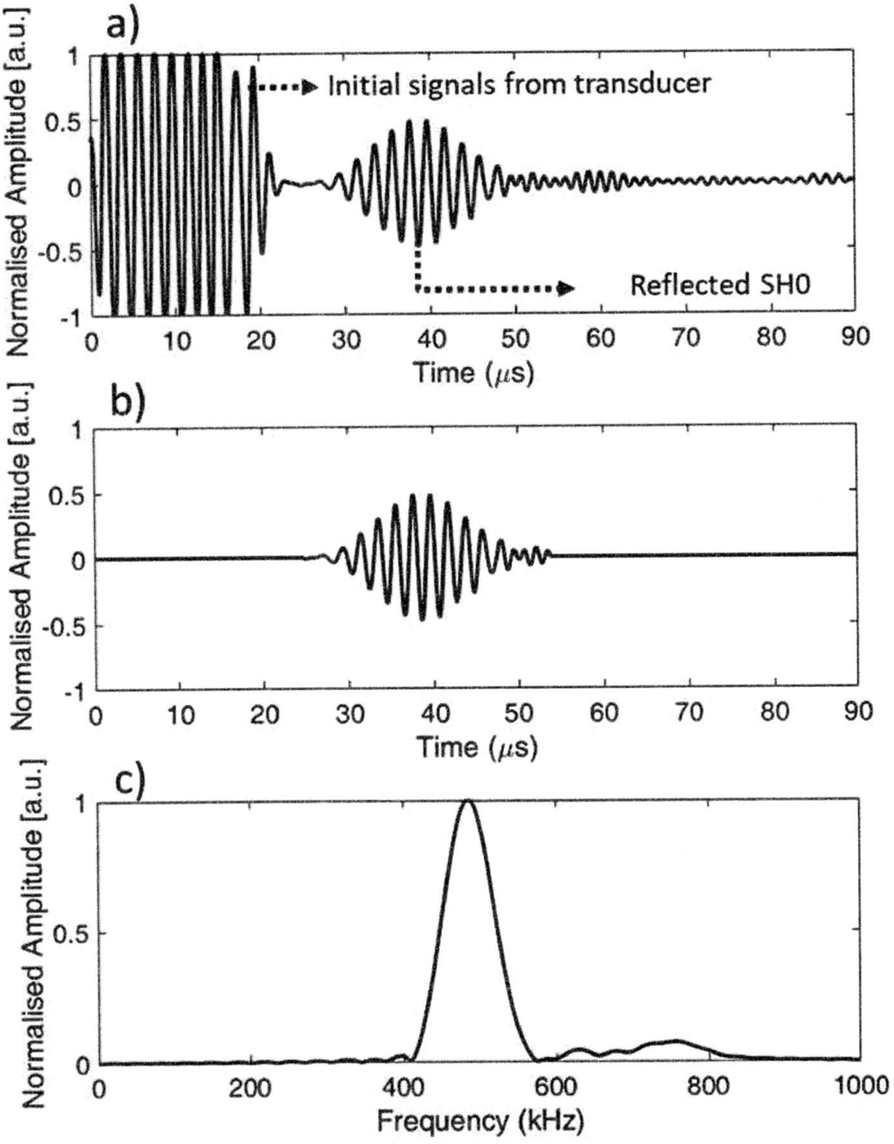

Fig. 6 **a** A-scan showing reflected SH0, **b** filtered A-scan showing reflected SH0 alone, **c** the frequency spectrum of the SH0 A-scan

facilitates rapid inspection for practical applications in the industries. The B-scan EMAT-based amplitude plots were found to be in agreement with respect to the digital X-ray imaging results. It is found that this approach is successful if the defect is continuously spread over a distance. However, since the wavelength of the EMAT-based SH0 mode used is of the order of 6.35 mm, it is likely that only gross defects

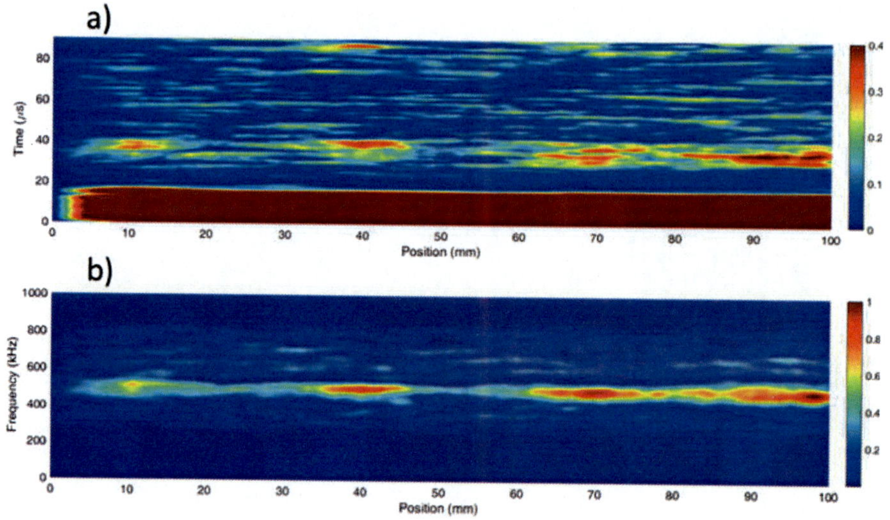

Fig. 7 **a** B-scan of Hilbert transformed time-domain signals, **b** B-scan of filtered frequency domain signals of SH0 alone

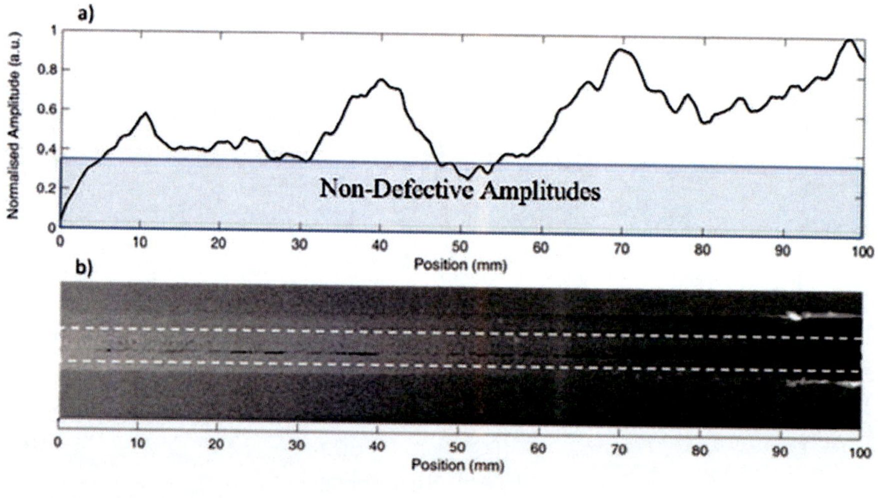

Fig. 8 **a** Amplitude of 485 kHz frequency at different scanning positions, **b** An X-ray image of the FSW portion inspected

such as clusters of porosity or tunnelling types are found using this approach. It is unlikely that defects smaller than 3 mm that are isolated in nature will not be resolved using this wavelength. Smaller wavelength EMAT probes may be required to further improve the resolution and sensitivity of this technique.

References

1. Kinchen DG, Martin L, Space M, Orleans N, Aldahir E (2002) NDE of friction stir welds in aerospace applications. Insp Trends 1–7
2. Leonard AJ, Lockyer SA (2003) Flaws in friction stir welds. In: 4th international symposium on friction stir welding, pp 14–16
3. Cao X, Jahazi M (2009) Effect of welding speed on the quality of friction stir welded butt joints of a magnesium alloy. Mater Des 30(6):2033–2042
4. Kim YG, Fujii H, Tsumura T, Komazaki T, Nakata K (2006) Three defect types in friction stir welding of aluminum die casting alloy. Mater Sci Eng A 415(1–2):250–254
5. Rosado LS, Santos TG, Piedade M, Ramos PM, Vilaça P (2010) Advanced technique for non-destructive testing of friction stir welding of metals. Measurement 43(8):1021–1030
6. Mandache C, Levesque D, Dubourg L, Gougeon P (2012) Non-destructive detection of lack of penetration defects in friction stir welds. Sci Technol Weld Join 17(4):295–303
7. Lamarre A, Dupuis O, Moles M (2004) Complete inspection of friction stir welds in aluminum using ultrasonic and eddy current arrays. In: 16th WCNDT, Monteral, Canada
8. Akinlabi E, Levy A, Akinlabi S (2012) Non-destructive testing of dissimilar friction stir welds. In: Proceedings of world congress, vol III, pp 4–8
9. Liu F, Liu S, Guo E, Li L (2008) Ultrasonic evaluation of friction stir welding. In: 17th World conference on nondestructive, pp 25–28
10. Rose JL (1995) Recent advances in guided wave NDE. In: International symposium on 1995 IEEE ultrasonic symposium proceedings, vol 1, pp 761–770
11. Alleyne DN, Cawley P (1992) The interaction of lamb waves with defects. IEEE Trans Ultrason Ferroelectr Freq Control 39(3):381–397
12. Carandente R (2011) Interaction between the fundamental torsional guided wave mode and complex defects in pipes, p 145
13. Rajagopal P, Lowe MJS (2008) Angular influence on the scattering of fundamental shear horizontal guided waves by a through-thickness crack in an isotropic plate. J Acoust Soc Am 124(4):2021–2030
14. Rajagopal P, Lowe MJS (2008) Scattering of the fundamental shear horizontal guided wave by a part-thickness crack in an isotropic plate. J Acoust Soc Am 124(5):2895–2904
15. Vasile CF, Thompson RB (1979) Excitation of horizontally polarized shear elastic waves by electromagnetic transducers with periodic permanent magnets. J Appl Phys 50(4):2583–2588
16. Hirao M, Ogi H (2003) EMATs for science and industry. Springer, US, Boston, MA
17. Huang S, Wang S (2017) Electromagnetic ultrasonic guided waves
18. Ribichini R, Cegla F, Nagy PB, Cawley P (2010) Quantitative modeling of the transduction of electromagnetic acoustic transducers operating on ferromagnetic media. IEEE Trans Ultrason Ferroelectr Freq Control 57(12):2808–2817
19. Dhayalan R, Balasubramaniam K (2010) A hybrid finite element model for simulation of electromagnetic acoustic transducer (EMAT) based plate waves. NDT E Int 43(6):519–526
20. Dhayalan R, Balasubramaniam K (2011) A two-stage finite element model of a meander coil electromagnetic acoustic transducer transmitter. Nondestruct Test Eval 26(02):101–118
21. Rose JL (2014) Ultrasonic guided waves in solid media. In: Ultrasonic guide waves solid media, pp 1–512
22. Suresh N, Balasubramaniam K (2019) Reflection study of SH0 mode with plate edge at different incident angles. In: AIP conference on proceedings, vol 2102, pp 0–8

Proper Orthogonal Decomposition-Based Coating Thickness Estimation in Quadratic Frequency Modulated Thermal Wave Imaging

G. T. Vesala, G. V. P. Chandra Sekhar Yadav, V. S. Ghali, B. Suresh, and R. B. Naik

1 Introduction

Elevated temperature operational conditions in aerospace and industrial applications of metals or metal substrates require proper protection for more reliable usage. Thermal barrier coatings, a thin layer of ceramic coatings sprayed on metal substrates with a typical thickness range of 100 μm to 2 mm to protect from elevated operational conditions [1, 2]. Either at the manufacturing phase or operating phase, they may undergo various deformations which leads to various flaw generations like uneven thickness, debonding, delaminations, and cracks [3]. A non-destructive, online monitoring, and quality control is necessary to identify these flaws to recommend for repair or replacement in order to enhance the reliability of materials. Over various available NDT techniques, active thermography (AT) is gaining interest due to its non-invasive, non-contact, whole field investigation capabilities. A controlled optical stimulus is used to heat up the object in AT, and the corresponding thermal variations created on the object surface is observed by an infrared imager. Further, these thermal variations are post processed and analyzed for surface or subsurface flaw detection and material characterization [4].

Short duration high peak power optical stimulus-based pulse thermography [PT] [5, 6] and a moderate peak power periodic excitation-based lock-in thermography [LIT] [7–9] are widely used techniques to detect thickness variations, delaminations, dis-bonds, and subsurface defects in TBC. The cooling phase of sample after the excitation is analyzed in PT, whereas the thermal response of periodic excitation in

G. T. Vesala · G. V. P. Chandra Sekhar Yadav · V. S. Ghali (✉) · B. Suresh
Infrared Imaging Center, Koneru Lakshmaiah Educational Foundation, Vaddeswram, Guntur, India
e-mail: gvs0raos@kluniversity.in

R. B. Naik
Naval Materials Research Laboratory, Ambernath (E), Thane, Maharashtra 421506, India

C. K. Mukhopadhyay and R. Mulaveesala (eds.), *Advances in Non-destructive Evaluation*, Lecture Notes in Mechanical Engineering,
https://doi.org/10.1007/978-981-16-0186-6_6

LIT is post processed by Fourier transform technique to analyze phase variations. Due to its robustness toward non-uniformity of thermal radiation, phase-based analysis is mostly recommended. Similar to LIT, FFT-based phase analysis is applied to PT named pulse phase thermography (PPT) [10]. The differences or similarities between PPT and LIT to estimate TBC thickness variations through empirical observation is presented by Ranjit et al in his consecutive works from numerical simulations and experimental investigations [8, 9, 11].

Recent advancements in AT using variants of frequency modulated stimulation proved to overcome the limitations associated with PT and LIT [12–14] that are high peak power excitation and repetitive experimentation, respectively. Frequency modulated thermal wave imaging (FMTWI) uses a band of low frequencies with moderate peak power stimulus which is able to provide defect detection and depth resolution [12]. On the other hand, variants of FMTWI probe more energy into the test sample at low frequencies that favors the detection of deeper defects [13, 14]. The recorded thermal response processed by pulse compression (PC) where it widens the bandwidth of incident wave to provide finer depth resolution, since the conventional FFT-based processing approach is inadequate to provide sufficient depth resolution [15]. The correlation coefficient contrast is used for anomaly detection, which improves signal-to-noise ratio of defect [16]. Further, the quantitative depth estimation is carried out by comparing the defect depths with corresponding peak delays of correlation profiles [17, 18].

In present article, a numerical simulation is carried out to model a thermal barrier coatings sample with uniform thickness variations, excited with quadratic version of frequency modulated stimulus (QFM) [14]. Then, the pulse compression is employed over the resultant thermal response, and coating thickness resolution is analyzed. Along with that, coating thickness estimation is carried out using peak delays of correlation profiles taken from each coating layer. A polynomial fitting is used to relate the TBC thickness variation with peak time delays, and further, the compressed response is applied to proper orthogonal decomposition (POD) [19, 20]. POD reduces the dimensionality keeping the system dynamics intact and enhances the detection criteria of coating thickness variations.

2 Methodology

In QFMTWI, a moderate peak power optical stimulus modulated by a band of low frequencies provided from a set of halogen lamps to the front surface of the object [13]. The experimental setup of QFMTWI is shown in Fig. 1a, and the expression for quadratic frequency modulated input stimulus is given by

$$Q(t) = Q_0 \sin\left(at + bt^3\right) \tag{1}$$

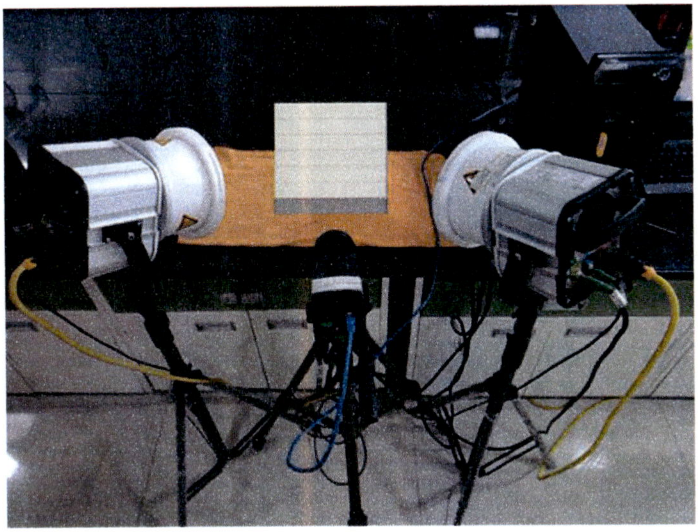

Fig. 1 Experimental setup for QFMTWI

where 'a' is the initial frequency, 'b' is bandwidth, and 'Q_0' is the intensity of the heat flux. The applied optical stimulus creates diffusive thermal waves into the subsurface layers of the sample. The one-dimensional heat diffusion equation is given by

$$\frac{\partial T}{\partial t} = \frac{1}{\propto}\frac{\partial^2 T}{\partial x^2} \tag{2}$$

where $\alpha = k/\rho c$ is the diffusion coefficient, and 'T' is temperature at depth 'x' corresponding to the time instant t. The corresponding thermal variations captured using an infrared camera. Further, Eq. 2 can be solved under boundary conditions with quadratic heat flux at front surface, the obtained thermal response in Laplacian domain is given by [7]

$$T(x, s) = \frac{T_0}{4k\sigma}\left(\frac{(1+R)(1+\mathrm{R}e^{-2\sigma l})}{1-\mathrm{R}^2 e^{-2\sigma l}}\right) \tag{3}$$

where $\sigma = (1 + j)/\mu$, 'T_0' is the intensity of thermal wave, and 'R' is the reflection coefficient between coatings and substrate given by $R = (1 - \mathcal{E}\text{ coating})/(1 - \mathcal{E}\text{ Substrate})$ since '$\mathcal{E}$' is the effusivity of the material and μ is the diffusion length of the thermal wave for QFM given by [13],

$$\mu = \sqrt{\frac{2\alpha}{\pi(a + 3bt^2)}} \tag{4}$$

where 'α' is thermal diffusivity computed from thermo-physical properties of the material. The thermal response in Eq. 3 is subjected to pulse compression by cross-correlating with a reference thermal profile given by

$$T_{\text{corr}} = X\text{corr}(T * T_{\text{ref}}) \tag{5}$$

Algorithm 1:

Initialize: Raw thermal response: $T(x, y, t)\ t = N/f$
Fitting and Mean removal: $T(x, y, N) = \text{linearfit}(T(x, y, N) - \text{mean}(T(x, yxN))$
Pulse compression: $T_{\text{corr}}(x, y, 2 * N - 1) = X\text{corr}(T(x, y, N) * T_{\text{ref}}(x, y, N))$
2-D Conversion: $X(x * y, N) \xleftarrow{\text{Reshape}} T_{\text{corr}}(x, y, N)$
Covariance computation: $C = X^{\text{T}}X$
Eigen Decomposition: $[V\ \ D] = \text{eig}(C)$
POD computation: for $i = 1 : m$
$\text{POD}(x * y, i) = {T(x * y, t) * V(t, i)}/{\sqrt{D_i}}$
end
3D reshape: $\text{POD}(x, y, m) \xleftarrow{\text{Reshape}} \text{POD}(x * y, m)$

This results in a correllogram cube by rearranging the compressed profiles in their respective spatial location. The correlation coefficient contrast is used to distinguish thickness variations from the correllogram, and the correlation peak delays are used to estimate the thickness of each layer. Though pulse compression improves depth resolution and minimizes noise, it required to analyze the total sequence for proper detection of thickness variations which is twice the length of original thermal response. An enhancement in detecting subsurface anomalies by projecting the pulse compressed sequence through a statistical approach is observed in [21]. Hence, the pulse compressed response is subjected to proper orthogonal decomposition (POD) algorithm [19, 20]. POD extracts the dominant modes in the pulse compressed thermograms sequence and improves the resolution of wedges. At first, the three dimensional pulse compressed thermal response $T(x, y, t)$ is reshaped to form 2D vector $T(x * y, t)$ with pixels along rows and their temporal variations along columns. Now, empirical correlation or covariance is calculated followed by Eigen decomposition. This produces Eigen vectors (V or V') or orthogonal basis vectors with corresponding energy distribution in a diagonal matrix as Eigen values (D). Then, projecting these orthogonal basis vectors onto data driven model gives the proper orthogonal modes; the process is given in algorithm 1. Since most of the energy resides in first few Eigen vectors, hence first 'm' Eigen vectors selected to project into data driven model to obtain POD modes [20]. Now, the obtained modes are in 2D form and again reshaped to get 3D form to visualize the observations as thermograms.

3 Material and Methods

A thermal barrier coatings sample is numerically simulated with a nickel-based super alloy substrate of dimensions 180 × 210 × 4 mm, an intermetallic bond coat of 0.1 mm placed between substrate and topcoat. The top coat is yttria stabilized zirconia (YSZ) with a varying thickness from 0.15 to 0.90 mm as shown in Fig. 2a, where typical TBC thickness ranging from 100 μm to 2 mm [2]. The intermetallic bond coat Ni–Cr-Al-Y having both metal and ceramic Yttria stabilized Zirconia (YSZ) thermal barrier coatings properties, hence provides maximum adhesion. Thermo-physical properties of the coatings sample are presented in Table 1 [8].

To calculate the temperature measurements with sufficient spatial resolution, fine tetrahedral mesh is adopted. The coatings side of the sample is excited by an optical heat flux with a peak of 2 kW for 100 s, which is modulated by a quadratic chirp with a frequency sweep of 0.01–0.1 Hz. The excitation bandwidth selected based on the thermo-physical properties of coatings sample and their diffusion length such that the time varying thermal diffusion lengths of excitation ranges from 4.4 to 0.086 mm from low to high frequencies, respectively. Also, from the literature, it is found to be a prominent excitation bandwidth to probe deep into the sample as well as preserve fine details of anomalies at high and low frequencies, respectively [18]. The resultant thermal response from sample surface is recorded at a rate of 25 frames per second.

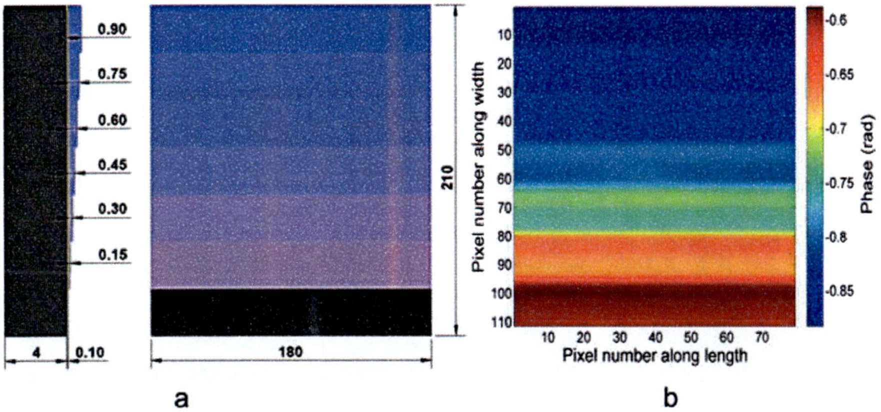

Fig. 2 **a** Layout of thermal barrier coatings sample side view on left and front view on right, **b** its phase response at 0.05 Hz

Table 1 Thermo-physical properties of TBC sample

Layers	Density (kg/m^3)	Conductivity (W/m K)	Heat capacity (J/kg K)
Substrate	8400	9.92	410
Bond coat	7380	3.3	755
Top coat	5600	0.9	505

4 Results and Discussion

The captured thermal response is linear fitted and mean removed to extract the dynamic thermal response. As like in PPT and LIT in [8, 9, 11], Fourier transform-based phase analysis is applied to verify the thickness variations of thermal barrier coatings sample. The corresponding phase variations observed at 0.05 Hz is given in Fig. 2b.

The limited frequency resolution of Fourier transform corresponding to non-stationary signals [18] limits the applicability of FFT-based phase analysis for present scenario. Hence, to study the coating thickness variation, pulse compression is employed over thermal response of each pixel in view. In pulse compression, the mean removed dynamic thermal response is cross-correlated with a pre-selected reference thermal profile from taken from the substrate region.

The corresponding normalized correlation coefficient is rearranged in respective pixel location to form correllogram cube. The depth resolution characteristics of pulse compression are analyzed for the coating thickness at different time instants. The coating thickness resolution is observed by analyzing correlation coefficient contrast is given in Fig. 3a–f for detecting each coating variation individually. But, the quantitative analysis is achieved by comparing the correlation peak delays from the auto-correlation peak of reference [17, 22]. Figure 4a shows the correlation peak delays

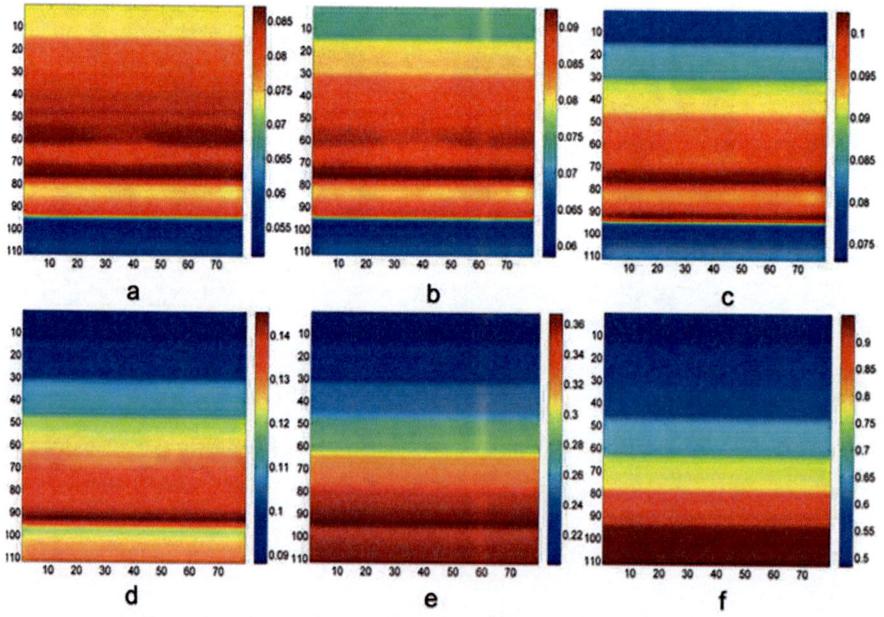

Fig. 3 Pulse compressed thermograms of different thicknesses detected, i.e., **a** 0.9 mm, **b** 0.75 mm, **c** 0.60 mm, **d** 0.45 mm, **e** 0.30 mm, and **f** 0.15 mm thickness at 10.57 s, 10.39 s, 10.07 s, 9.18 s, 6.28 s, and 0.93 s, respectively. Color bar presents the normalized correlation coefficient values

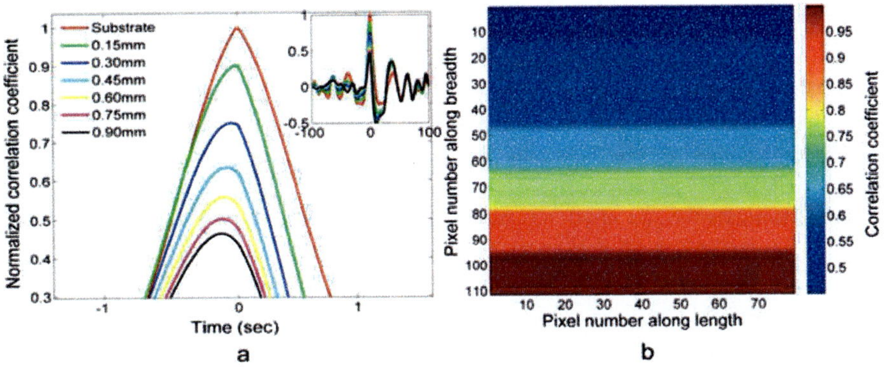

Fig. 4 **a** Pulse compression profiles of substrate and coating layers, **b** corresponding correllogram at 0 s

of TBC layers to that of reference profile peak from substrate region; Fig. 4b represents the correllogram observed at center of the compressed sequence where each coating thickness is discriminated efficiently by corresponding normalized correlation coefficient. Further, the polynomial relation is drawn in Fig. 5 to empirically fit the coating thickness variations with corresponding correlation peak delays [22].

$$L = a_1 * t^3 + a_2 * t^2 + a_3 * t + a_4 \tag{6}$$

where 'L' is the thickness of the top coat thickness, 't' is the peak time delay of detection, and a_1, a_2, a_3, and a_4 are quadratic coefficients given by 1.1484, −0.9646, 0.6750, and 0.1250, respectively. The detectability or discrimination of each thickness is enhanced by employing proper orthogonal decomposition (POD) to compressed sequence [21], which also reduces dimensionality of the system.

Fig. 5 Correlation peak delays Vs layer thickness of TBC coating

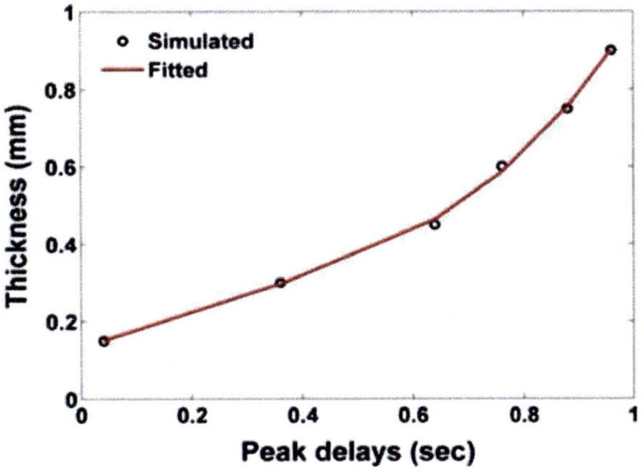

The flow chart for PC-POD is given in Fig. 6. The three-dimensional pulse compressed thermal response is reshaped into two-dimensional data set as time variations along columns and pixel numbers along rows. This 2D data set is subjected to empirical correlation or covariance calculation followed by Eigen decomposition. For the covariance vector, Eigen decomposition generates orthogonal basis vectors or dominant modes known as Eigen vectors and their respective energy variations in a diagonal matrix named as Eigen values.

Since, most of the energy resides in first few Eigen vectors correspond to maximum variance of the data. First 6 Eigen vectors selected from the Eigen decomposition are given in Fig. 7a. Projecting the selected dominant modes into data driven model results in POD modes. The 2D POD vector is again reshaped to 3D by rearranging the

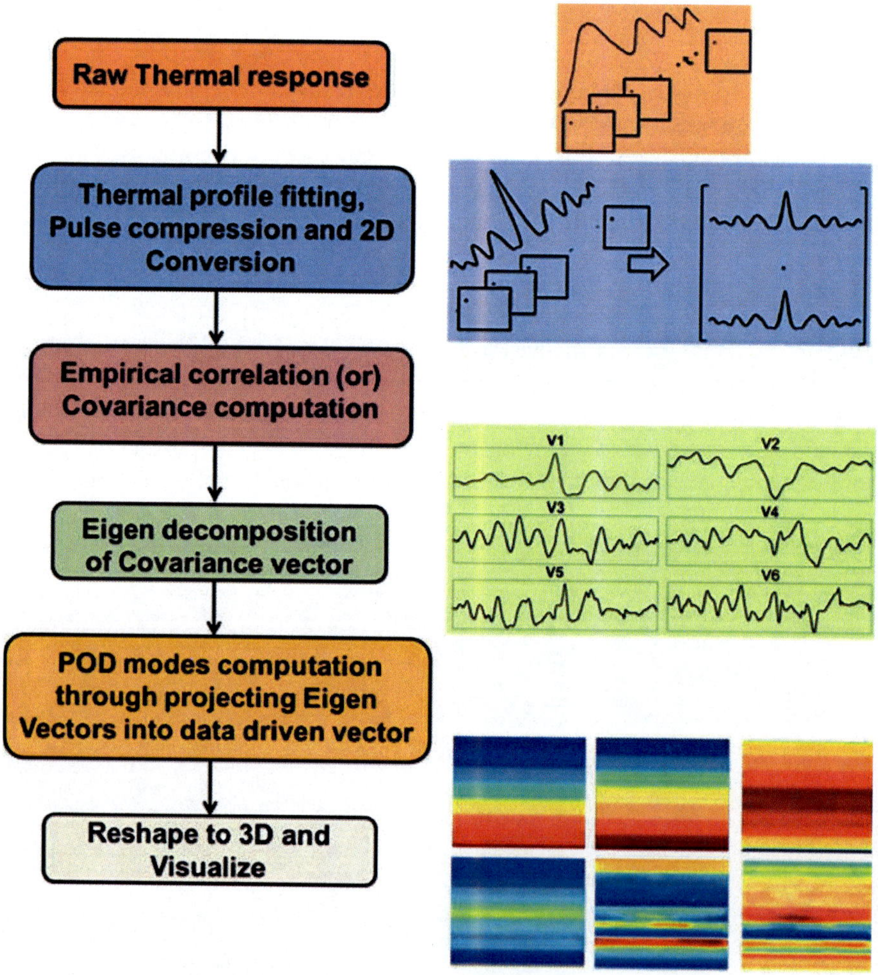

Fig. 6 Flowchart of QFM-PC-POD

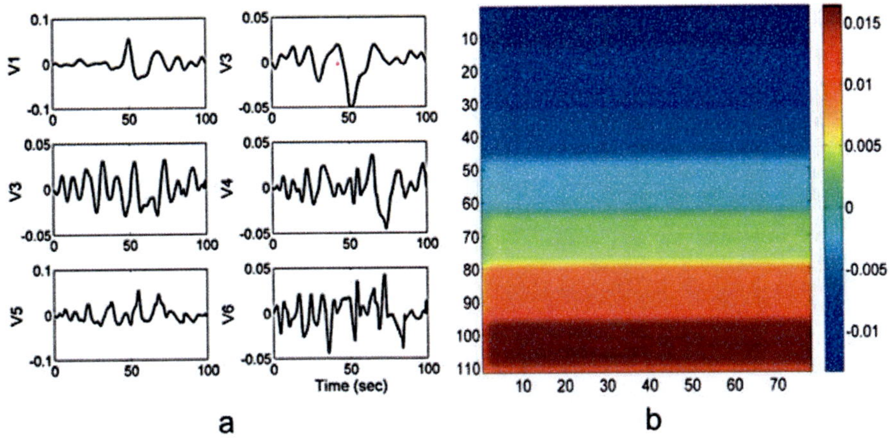

Fig. 7 **a** First 6 dominant modes of pulse compressed thermal sequence and **b** 2nd POD mode

pixels according to their corresponding spatial locations and analyzed. It is observed that 2nd mode of reconstructed POD sequence provides enhanced detection of thickness variations as presented in Fig. 7b compared to that of previous detection of pulse compression and FFT phase from Figs. 4b and 2b, respectively. It is also observed that the dimensionality of the system also reduced from thousands of frames to units count with enhanced detectability.

5 Conclusion

The present article focused on the study and estimation of thickness variation of a thermal barrier coatings sample. A numerical simulation is carried out by applying QFM heat flux over the coating side of TBC sample, and the recorded thermal response is analyzed with the pulse compression technique. An empirical correlation is made between thickness variations to peak time delays of correlation profiles. From the relation, it is concluded that the polynomial expression estimates coating thickness with less deviation. It is also concluded that the application of POD over pulse compression response reduces dimensionality and enhances the detection of coating thickness variations.

Acknowledgements This work is supported by Naval Research Board, India under the grant no. NRB-423/MAT/18-19.

References

1. Zhu D, Miller RA (2000) Thermal-barrier coatings for advanced gas-turbine engines. MRS Bull. https://doi.org/10.1557/mrs2000.123
2. Keiteb AS, Saion E, Zakaria A, Soltani N (2016) Structural and optical properties of zirconia nanoparticles by thermal treatment synthesis. J Nano-Mater. https://doi.org/10.1155/2016/1913609
3. Thoules MD (1991) Cracking and delamination of coatings. J Vac Sci Technol A 9(4). https://doi.org/10.1116/1.577265
4. Maldague XPV (2001) Theory and practice of infrared technology for non-destructive testing. Wiley, New York
5. Švantner M, Muzika L, Houdková S (2019) Quantitative inspection of coatings thickness by time-power transformation flash pulse thermographic method. In: Proceedings, pp 27–32. https://doi.org/10.3390/proceedings2019027032
6. Polo GB, Sergio Marinetti GG, Vladimir PV, Federico C, Daniele R (2003) Inspecting thermal barrier coatings by IR thermography. In: Proceedings of SPIE 5073, Thermosence, vol XXV. https://doi.org/10.1117/12.486019
7. Zhang J-Y, Meng X-B, Ma Y-C (2016) A new measurement method of coatings thickness based on lock-in thermography. Infrared Phys Technol 76:655–660. https://doi.org/10.1016/j.infrared.2016.04.028
8. Shrestha R, Kim W (2017) Evaluation of coating thickness by thermal wave imaging: a comparative study of pulsed and lock-in infrared thermography—part I: simulation. Infrared Phys Technol 83:124–131. https://doi.org/10.1016/j.infrared.2017.04.016
9. Shrestha R, Kim W (2018) Evaluation of coating thickness by thermal wave imaging: a comparative study of pulsed and lock-in infrared thermography—part II: experimentation. Infrared Phys Tech 92:24–29. https://doi.org/10.1016/j.infrared.2018.05.001
10. Maldague X, Marinetti S (1996) Pulse phase infrared thermography. J Appl Phys 79(5):2694–2698. https://doi.org/10.1063/1.362662.doi:10.1063/1.362662
11. Ranjit S, Chung Y, Kim W (2016) Thermal behavior variations in coating thickness using pulse phase thermography. J Korean Soc Nondestruct Test 36(4):259–265. https://doi.org/10.7779/JKSNT.2016.36.4.259
12. Mulaveesala R, Tuli S (2006) Theory of frequency modulated thermal wave imaging for nondestructive subsurface defect detection. Appl Phys Lett 89:191913. https://doi.org/10.1063/1.2382738
13. Mulaveesala R, Tuli S (2006) Phase sensitive digitized frequency modulated thermal wave imaging and pulse compression for NDE applications. Thermosense XXVIII 6205:620515. https://doi.org/10.1117/12.669530
14. Subbarao GV, Mulaveesala R (2012) Quadratic frequency modulated thermal wave imaging for non-destructive testing. Prog Electromagnet Res M 26:11–22. https://doi.org/10.2528/PIERM12062101
15. Suresh B, Subhani SK, Vijayalakshmi A, Vardhan VH, Ghali VS (2007) Chirp Z transform based enhanced frequency resolution for depth resolvable non-stationary thermal wave imaging. Rev Sci Instrum 88:014901. https://doi.org/10.1063/1.4973192
16. Ghali VS, Mulaveesala R (2011) Comparative data processing approaches for thermal wave imaging techniques for non-destructive testing. Sens Imaging Int J 12(1–2):15–33. https://doi.org/10.1007/s11220-011-0059-0
17. Ghali VS, Mulaveesala R (2010) Frequency modulated thermal wave imaging techniques for non-destructive testing. Insight-Non-Destruct Test Cond Monitor 52(9):475–480. https://doi.org/10.2528/PIERM12062101
18. Subhani S, Chandra Sekhar Yadav GVP, Ghali VS (2020) Defect characterization using pulse compression-based quadratic frequency modulated thermal wave imaging. IET Sci Measure Technol 14(2):165–172. https://doi.org/10.1049/iet-smt.2019.0118

19. Garg S, Shama BR, Cohen K, Kumar M (2013) A proper orthogonal decomposition based algorithm for smoke filtering in videos. In: 2013 American control conference (ACC), Washington, DC, USA. https://doi.org/10.1109/ACC.2013.6580377
20. Xiao T, Li X-D, Wang S (2019) Dominant-modes-based sliding-mode observer for estimation of temperature distribution in rapid thermal processing system. IEEE Trans Ind Inf 15(5). https://doi.org/10.1109/TII.2018.2871802
21. Kaur K, Mulaveesala R (2019) Experimental investigation on noise rejection capabilities of pulse compression favourable frequency-modulated thermal wave imaging. Electron Lett 55(6):352–353
22. Mulaveesala R, Somayajulu Vaddi J, Singh P (2008) Pulse compression approach to infrared nondestructive characterization. Rev Sci Instrum 79:094901. https://doi.org/10.1063/1.2976673

High Temperature Eddy Current Sensor for Real-Time Structural Health Monitoring of Critical Engineering Components

Chandan Dutta, Tarun K. Das, Alok Kumar, Jayendra Kumar, and S. Palit Sagar

1 Introduction

Structural health monitoring (SHM) of critical engineering components is of paramount importance nowadays to the industries. The hostile operating conditions such as temperature and pressure, contraction/expansion, vibrations, etc., may lead them to suffer from creep, thermo-mechanical fatigue, environmental attack like oxidation and hot corrosion, etc. [1, 2] Consequently, they are prone to get damaged and their replacement and maintenance costs are too high. Non-destructive testing (NDT) techniques are suggested to be applied for the structural health monitoring or inspection of the components. Several NDT techniques such as acoustic emission (AE), eddy current (EC), holographic inter-ferometry, laser ultrasonic, guided wave testing (GWT), and infrared thermography (IR) can either be used for the inspection or the monitoring of high-temperature structures. These techniques have been reported for operating at temperatures up to approximately 300 °C, though with various shortcomings [3]. Some of the drawbacks are qualitative results, sensitivity to noise, laboratorial utilisation and the need for coupling media. AE is a passive NDT technique that has been widely deployed in structural health monitoring (SHM); it monitors the elastic waves generated after the initiation or propagation of a crack [4]. Nevertheless, AE is sensitive to noise and gives qualitative results. Eddy current testing (ECT) is a non-contact technique that can be employed for the inspection of any electrically conductive material; however, it is subject to the skin effect, leading

C. Dutta (✉) · T. K. Das · A. Kumar · S. Palit Sagar
CSIR-National Metallurgical Laboratory, Jamshedpur, India
e-mail: dutta.chandan93@gmail.com

C. Dutta · J. Kumar
National Institute of Technology, Jamshedpur, India

C. K. Mukhopadhyay and R. Mulaveesala (eds.), *Advances in Non-destructive Evaluation*, Lecture Notes in Mechanical Engineering,
https://doi.org/10.1007/978-981-16-0186-6_7

it to be mainly used for the detection of surface and subsurface defects [5]. Temperature effecting eddy current (EC) probe used in non-destructive testing (NDT) especially for oxide scale thickness measurement has been reported in [6]. Though the temperature was limited to maximum 30.6 °C, it shows significant error in thickness measurement which was in the order of several microns. Several compensation techniques to compensate error due to exponential hysteresis characteristics of high-temperature eddy current-based displacement sensor up to 350 °C has been proposed by several authors [7]. This eventually strengthens the candidature of eddy current sensor over other sensors for health monitoring at high-temperature environment. Holographic inter-ferometry can give detailed results and be used for the detection of small defects, but it is mainly used for laboratorial tests as its setup is complicating, and it is sensitive to vibrations [8]. Laser ultrasonic has a small and adjustable footprint; therefore, it can be used for the inspection of irregular surfaces and samples of small and complex geometry. It induces high-frequency ultrasound, and thus, very small defects can be detected as well. However, its setup is complicated, and it is mainly used for laboratorial tests. GWT is used for the inspection or monitoring of large structures; mainly, piezoelectric transducers are employed upon the structure being tested exciting/receiving guided waves [9]. Piezoelectric transducers require direct access to the specimen and a coupling medium (usually a water-based gel), and their response cannot travel through a vacuum. IR can be used in this application mainly for overheating identification. However, the length of the components makes this technique practically inefficient. The camera needs to scan the whole length of the components which is time-consuming and may not be possible while the component is operating [10]. Furthermore, the high temperatures, the access, size and structural complexity of the specimen can limit the number of NDT techniques that can be efficiently employed. Thus, there is a specific need for structural health monitoring (SHM) of critical components at operating conditions in ultra-supercritical power generating plants to enable advanced decision making tools. Various degradations such as cracking, creep damage, fatigue, loss of thermal stability (precipitation of undesirable phases) and environmental attack (oxidation and hot corrosion) occur during extended exposure at elevated temperatures. Assessing and monitoring of such degradation during operation at high temperature need to overcome the present shortcoming of available sensors used in NDE. The present paper addresses this issue by developing application specific high-temperature sensors for real-time condition monitoring of the components operated at high temperature. The goal is to provide superior performance for in situ material condition monitoring (material degradation, flaw detection, stress relaxation and/or creep monitoring) and through-wall temperature measurement.

2 Finite Element Method (FEM)-Based Simulation

Prior to the fabrication of actual sensor, the same has been simulated in ComSol Multi-Physics simulation software.

The intrinsic electrical parameters of the cascaded Archimedean spiral coil are computed in FEM simulation model and listed in Table 2. The governing equations associated with the simulation model follow analytical expressions derived through Maxwell equations for time varying magnetic field. Magnetic field formulation physics and electric current physics present in the AC/DC module of Comsol Multi-Physics have been used to develop the above-mentioned simulation model. The design model with single layer Archimedean spiral coil is shown in Fig. 1.

$$B = \nabla \times A \tag{1}$$

$$E = -j\omega A \tag{2}$$

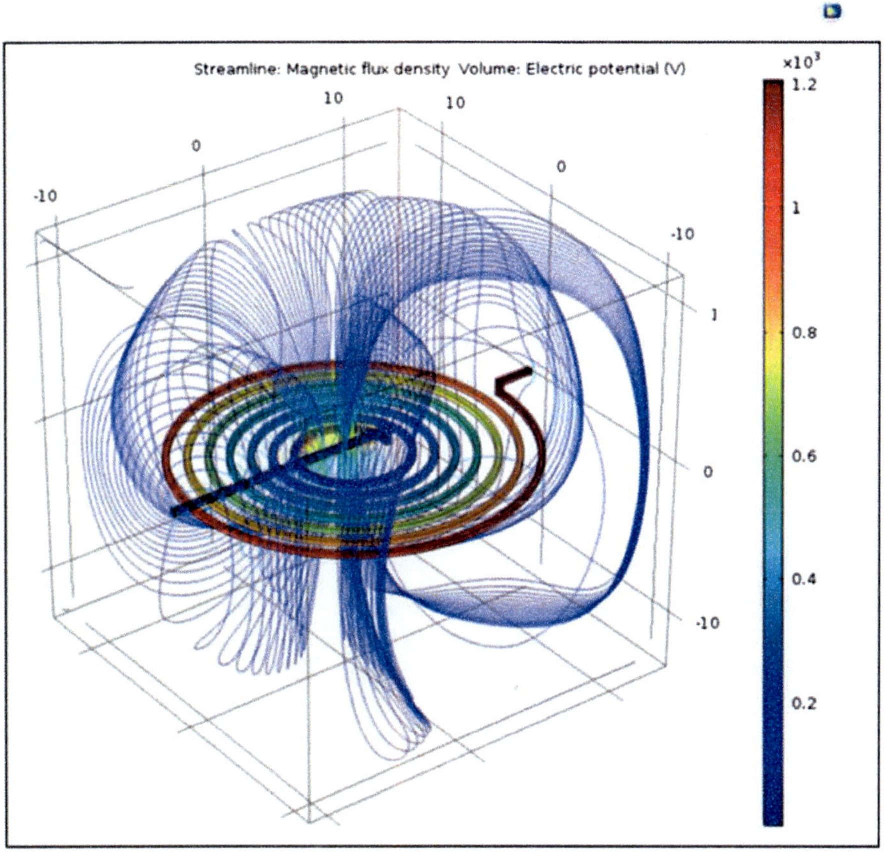

Fig. 1 Single layer Archimedean spiral coil simulation model

$$J = \sigma E + j\omega D + J_{\mathrm{e}} \tag{3}$$

where

(1) B: Magnetic flux density, (T) or $\left(\mathrm{Wb/m^2}\right)$
(2) E: Electric field intensity, (V/m)
(3) A: Magnetic vector potential, (Wb/m)
(4) σ: Electrical conductivity, (MS/m)
(5) J: Electric current density, $\left(\mathrm{A/m^2}\right)$
(6) J_{e}: Electric current density due to eddy current, $\left(\mathrm{A/m^2}\right)$
(7) ω: Angular frequency, (radian/s)
(8) D: Electric displacement field $\left(\mathrm{C/m^2}\right)$.

The Archimedean spiral coil is made using parametric curve option available in geometry section of Comsol Multi-Physics simulation software. The results obtained after sensor optimisation have been discussed in Sect. 4.

3 Experimental Procedure

The present work highlights to the development of eddy current-based non-destructive sensor that can operate at an elevated temperature up to 300 °C. In the proposed work, we have tried to study the effects on high temperature over ECT sensor and to characterise the intrinsic behaviour of the ECT sensor for different test conditions. The strength of the oscillatory magnetic field which is the only reason for the generation of eddy current within the specimen under test depends on the inductance of the sensing coil. Eddy current probes can be of several types depending upon its mode of operation. In generic way, probes are classified as transmit/receive and absolute sensing coil. The transmit/receive coil consists of one coil unit for generating primary magnetic field knows as excitation coil which generates eddy current in the specimen and one or more pickup coils (both in absolute or in differential modes) which mainly sense the secondary magnetic field which are generated due to the eddy current flowing within the specimen. The other type of sensing coil is purely absolute coil which generates both primary magnetic field and senses the secondary magnetic field which is reflected as impedance change of the coil due to the presence of abnormality within the specimen under test [11]. The secondary magnetic field can be sensed through many other types of magnetic sensor which meets the frequency requirement of the eddy current testing. The commonly used magnetic sensors along with ECT coils are hall sensor, giant magneto resistive sensor (GMR). The only problem with such sensing technology is its operation in high temperature (>250 °C). In the present scenario, eddy current systems come with different configurations and size, as single sensor as well as multiple sensor arrays. The systems are built mostly with one or more coils and electronics. Most of the eddy current sensors are made using copper wire with enamel coating wounded around or inside ferrite core. Alternatively, coils are also built with printed circuit technology in which spiral coils are printed with one layer on the substrate. The temperature ranges of these sensors are limited by enamel coating around the copper wire. This limits the usage of these sensors to 250 °C. Another limit is the ferrite core which loses its magnetic properties at the so-called Curie temperature, which is also in the range of 250 °C for most of the ferrites. Currently, there are no such ECT sensors known which are stable and safe beyond the aforementioned temperature level.

3.1 Material and Method

Printed circuit technology with Archimedean spiral coil printed on top of FR4 glass epoxy substrate shows very good thermal resistance at an elevated temperature up to 300 °C. The spiral coils are micro-machined using computerised numeric control (CNC)-based engraving tools on top of single side copper cladded FR4 grade glass epoxy board. The number of the spiral coils for efficient performance is optimised by developing a numerical model in computer simulation software based on finite element method (FEM). With limited machining tool resolution, 07 turn spiral coil on top of 15 mm × 14.5 mm FR4 glass epoxy substrate has been fabricated. The track width of the conductor in each turn is kept around 300 μm. The coils terminate with two terminals for electrical connection as shown in Fig. 2. Eagle cad-soft printed circuit board (PCB) designing software has been used to precisely design the Archimedean spiral coil of desired structural dimensions.

In the above shown figure, Archimedean spiral coils with the following dimensions shown in Table 1 have been designed in Eagle Cad-Soft PCB designing software. These coils have been fabricated on top of FR4 grade single side copper cladded PCB board. Prior to fabrication and post processing/cleaning, the coils can be bifurcated in separate unit for proper housing.

The structural design of the above fabricated sensor is same as that has been optimised in the simulation model to estimate the value of series resistance and inductance of the coil unit.

3.2 Sensor Characterisation

The micro-machined Archimedean spiral coil is then characterised by measuring its passive electrical parameters such as series resistance R_S, series inductance L_S, using precision impedance analyser. It is very important to stabilise these electrical parameters of the proposed ECT sensor at varying surrounding conditions, especially at high temperature. The measurement of the proposed sensor has been carried out using 4294A Agilent precision impedance analyser for a single tone sinusoidal excitation signal of 50 and 100 kHz. The measurement of the desired electrical parameters has been done at room temperature and also at several elevated temperature points.

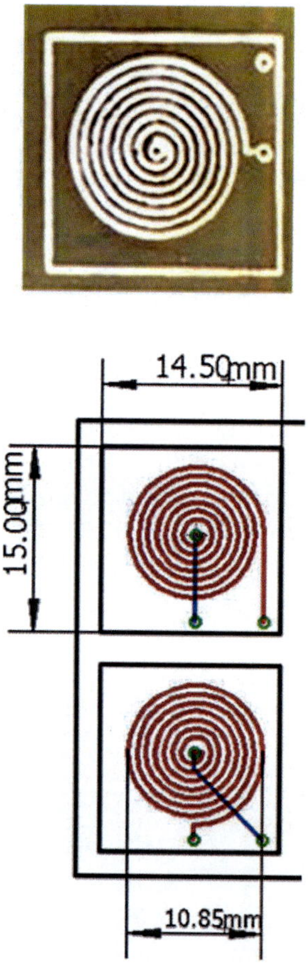

a. CAD drawing of spiral coil

b. Micro-machined spiral coil

Fig. 2 Single layer Archimedean spiral coil

Table 1 Dimension of each sensor unit

Length (mm)	Breadth (mm)	Diameter of coil (mm)	Conductor track width (mm)	Number of turns
15	14.50	10.85	0.300	7

The high-temperature requirement has been met through resistive heating tubular furnace. Sensor behaviour at high temperature is discussed in Sect. 4 of this article.

4 Result and Discussion

The validation of the CNC-based micro machining codes with its highest possible resolution has been checked by developing spiral coils with maximum possible number of turns and minimum pitch/track width shown in Fig. 2. Proper terminations for electrical contact are made for sensor characterisation using 30 SWG single stand copper wires. The series inductances as well as series resistance for a single coil shown above are very low in terms of nH and mΩ. The same has been verified with a single layer coil simulation model developed in ComSol Multi-Physics. The computed electrical parameters and experimentally measured electrical parameters of an optimised single layer spiral coil obtained from simulation model and actual fabricated coil are shown in Tables 2 and 3, respectively. In order to increase the value of series inductance and series resistance, so as to get significant potential differences across the spiral coil for eddy current testing, further models has been developed with spiral coils cascaded as multi-layers. The model has been developed with an increment of one layer, and series inductance/resistance has been calculated against fixed excitation current of 5 mA. The potential difference across the terminals of the sensor coil unit is also evaluated for each subsequent model. The simulation results along with experimentally measured values is plotted and shown in Fig. 4.

Table 2 Simulation results for single layer spiral coil

S. No.	No. of layers	Excitation frequency, f	Series inductance, L_s	Series resistance, R_s
	Unit	kHz	μH	Ω
1	1	50	0.4511	2.357

Table 3 Actual ECT sensor characteristics

S. No.	No. of layers	Excitation frequency, f	Series inductance, L_s	Series resistance, R_s
	Unit	kHz	μH	Ω
1	1	50	0.4535	0.252
2		100	0.4594	0.251

Figure 3 shown below shows the simulation model of cascaded spiral coil with 05 layers connected in series with consecutive layer thickness of 300 μm.

The characteristics of the actual sensor with one layer spiral coil over FR4 glass epoxy board are given in Table 3.

The FEM simulation results makes it very clear that cascading spiral coil layer by layers into a single unit with all coils in series combination increases the accumulative series inductance into manifold. The results obtained through simulation were then used to develop actual coils, and the same was then characterised using Agilent 4294A impedance analyser. The electrical parameter comprising series inductance (L_s) and series resistance (R_s) of the coils were evaluated at 50 kHz. This eventually increases the total impedance of the ECT coil, which makes instrumentation of the proposed sensor possible. The strength of the acquired signal across the sensor unit

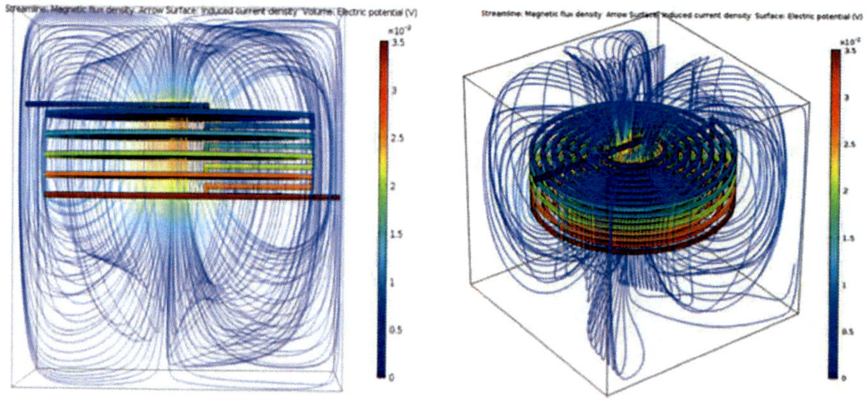

a.Cascaded spiral coil (Side View) **b.**Cascaded spiral coil (Isometric View)

Fig. 3 05 layers Archimedean spiral coil simulation model

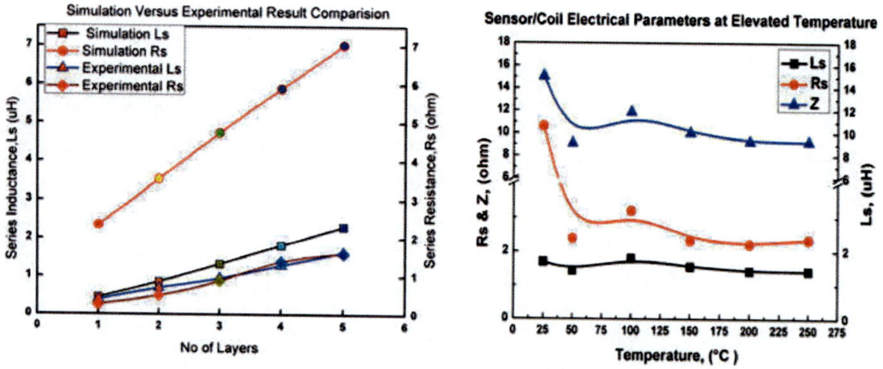

a b

Fig. 4 (**a**) Simulation versus experimental sensor characteristics, (**b**) actual sensor characteristics at different temperature condition increases and eventually the signal-to-noise ratio (SNR) improves. It also allows proper selection of specific valued passive components for post processing of the impedance change due to small perturbations in eddy current resulting due to abnormalities present in the service exposed specimens. The characteristic can further optimised by reducing the gap between consecutive layers of spiral coils and by increasing the compactness or fill factor of each coil. Apart from sensor characterisation at ambient temperature, the intended purpose of the proposed sensor is to operate it at high temperature for structural health monitoring of the serviced exposed engineering components. In order to ensure robustness against high temperature with repeatable signal response and sensitivity against intrinsic specimen abnormalities, the sensor characteristics should not change as surrounding temperature rises. The characterisation of the proposed sensor unit has been carried out at several elevated temperature points using resistive heating tubular furnace. Measurements were taken at every 50 °C intervals up to 250 °C. The results obtained were shown in Fig. 4.

5 Conclusion

An attempt has been made to develop eddy current-based sensor that is capable of operating at harsh industrial environments such as high temperature and pressure for real-time structural health monitoring (SHM) of service exposed power plant alloys. The capabilities of Archimedean spiral coils as eddy current sensor for structural health monitoring at high temperature up to 250 °C has been established. A numerical model for optimising the structural characteristics of the proposed sensor has been developed using FEM-based simulation software. The results obtained as sensor characteristics at different elevated surrounding temperature are almost constant. This proves the robustness of the proposed sensor against high temperature. Even it has been established through simulation and actual experimentation that by cascading spiral coil layer by layer improves the passive electrical parameters such as series resistance and series inductance of the sensor which is highly desirable for the intended application of these sensors. These observations will give an insight to operate such eddy current sensors at an elevated temperature, which can be more precisely achieved through fabricating these sensors using low temperature co-fired ceramic (LTCC) technology. LTCC technology offers precise development of such planner coils which can be used for radio frequency (RF) applications, eddy current-based non-destructive testing (EC-NDT), proximity measurements in harsh industrial environments and high surrounding temperatures up to 1000 °C.

Acknowledgements Authors are grateful to the Director, CSIR-National Metallurgical Laboratory, Jamshedpur for his kind permission to publish the work.

References

1. Papaelias M, Cheng L, Kogia M et al (2020) Inspection and structural health monitoring techniques for concentrated solar power plants. 85:1178–1191. https://doi.org/10.1016/j.renene.2015.07.090
2. Kogia M, Cheng L, Mohimi A et al (2016) Electromagnetic acoustic transducers applied to high temperature plates for potential use in the solar thermal industry. https://doi.org/10.3390/app5041715
3. Saillant J, Marlier R, Baqu F (2019) Ultrasonic transducer for non-destructive testing of structures immersed in liquid sodium at 200 °C. Sensors 1–13. https://doi.org/10.3390/s19194156
4. Beattie G (1983) Acoustic emission principles and instrumentation. J Acoust Emiss 2:95–128
5. Hagemaier DJ (1990) Fundamentals of eddy current testing
6. Beck FR, Lind RP, Smith JA et al (2018) Research in nondestructive evaluation temperature sensitivity study of eddy current and digital gauge probes for oxide measurement temperature sensitivity study of eddy current and digital gauge probes for oxide measurement. Res Nondestruct Eval 00:1–16. https://doi.org/10.1080/09349847.2018.1557770
7. Zheng S, Liu X, Zhang Y et al (2019) Temperature drift compensation for exponential hysteresis characteristics of high-temperature eddy current displacement sensors. 1748. https://doi.org/10.1109/JSEN.2019.2933347
8. Malmo JT, Jøkberg OJ, Slettemoen GA (1998) Testing at very high temperatures by TV holography (ESPI). Exp Mech 315–321
9. Silk MG, Nton KFBAI (1979) The propagation in metal tubing of ultrasonic wave modes equivalent to lamb waves. 11–19
10. Pfa M, Lu E, Pistor P (2007) Infrared temperature measurements on solar trough absorber tubes. 81:629–635. https://doi.org/10.1016/j.solener.2006.08.016
11. Garcí J (2011) Non-destructive techniques based on eddy current testing. Sensors 2525–2565. https://doi.org/10.3390/s110302525

Damage Assessment for a Sandwich-Like Panel Using Experimental and Numerical Analysis of Guided Waves

Kaleeswaran Balasubramaniam, Piotr Fiborek, Shirsendu Sikdar, and Pawel H. Malinowski

1 Introduction

Aluminum Nomex sandwich core structure (SCS) materials are mainly used in the automobile and aerospace industry for its high strength to weight ratio. Evaluation of these materials is necessary from time to time to prevent disasters to the entire structures. Barely visible impact damage (BVID) is a type of damage that occurs due to some impact drops or mishandling of tools leading to a small or big indentation in the materials. BVIDs on aluminum or steel creates lesser damage to the structure, but on fiber-reinforced polymers like in glass fiber-reinforced polymer (GFRP) or honeycomb structures cause changes to core/core crush or failure of the matrix, etc. Laser Doppler Vibrometer (LDV) is an advanced non destructive testing (NDT) tool used in the industry to detect damage in the early stage. LDV is reliable, handles large materials, and detects the damage sensitive regions.

The lamb waves or guided waves help to evaluate such defects as they can propagate to long distances and its excitation be achieved by a piezo zirconium titanate (PZT) actuator. These lamb waves get reflected from discontinuities and thus help in clear visualization of the damaged region. The usage of lamb waves in the FRP is well known and mentioned in many papers [1], and similarly usage of LDV methods

K. Balasubramaniam (✉) · P. Fiborek · S. Sikdar · P. H. Malinowski
Institute of Fluid Flow Machinery, Department of Mechanics of Intelligent Structures, Polish Academy of Sciences, Fiszera 14 Street, 80-231, Gdansk, Poland
e-mail: kaleeswaranb@imp.gda.pl

P. Fiborek
e-mail: pfiborek@imp.gda.pl

S. Sikdar
e-mail: ssikdar@imp.gda.pl

P. H. Malinowski
e-mail: pmalinowski@imp.gda.pl

C. K. Mukhopadhyay and R. Mulaveesala (eds.), *Advances in Non-destructive Evaluation*, Lecture Notes in Mechanical Engineering,
https://doi.org/10.1007/978-981-16-0186-6_8

in carbon fiber reinforced polymer (CFRP), aluminum is studied in [2]. When lamb waves interact with the damage region, it gets reflected, mode change occurs, and interference in the wave field occurs. The lamb waves occur in the form of asymmetric and symmetric waves. Thus, with the help of LDV and signal processing methods, the analysis of scattered waves is done. LDV works on the principle of Doppler effect and observation of the Doppler shift in frequency. LDV uses either Nd:YAG laser or helium–neon (He–Ne) laser. LDV finds its application mainly in detecting BVID [3], delamination [4], and debonding analysis [5]. The author [6] talks about honeycomb structures and BVID's. The impact test on GFRP [7] is studied and the use of LDV [8] on a composite plate to analyze wave patterns. The spectral element method (SEM), which was developed for the numerical solution of the laminar fluid flow by Patera [9], combines the flexibility of finite element method (FEM) and fast convergence of spectral methods in the frequency domain. This SEM technique was successfully implemented in the field of modeling the phenomena of elastic wave propagation [10–13].

In this research paper, we have used LDV in analyzing the BVID in a semi-sandwich panel with Nomex core and aluminum skin. The damage caused by impacts at 4 and 10 J is studied. In one region, a small portion of Nomex core is completely removed manually, BVID is studied, and in another region, an indentation with a Nomex core protrusion is studied. The results obtained from LDV are compared with the numerical studies to visualize the damage region. The numerical approach is performed using the spectral element method (SEM) code. The gathered data is processed with the elliptical signal processing method for damage localization with various sensor network arrangements.

2 Experimental and Numerical Process

The investigated object is a structure ($50 \times 50 \times 0.1$ cm^3) with aluminum skin and a Nomex honeycomb core at the back. The BVID is made by a low-impact equivalent to 4 and 10 J from the Nomex core side. On one side, a small portion of Nomex cells are manually removed to create a BVID, and another BVID is made with the crushing of Nomex cells. The device used to create the BVID is a circular impactor of 10 cm diameter. The specimen is shown in Fig. 1a. The largest diameter BVID is zoomed in Fig. 1b, and the Nomex attached to the facet is also shown in Fig. 1c with an area, where the BVID is made.

The following damage cases are in the paper:

1. BVID's of the diameter 1.4 and 1.8 cm.
2. Hole of 0.5 cm diameter.

The SLDV is an NDT device largely used in the industries for analyzing the vibrations of the structures. The SLDV instrument used is PSV 400 from Polytec© GmbH. The laser vibrometry uses the Doppler effect for non-contact optical vibration analysis. This uses light as a sensor and thus a contactless experimental process. This

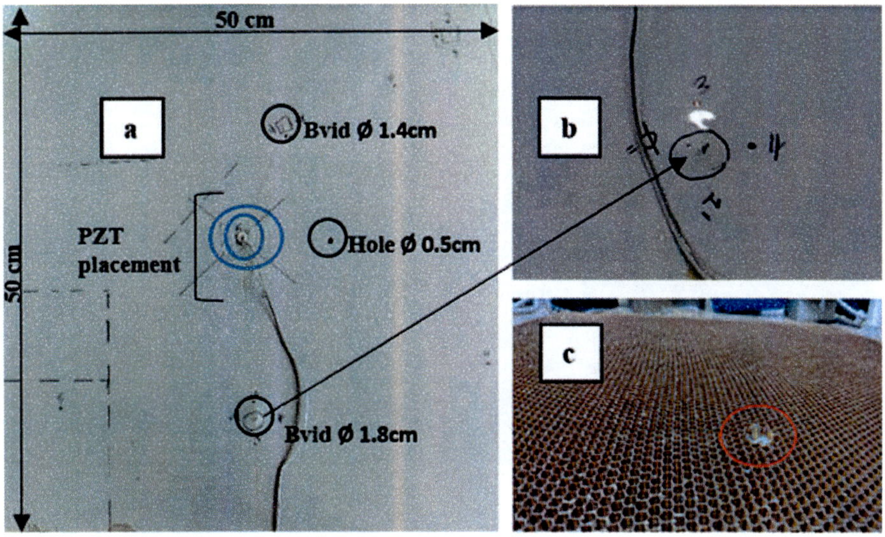

Fig. 1 **a** SCS plate with damages, **b** zoomed portion of BVID of $\emptyset = 1.8$ cm, and **c** Nomex core attached at the backside of the plate and region, where BVID is made

gives real-time measurements on temporal signatures. It works on the frequency of laser shift between the reference beam and the incident beam from the laser. The interferometer inside the laser head analyses this shift by propagating them to the photodetector. Thus, depending on the velocity, the backscattered light gets changed in frequency and phase.

Thus, the Doppler shift in frequency is analyzed based on the modulated detected output signals. Thus, after performing inbuilt signal processing techniques, the velocity and displacement of the structure surface are detected. The laboratory setup of LDV is shown in Fig. 2. The PZT's are bonded with bonding glue to the plate. A sine tone bursts signal with 10 cycles and with an amplification value of 16 V_{PP}, and gain of 20 (16×20) produces the excitation. The total number of grid points and frequency used in the experiment is summarized as follows in Table 1.

For the spectral element method (SEM), a hole of Ø 0.5 cm is studied. The engineering constants used for the analysis are highlighted in Table 2. The parameters used in SEM are explained in Tables 3, 4, and 5.

Dielectric, electromechanical, and mechanical properties used for PZT (Sonox P502) is shown in Tables 3 and 4, respectively, where d_{33}, d_{31}, d_{15} are charge constants, $\frac{\varepsilon_{33}^T}{\varepsilon_0}$, $\frac{\varepsilon_{11}^T}{\varepsilon_0}$ relative permittivity values, $S_{11}{}^E$, $S_{33}{}^E$ are elastic compliance values, ρ represents density value, and η_{13} represents the Poisson ratio.

The SEM technique, similarly to FEM, divides modeled domain into finite elements, imposing the arbitrary boundary conditions and external forces in the particular nodes. Distribution of the nodes in the element is the main difference between both methods, while the nodes in SEM are non–uniformly distributed. They

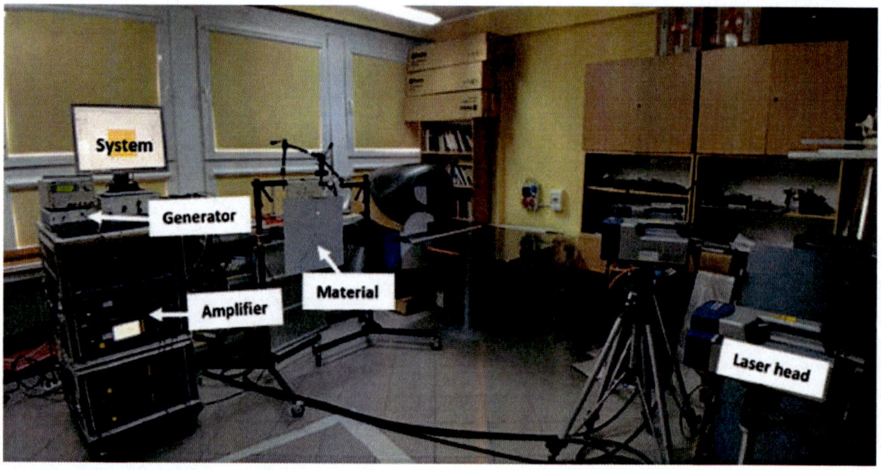

Fig. 2 LDV setup

Table 1 Experimental table with measurements performed in the LDV

No. of cycles	Frequency (kHz)	VPP	Grid points	Time increment (μs)
10	150	16	411 × 415	0.78125

Table 2 Engineering constants used in the numerical study

Material	Young's modulus (GPa)	Poisson ratio	Density (kg/m^3)
Nomex	9	0.30	1384
Aluminum	68	0.33	2600

Table 3 Electromechanical and dielectric properties of PZT

d_{33} (C/N)	d_{31} (C/N)	d_{15} (C/N)	$\frac{\varepsilon_{33}^T}{\varepsilon_0}$	$\frac{\varepsilon_{11}^T}{\varepsilon_0}$
440E−12	−185E−12	560E−12	1850	1950

Table 4 Mechanical properties of PZT and other values

$S_{11}{}^E$ (m^2/N)	$S_{33}{}^E$ (m^2/N)	η_{13}	ρ (kg/m^3)
18.5e−12	20.7e−12	0.44	7740

Table 5 Numerical table with values used

No. of cycles	Frequency (kHz)	Grid points	Time increment (μs)
10	150	251 × 251	6.1e−3

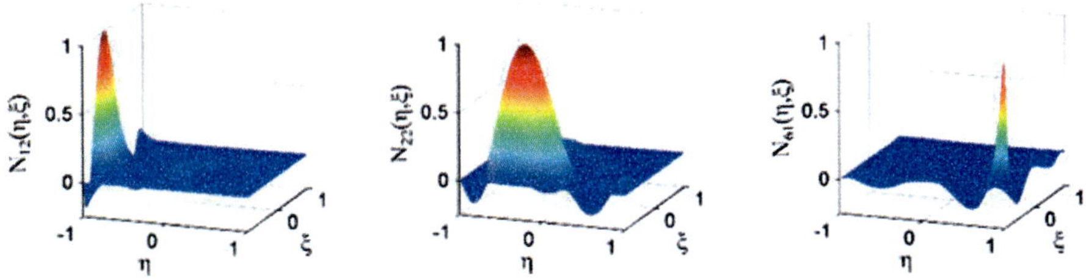

Fig. 3 Two-dimensional spectral shape function

are coincident with the coordinates of the Gauss–Lobatto–Legendre (GLL) integration points. Figure 3 and Eq. 1 depicts examples of 2D orthogonal shape functions which are constructed as a tensor product of $N_{\mathrm{m}}(\xi)$ and $N_{\mathrm{n}}(\eta)$.

$$N_{mn}(\hat{\imath}, \text{ç}) = N_m(\hat{\imath})\ N_n(\text{ç}),\ \text{for } m = 1\ldots\ p,\ n = 1\ldots r \qquad (1)$$

As a consequence of the orthogonality of shape functions as well as the application of GLL quadrature, the mass matrix is diagonal in the formulation. This property significantly reduces computation time for solving the equation of motion due to avoiding mass matrix inversion.

3 Results and Discussion

3.1 Analysis of the Results from the LDV

Velocity plots obtained directly from the LDV is analyzed. The lamb wave propagation of the specimen is visualized in Fig. 4. Figure 4a analyzed with 150 kHz frequency shows the propagation of the waves in the form of circles. The waves

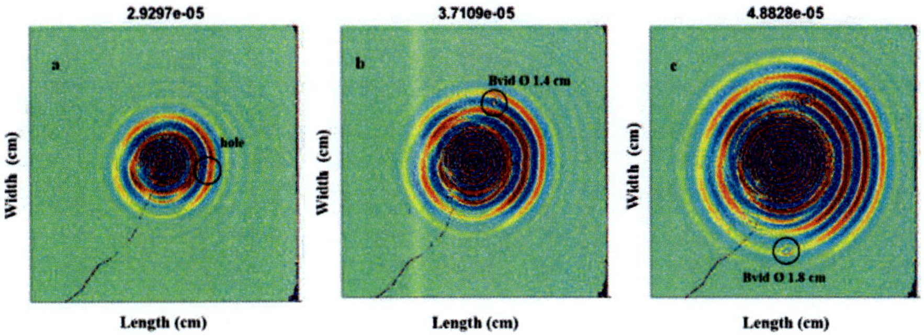

Fig. 4 **a** A hole of Ø 0.5 cm detected at 29 μs, **b** BVID of Ø 1.4 cm detected at 37 μs, and **c** BVID of Ø 1.8 cm detected at 48 μs

Table 6 Time where the damage gets detected

Frequency (kHz)	Hole (μs)	BVID1 (μs)	BVID2 (μs)
150	29	37	48

reach the damage region (hole) and get distracted, thus showing the location and damage shape. It shows the hole region perfectly with lamb waves getting scattered.

The experiment is carried out with a frequency range of 10, 15, 20, 25, 50, 100, 150, 200 kHz, and it is found that the 150 kHz frequency provides good visualization of results. The amplitude of the S0 mode is much lower than A0 mode; therefore, it is barely noticeable on the full field of the wave measured by the LDV. Thus, mostly, the antisymmetric A0 mode waves are clearly visible and marked in all the figures, respectively.

When the waves reach the damage zone, it gets reflected and mode change occurs. The important aspect is to visualize the wave reflections after it touches the impact damage zone. Thus, LDV helps in quality analysis of the material without destructing it. Figure 4b shows the BVID Ø 1.4 cm, getting detected. The waves get scattered at the damage region (marked with black circles), and mode change occurs. Figure 4c shows the wave pattern analysis and detection of BVID of Ø 1.8 cm.

The LDV measures the out of plane displacement and helps in the early inspection. As shown in Fig. 4, one can visualize the *S* waves reaching the damage region and its conversion. This shows a clear visualization of the damage. The time for the waves to reach the damage zones is summarized as follows in Table 6.

3.2 *Analysis of the Results from Numerical Studies*

In this section, the numerical results from the SEM are shown. Thus, the main idea is to check the experimental results visualized with the numerical studies and to put a proper comparison of the results. For this spectral elemental study, a hole Ø 0.5 cm section is selected to be analyzed. Figure 5a shows the mesh arrangement and PZT position in the SEM approach; Fig. 5b shows the result obtained with the SEM numerical study.

Thus, a comparative study of numerical analysis and experimental results has shown that the hole gets detected at a time period of 29 μs in both cases as shown in Figs. 4a and 5b, respectively. Thus, by comparing the experimental and numerical figures, 4 and 5 reveal the wave field of the lamb waves and its change due to the damage zones, respectively.

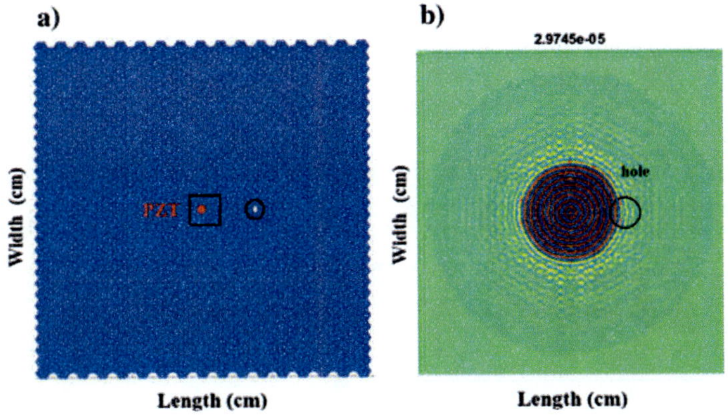

Fig. 5 **a** Mesh arrangement and PZT position in SEM, **b** hole of Ø 0.5 cm detected at 29 μs for 150 kHz excitation

3.3 Analysis of the Results Using an Elliptical-Based Signal Processing Method

The numerical data from SEM is used to locate the damaged region. This section focuses on the application of the elliptical-based signal processing method in localizing the damage [14]. The concept behind is the intersection of the ellipses and at the region where a large number of ellipses intersects, thereby identifying the region of damage with higher energy. Optimum sensor positioning and damage localization are considered. Three cases with different sensor positioning are shown in Table 7. Cases 1 and 3 are taken from research on the optimal sensor placement assisted with a genetic algorithm [15].

For the elliptical-based method to function, a velocity profile is to be determined because the velocity can be angle-dependent. Thus, the velocity of the waves is determined. The steps performed to obtain the velocity profile are summarized below and with Eq. 2.

The scanning points are placed at equal distances, and thus, the velocity of the waves is calculated based on the formulae shown below.

$$\text{Velocity} = \frac{\text{delx}}{\text{delt}} \tag{2}$$

Table 7 Case models for elliptical approach with experimental and numerical data

Case	Test	Sensor placement	Frequency (kHz)	PZT	No. of sensing points
1	Experimental	Quadrant I, II, III, IV	150	Center	8
2	Numerical	Quadrant II	150	Center	6
3	Numerical	Quadrant I, II, III, IV	150	Center	8

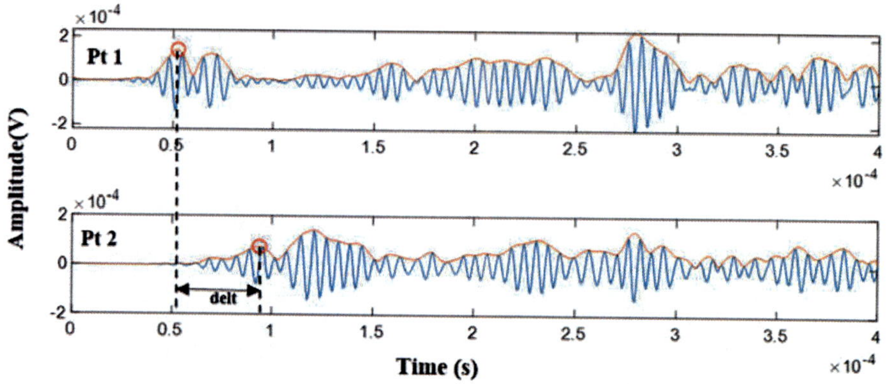

Fig. 6 Delt calculation between two points

where delx is the distance between the scanning points, and delt is the distance between the max peak value of the first wave packet of one point to the similar max peak value of another point. The calculation of delt between pt1 and pt2 of case 2 signals is shown in Fig. 6. The velocity profile is calculated by taking 0°, 90° scanning points of case 2. It is then used to calculate the entire 360° values and marked with a polar plot in Fig. 7 indicating an almost perfect isotropic profile. These velocity profile values obtained are used for case 1 and 3 calculations, respectively.

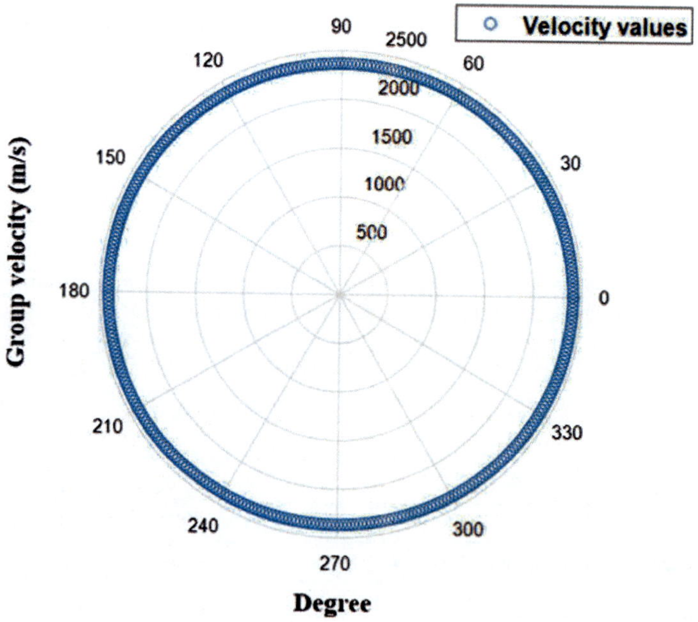

Fig. 7 Velocity profile obtained

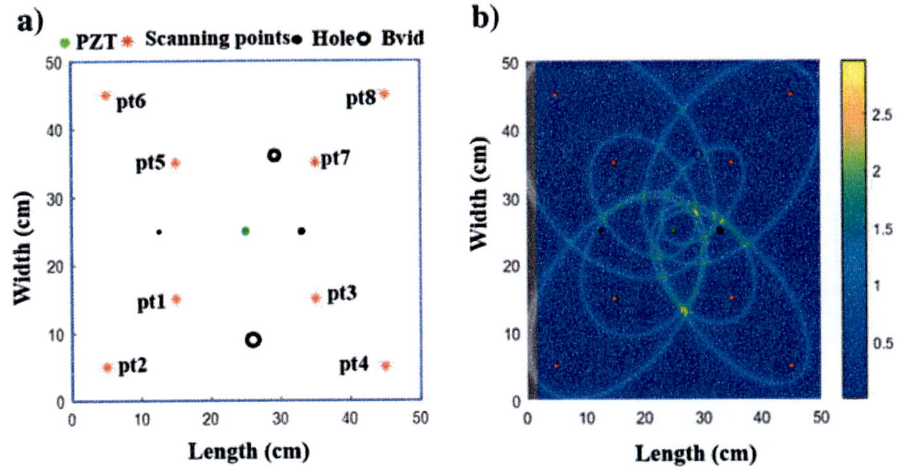

Fig. 8 Experimental case: **a** case 1 quadrant I–IV arrangement of sensors, **b** elliptical method on case 1

The elliptical method is applied to the complete experimental case 1 with 2 BVIDs and hole as shown in Fig. 8a. The results tried to identify most of the damage locations but in some cases, it could not due to multiple damage locations and thereby causing multiple reflections as shown in Fig. 8b. Thus, the research is later shifted to the data obtained from a numerical study which was simpler due to only one damage and gave good clarity of results as shown in Fig. 9.

Figure 9 shows the calculation performed and the damage located, respectively. Figure 9a, b shows the sensor network placements in two cases. The analysis of the results from Fig. 9 shows the region of higher energy, where ellipses get coincident more. In Fig. 9c, the damage hole region is properly detected with the help of intersecting ellipses. The higher energy is due to wave reflection from the damage. This result is obtained by analyzing results from various peak values extracted from the time signals.

In Fig. 9c, the scanning sensor points are placed at an interval of 10 cm and located at 0°, 45°, and 90°, respectively. The highest reflection is obtained at a distance of $(31.9, 26)\ cm^2$, whereas the damage is at $(33, 25)\ cm^2$. Thus, the high energy reflection from the damage is captured at a distance of 1.49 cm from the actual damage location.

In a similar way for Fig. 9b, the scanning points are placed at all quadrants of a square plate (according to [15]). This resulted in more convergence of ellipses and more scanning sensor points as shown in Fig. 9d. The highest convergence of ellipses reflection is obtained at a distance of 4.1 cm from the damage location. The damage location is at (33, 25) cm^2, and the maximum reflection is obtained at (29.8, 27.8) cm^2. Case 2 analysis gave more precise results when compared to case 3.

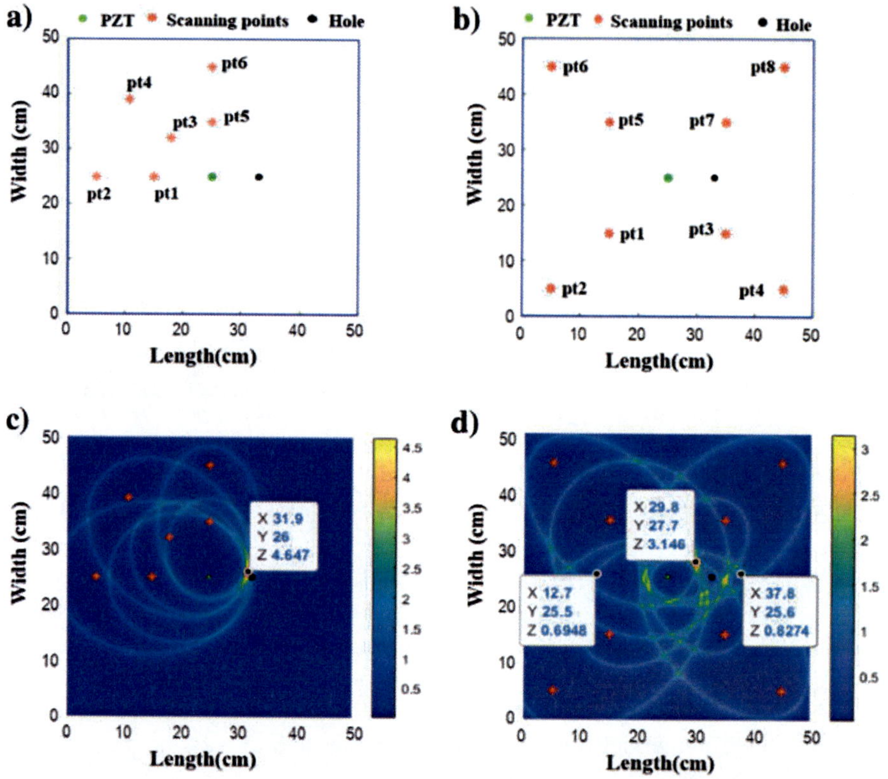

Fig. 9 Numerical cases: **a** case 2 quadrant II arrangement of sensors, **b** case 3 quadrant I–IV arrangement of sensors, **c** elliptical method on case 2, and **d** elliptical method on case 3

4 Conclusion

The research purposed is about the NDT inspection of barely visible impact damage (BVID) in the aluminum Nomex SCS plate using guided wave propagation. The damage regions are successfully analyzed with the help of the A0 wave mode. The lamb waves interactions and impact damages produced are visualized and are detected. The frequency of excitation at 150 kHz is chosen after studying different frequency ranges. The lower excitation frequencies (10, 15, and 50 kHz) were studied, and it failed to capture the damage results. Thus, further increase in the frequency like 100, 150, 175, and 200 kHz gave good quality results. The results from the LDV gave a good appropriate identification of the damage location. This gave a good analysis of wave pattern study and mode changes in the damage region. The lamb wave propagation and reflection from damages are identified with LDV and shown in Fig. 4. The LDV results identify the two different BVID damages and the small hole with the help of guided waves.

The spectral elemental code is used to analyze similar conditions numerically. The results obtained from it identified the damage at a similar time period as the experimental case. The elliptical signal processing technique works on the methodology of visualization of the wave reflection site. The interaction of the wave with damage creates higher energy reflected zone. In this method, three different cases are studied as shown in Table 7. Case 1 corresponds to the result obtained from experimental studies (LDV), and cases 2 and 3 are from SEM calculations. The sensor network positions are different for all three cases. Case 2 provided a better result in identifying the damage location. Thus, by using the elliptical-based signal processing technique, the data obtained from LDV and SEM is again used to locate the damaged region. Thus, overall a case study of experimental, numerical, and analytical studies is carried out to understand the wave propagation phenomenon and to locate the damage with sensor network arrangements.

Acknowledgments The authors would like to acknowledge the support funding provided by the National Science Center, Poland under SONATA BIS project titled: Study of piezoelectric sensors placement and their interaction with structural elements (2016/22/E/ST8/00068).

The research was also funded by the Polish National Science Center under grant agreement no UMO-2018/31/N/ST8/02865 in the frame of PRELUDIUM project entitled: Model-assisted damage identification function for Structural Health Monitoring of composite structures under a varied environmental condition.

The authors also acknowledge Task-CI for allowing the use of computational resources.

References

1. Guo N, Cawley P (1993) The interaction of Lamb waves with delaminations in composite laminates. J Acoust Soc Am 94:2240–2246
2. Toyama N, Yamamoto T, Urabe K, Tsuda H (2017) Ultrasonic inspection of adhesively bonded CFRP/aluminum joints using pulsed laser scanning. Adv Compos Mater
3. Diamanti K, Hodgkinson JM, Soutis C (2004) Detection of low-velocity impact damage in composite plates using Lamb waves. Struct Health Monit 3:33–41
4. Hayashi T, Kawashima K (2002) Multiple reflections of Lamb waves at a delamination. Ultrasonics 40:193–197
5. Park B, An YK, Sohn H (2014) Visualization of hidden delamination and debonding in composites through noncontact laser ultrasonic scanning. Compos Sci Technol 100:10–18
6. Sikdar S, Kudela P, Radzienski M, Kundu A (2018) Online detection of barely visible low-speed impact damage in 3D core sandwich composite structure. J Compos Struct 185:646–655
7. Philbert M, Soutis C, Gresil M, Yao K (2018) Damage detection in a composite T-joint using guided lamb waves. Aerospace MDPI 40:1–13
8. Ruzzene M, Joeng SM, Michaels TE, Michaels JE, Mi B Simulation and measurement of ultrasonic waves in elastic plates using laser vibrometry
9. Patera AT (1984) A spectral element method for fluid dynamics: laminar flow in a channel expansion. J Comput Phys 54:468–488
10. Kudela P, Zak A, Krawczuk M, Ostachowicz W (2007) Modelling of wave propagation in composite plates using the time domains spectral element method. J Sound Vib 302:728–745
11. Kudela P, Ostachowicz W (2009) 3D time-domain spectral elements for stress waves modelling. J Phys Conf IOP publishing 181:1–9

12. Ha S, Chang FK (2009) Optimizing a spectral element for modeling PZT-induced lamb wave propagation in thin plates. Smart Mater Struct 19:1–12
13. Ge L, Wang X, Wang F (2014) Accurate modeling of PZT induced lamb wave propagation in structures by using a novel spectral finite element method. Smart Mater Struct 23:1–15
14. Fendzi C, Mechbal N, Rebillat M, Guskov M, Coffignal G (2015) A general bayesian framework for ellipse-based and hyperbola-based damage localisation in anisotropic composite plates. J Intell Mater Syst Struct 1–32
15. Soman R, Malinowski P (2019) A real-valued genetic algorithm for optimization of sensor placement for guided waves based structural health monitoring. J Sens

Development of Sectored External Pipeline Inspection Gauge for Health Monitoring of Industrial Carbon Steel Pipelines

Shilpi Saha, Debmalya Mukherjee, Y. Chandra, S. C. Ramrane, S. K. Lahiri, and P. P. Marathe

1 Introduction

Magnetic Flux Leakage technique (MFL) is one of the most widely used technologies for Non-Destructive Evaluation (NDE) of buried Carbon Steel (CS) pipelines. This technology is extensively used by the petroleum industries for periodic health assessment of its large network of buried pipelines. An Instrumented Pipeline Inspection Gauge (IPIG) is used for the purpose of In-Line Inspection (ILI) of these pipelines, wherein the tool is inserted at one end of the pipe, and it traverses within the pipe along with petroleum cargo for hundreds of kilometres and is retrieved at the other end. An IPIG based on MFL technology contains an array of circumferentially disposed permanent magnet assembly and hall-effect sensors which senses the axial and/or radial component of the leakage flux. These leakage flux signals are thereafter mapped to pipeline defects (cracks, metal-loss, corrosion, etc.) and features (weld, flange, etc.) based on empirical non-linear models. Further details on MFL-IPIG and its construction can be found in [1]. It is to be noted that some IPIGs also employ other NDE technologies like Ultrasonic (UT) [2] and Eddy Current (EC) [3] for pipeline health assessment, each technique having its unique merits and shortcomings.

Pipelines that can be inspected by IPIG are called piggable pipelines. They must have specially designed launching and receiving barrels for tool launch and retrieval and must maintain a minimum flow across the pipe length to generate sufficient motive force for the tool to move. Also, the pipe must conform to stringent guidelines on curvature and bends laid down by Pipeline Operators Forum (POF) [4]. Most pipelines used by other heavy industries like chemical refineries, process and power plants (conventional and nuclear), etc. are unpiggable. Full length, full periphery inspection of these pipelines is seldom carried out in their operational lifetime due to

S. Saha (✉) · D. Mukherjee · Y. Chandra · S. C. Ramrane · S. K. Lahiri · P. P. Marathe
Control Instrumentation Division, Bhabha Atomic Research Centre (BARC), Mumbai, India
e-mail: shilpis@barc.gov.in

C. K. Mukhopadhyay and R. Mulaveesala (eds.), *Advances in Non-destructive Evaluation*, Lecture Notes in Mechanical Engineering,
https://doi.org/10.1007/978-981-16-0186-6_9

lack of any standardized NDE methodology or available gauges. Generally, inspection of these pipelines is carried out with spot measurements at specific places that are more prone to failure, leaving out most of its length. This sometimes lead to early failure or premature decommissioning of such pipelines due to lack of thorough health assessment.

BARC has successfully designed and developed a fleet of inline inspection tools (MFL-IPIGs) of various sizes for health monitoring of petroleum pipelines. These tools, developed for Indian Oil Corporation (IOCL), are currently undertaking commercial ILI services across major buried CS pipelines in the country. Using the already developed expertise in ILI technology, an external MFL tool is designed and developed, to address the health assessment challenges of unpiggable pipelines, for its comprehensive and full periphery ISI. Research for tools with similar objectives (inspection of unpiggable pipelines) is being carried out all over the world, and there are a few prototype/products available for commercial use. Popular technologies used by these tools are imaging, conventional UT and Long-Range UT (LRUT). Advances in inspection of unpiggable pipelines have primarily focused on the development of autonomous robots capable of high-resolution imaging for health assessment [5]. For the inspection of CS pipelines, use of robots with magnetic wheels is a popular choice [6] as it adheres to pipe surface while negotiating sharp bends or vertical drop/climb. Visual inspection robots can be customized for inspecting the pipe from within (giving full periphery image of the inside pipe-wall) or externally, which provides sectored image of the pipe. These tools, however, are not without some shortcomings. An inspection robot launched from within cannot be used for ISI for liquid cargo pipelines. In most CS pipelines, there is a coat of paint exterior to the pipe-wall, rendering visual inspection ineffective, when done externally. Also, given that cracks, corrosion and other pipeline defects can be subsurface or occur on the opposite wall with respect to where the inspection tool is moving; visual inspection will not be comprehensive.

UT-based inspection gauges are also available and provide a detailed measurement of pipe wall thickness. 'Pipe crawler' by 'Inspector Systems' [7] uses UT technology for inspection of few 100 m of pipeline from within, negotiating sharp bends (1.5D). While this and similar other UT-based internal tools are very effective for pipeline crack or corrosion detection, it needs access to interior of the pipe. This would not be possible in nuclear (or conventional) power plants and hazardous chemical industries. An alternate solution is a crawler for thickness measurement and external corrosion mapping of pipelines by IMG ultrasonic [8]. However, conventional UT-based tools need a couplant between the UT probes and material under inspection. Also, poor surface finish of the material under inspection adversely affects the quality of UT inspection [9]. This is one of the major drawbacks of UT against MFL with respect to pipeline inspection.

LRUT, also known as Guided Wave UT (GWUT), is one of the most effective solutions currently used by the industry for external inspection of unpiggable pipeline. It involves deploying a ring of UT sensors (transducer assembly collar) across the exterior of pipe wall and generating low-frequency mechanical waves on both sides. Any pipe wall anomaly will reflect back some of the wave, which is then analysed for sizing and distance. Each collar assembly can inspect up to 90 m of pipe on either side and is very effective against pipes that are difficult to access. LRUT is, however, known to miss isolated pits within the pipe and is used primarily as a screening tool due to its poor defect sizing capability. Also, LRUT has lower efficiency in defect sizing when there are sharp bends in the piping layout.

Use of MFL technology for inspection of unpiggable pipeline has been reported by Korean Gas Corporation (KOGAS) in 2016 [10] using a self-propelled in-pipe robot for gas pipelines. The present development of SEPIG-10 can be seen as an attempt to use MFL technology for inspection of unpiggable pipelines externally. One of the major challenges in this pursuit is to make the tool light, compact and manoeuvrable by hand, as it involves use of strong permanent magnets that generate large attractive force with the pipe wall. With the domain knowledge and expertise acquired from development of IPIGs, the magnetic circuit and sensor arrangement were re-engineered to suit the application. Multiple iterations of Finite Element (FE) analysis were carried out to optimize the tool design. Initially, a full periphery external PIG (EPIG-6) was designed and developed for 6″ nominal bore (NB) CS pipeline [11]. After its multiple runs in test bench and a successful field trial of a 250 m long discharge pipeline at BARC, an EPIG was designed for 10" NB CS pipelines. Due to large diameter constraints, as well as lack of back wall access of the target pipeline under inspection, the tool is re-engineered for a smaller size and would perform sectorial inspection of the pipe. Unlike conventional MFL tools, where the permanent magnets are in contact with the pipe surface through CS brushes, this MFL-EPIG has rollers maintaining an air gap between magnetic circuit and pipe surface, thereby minimizing frictional drag during inspection and enabling movement by hand/winching mechanism. Also, this arrangement will not peel off the external protective paint coating of the pipelines. Use of MFL technology ensures precise defect sizing conforming to the requirements and specifications laid down by POF [12].

The paper is structured as follows. Section 2 covers basic understanding of MFL technology with respect to pipeline inspection, tool development with optimized design based on FE analysis and MFL signal analysis for defect detection and characterization. Section 3 deals with the test bench results. Section 4 concludes the paper highlighting the major advantages of MFL-EPIG and future scope of work.

2 EPIG Development Overview

2.1 MFL Basics

An MFL-PIG contains a strong permanent magnet assembly that magnetizes the CS pipe wall to near saturation. An array of circumferentially disposed hall-effect sensors located at magnetic neutral plane senses the leakage flux from pipe wall. In a defect free region, a background leakage flux is sensed. However, any metal loss

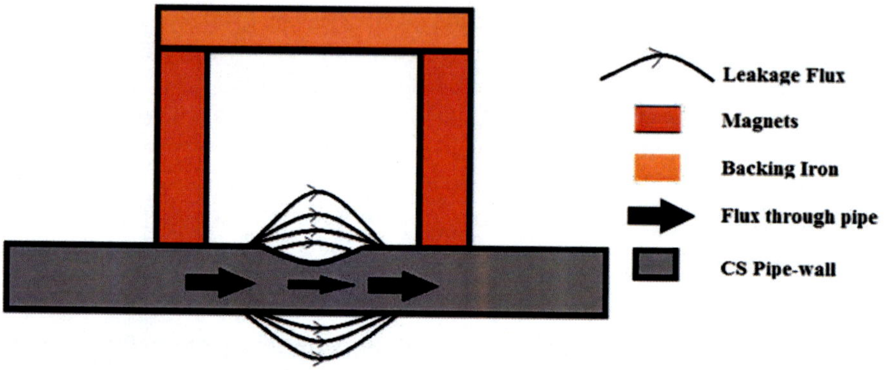

Fig. 1 Axial cross section of magnetized CS pipe wall with defect and leakage flux

on or within the pipe wall will act as a region of increased magnetic reluctance and cause a local disturbance in the distribution of magnetic flux. Thus, the density of leakage flux will increase locally in the vicinity of a defect (Fig. 1). The information in the sensed magnetic flux leakage (MFL) signal is analysed for defect detection and its characterization.

2.2 *Tool Design*

The MFL-EPIG consists of permanent magnet assembly with hall-effect probes at its magnetic neutral plane. These permanent magnets are required to magnetize the pipe wall to near saturation (optimum magnetization). For sub-optimal magnetization, leakage flux around metal loss defects would be low as the pipe wall can still accommodate more flux within itself. If pipe is magnetized beyond optimum level, then background leakage flux (noise signal) would increase leading to lower Signal-to-Noise Ratio (SNR). As stated earlier, the EPIG envisaged for 10″ NB pipeline is sectored because (1) the target pipes for inspection is too close to building wall and does not provide sufficient access and (2) a full periphery tool would be too bulky to manoeuvre. Therefore, unlike IPIG, where magnets are disposed uniformly about the circumference of the pipe leading to a near uniform pipe magnetization, SEPIG-10 has localized magnetization near the pipe sector under inspection. The magnetic assembly needs to be designed to achieve an optimum magnetization level, thus maintaining its desired signal integrity, without significant increase in frictional drag of the tool.

Magnetic Circuit Study of EPIG FE analysis to understand the effects of different magnetic circuits on CS pipe wall is a popular choice among IPIG designers due to the limitations of closed form solutions. In the present study, a sectored tool is designed and tested for optimum level of magnetization on the pipe wall. The permanent magnets used are NdFeB, and the backing iron is designed with SOFTMAG 48B.

The choice of these materials is based on the IPIG designs developed earlier. To minimize frictional force, a non-contact magnetic circuit concept was adopted. A constant air gap is maintained between the magnets and pipe surface by incorporating steel rollers for smooth movement of the tool over pipe wall. This is a trade-off between signal strength at defect locations (due to lower level of magnetization) and ease of inspection. Magnetostatic analysis of various designs was carried out using standard FE analysis software, few of which are discussed in the following paragraph, to arrive at an optimal magnetic circuit design for the tool.

SEPIG-10 was initially designed as two backing iron (10 mm thick) modules 35.5° apart with two sectors in each module (Fig. 2a). Each module consists of four sets of NdFeB magnets with 2 mm lift-off from pipe surface. This yielded a magnetization of 55% of optimum saturation level at the magnetic neutral plane for the pipe section under inspection. Redesigning the system to a single module with four sectors, 7.5 mm lift-off and 10 mm backing iron thickness increased the magnetization level to 75% saturation (Fig. 2b). Again, doubling the backing iron thickness (20 mm) resulted in a magnetization level of 91.6% (Fig. 2c). The magnetization level of CS pipe for each design is shown in Fig. 3.

It is apparent from the above studies that the size of backing iron, magnets and lift-off parameters affect the magnetization level inside the pipe and hence are typical design parameters to be tuned for optimal magnetization level at the zone of inspection. Based on the FE analysis results from multiple iterations, a single module with 20 mm thick backing iron design was adopted for SEPIG-10 (Fig. 4).

Besides having a set of rollers at the front and back for axial scanning of the pipe, the tool has an additional set of spring loaded rollers on either side for circumferential movement for periphery coverage.

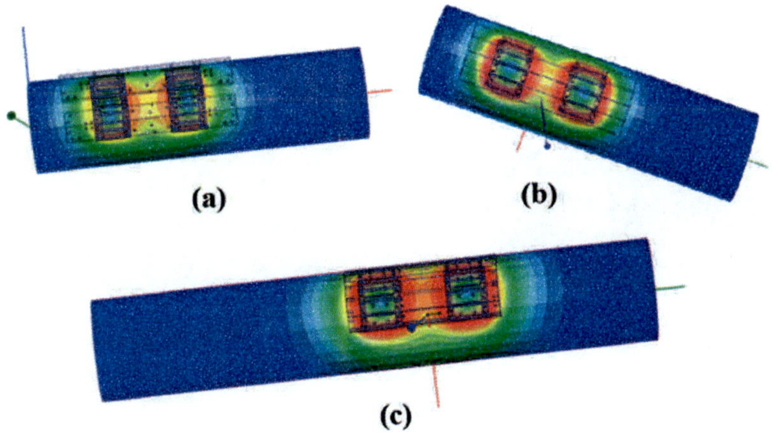

Fig. 2 FE analysis of various SEPIG-10 designs to study magnetization of CS pipe

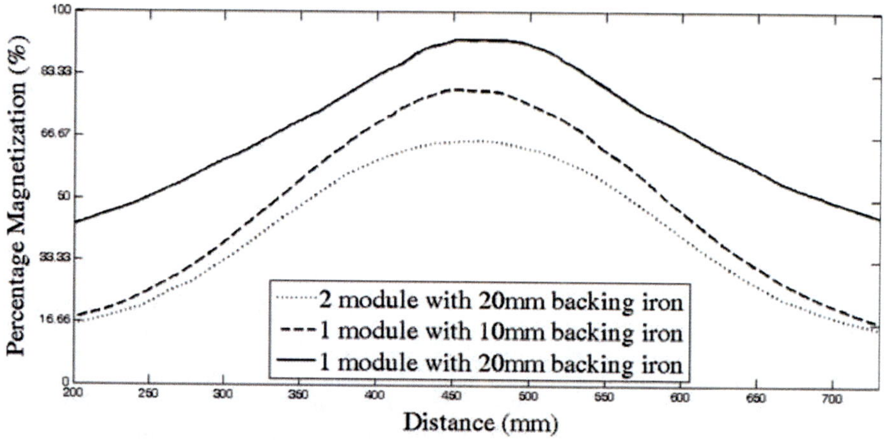

Fig. 3 Magnetization level on CS pipe for each design discussed in Sect. 2.2

Fig. 4 SEPIG-10 tool for MFL inspection of pipeline

2.3 *Acquisition and Processing System Design*

The sensors of the tool are connected to a portable data acquisition system (DAS) with inbuilt battery. The DAS consists of an in-house developed 32 channel FPGA-based electronics with onboard flash memory as storage option. The system can be connected to a laptop and C-scan (heat map) of the pipe surface can be viewed in real time along with data logging. Data is collected per sensor at the rate of 1 sample per mm when moving at a speed of 1 cm/s. The online C-Scan gives a clear idea of pipe sections having corrosion defects. These defects are then sized for its dimension estimation (length, width and depth) using an offline post-processing software.

2.4 *Defect Characterization and Signal Analysis*

Metal-loss defect characterization involves accurate sizing from MFL signals. The leakage flux density is primarily affected by percentage wall loss (%WL), defect length (along the direction of magnetization) and width (perpendicular to the direction of magnetization).

Pre-Processing of Raw Data Using Wavelet Transform (WT) MFL data acquired is subjected to baseline equalization and de-noising. Given the nature of pipeline defects and features are not deterministic (randomly oriented with random dimensions), the signals need to be addressed in a stochastic non-stationary framework. Also, the noise in these signals could be attributed to variety of factors like sensor bounce (vertical), sensor movement across magnetic neutral plane (horizontal), jerky movement of IPIG, etc. Thus, MFL signals are non-stationary in nature [13]. As the defect signal shape, magnitude and its extent are of paramount importance in determining defect parameters as well as its severity, the presence of unwanted noise can undermine a critical defect. As apriori knowledge about signal spectra is needed for standard de-noising filters, which is not available for such non-stationary signals, it is de-noised using discrete wavelet transform (DWT).

Wavelet transform projects signals onto a set of finite length basis functions, derived from a single mother wavelet, and have the form $\psi_{(a,\tau)}(t)$, where ψ is the mother wavelet, a is the scaling parameter, and τ is the translation parameter (Eq. 1).

$$\psi_{(a,\tau)}(t) = \psi\left(\frac{t-\tau}{a}\right) \tag{1}$$

For DWT, the scaling parameter a is changed in values of 2^n, $n \in \mathbb{Z}$ and can be interpreted as a filter bank of constant Q band-pass filters. Each of these filters has varying bandwidth and centre frequency. Short-lived wavelets (higher scales) would project high frequency component of the signals with better time resolution and vice versa. Equation (2) shows the wavelet transform of an arbitrary signal, $x(t)$.

$$y(\tau, a) = \frac{1}{\sqrt{|a|}} \int x(t)\psi^*\left(\frac{t-\tau}{a}\right)\mathrm{d}t \tag{2}$$

$y(\tau)$'s are the wavelet coefficients for each discrete scale; a and ψ^* are the complex conjugate of ψ.

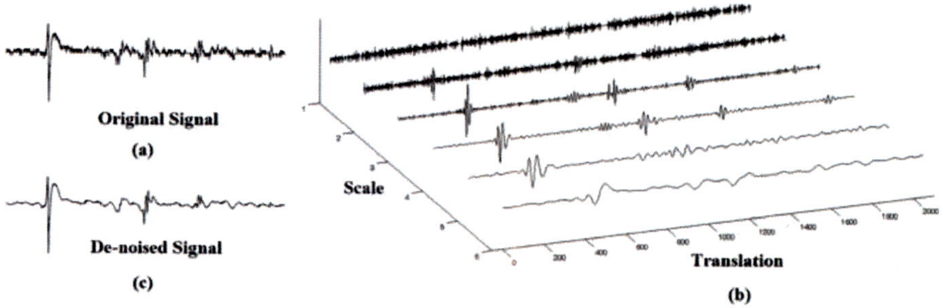

Fig. 5 Wavelet de-noising of MFL signal and its decomposition into scales

The noise in MFL signals could be attributed to variety of reasons like sensor bounce (vertical), movement of the sensor across the magnetic neutral plane (horizontal), jerky movement of EPIG, etc. As the signal shape, its extent and magnitude are of paramount importance with respect to defect sizing, any unwanted noise can affect the signal analysis leading to incorrect defect sizing.

The MFL signals are subjected to DWT by bior-2.2 into 6 scales. The choice of this wavelet was based on its peak preserving property, as peak to peak value of MFL signal is representative of severity of the defect [14]. Each scale represents the signal and noise coefficients projected on wavelet corresponding to its bandwidth. It can be seen in Fig. 5b that the first two scales for the MFL signal shown in Fig. 5a mostly contain noise projections while the defect features are captured by other scales. In every scale, these noise components are suppressed with an appropriate threshold [15]. The coefficients are thereafter converted back to time domain (Fig. 5c) with Inverse Wavelet Transform (IDWT).

Characterization or Sizing of Defects The relationship between signal and defect parameters is nonlinear and ill-posed inverse mapping problem [16]. After analysis of large datasets obtained from IPIG development, an empirical relationship has been established between the defect parameters and recorded signal. Other techniques for data analysis have also been reported, as like in [17] defect MFL signals are analysed using Wavelet basis function neural network (WNN) and multi-sensor data fusion [18]. Defect characterization amounts to estimation of three key parameters: length, width and %WL from the signal. Hall-effect sensors record either the radial or axial component of leakage flux. This tool is designed to record the radial component of the same. In this component, the defect signal is bipolar in nature and is characterized by three primary parameters—peak to peak span (samples), number of sensors seeing the defect and maximum peak to peak flux density $B_{\max}$ (Fig. 6). The difference in samples between positive and negative signal peak is called span and can be directly related with the length of a defect. Width and %WL are estimated based on nonlinear models developed empirically from the defect signals generated by IPIGs on commercial pipelines. This is discussed in details in [19]. Width is estimated as an function of number of sensors (N), sensor circumferential pitch and $B_{\max}$ while %WL is a function of $B_{\max}$, the ratio of estimates of width and length and the area

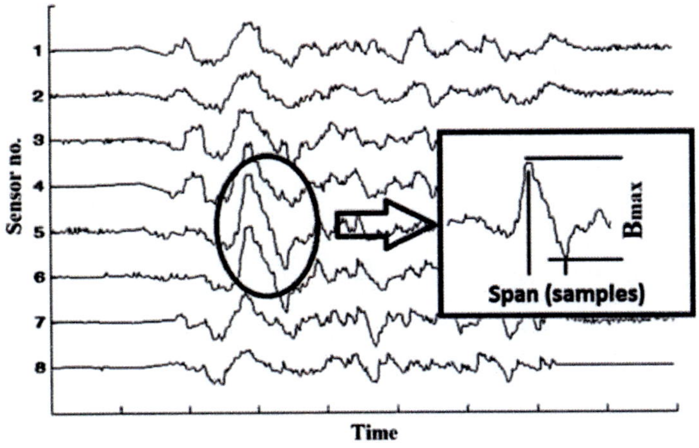

Fig. 6 A typical radial MFL signal and its parameters used for defect characterization

of the metal loss defect. This model has been optimized for EPIG signals to account for non-uniform reduced pipeline magnetization.

3 Test Bench Trials

An array of faults were created (both internal and external) on test bench (standard 2.5 m long 10 in. NB CS pipe), and multiple trial runs of the tool were carried out to calibrate the empirical model for defect characterization and generate a data bank for the same. The tool was pulled with a battery operated winching machine for uniform speed of inspection (1 cm/s). The entire arrangement is shown in Fig. 7. Two set of defects (three external and three internal) having length and width of 18 mm × 18 mm and depth of 10%WL, 25%WL and 35%WL were created on a sectored pipe and welded on the test bench. Fig 8 shows the radial MFL signal of defects recorded by SPEIG-10. It can be seen in Fig. 8a that weld signature has been picked by all the sensors at the same time instant. The C-Scan of MFL signal (Fig. 8a) is shown in Fig. 9a. For better visualization of defects, the weld signatures are suppressed (linear interpolation between two data points on either side of weld signature having similar magnitude), and the signal is de-noised using DWT. The enhanced C-Scan is shown in Fig. 9b. These defect signals were thereafter calibrated against the known defect dimensions by altering the constants of the empirical nonlinear map discussed earlier. Multiple trial runs were carried out for repeatability and consistent results.

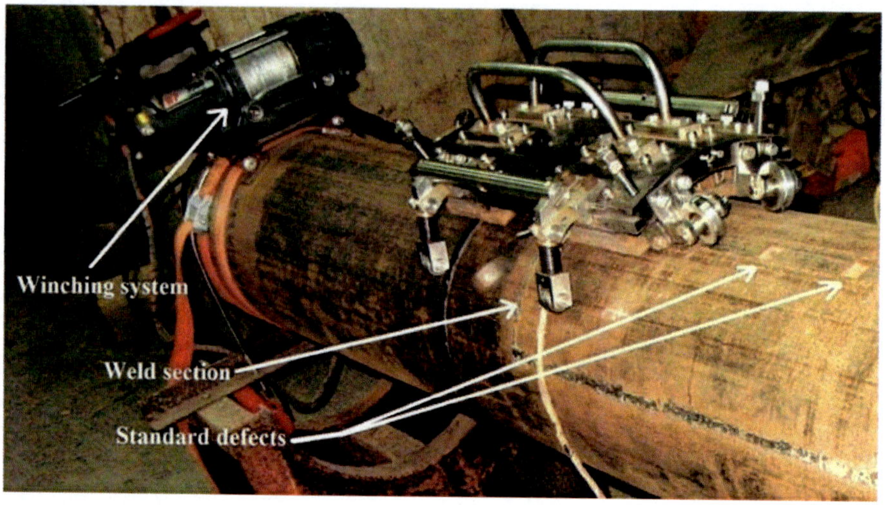

Fig. 7 EPIG tool and test rig arrangement

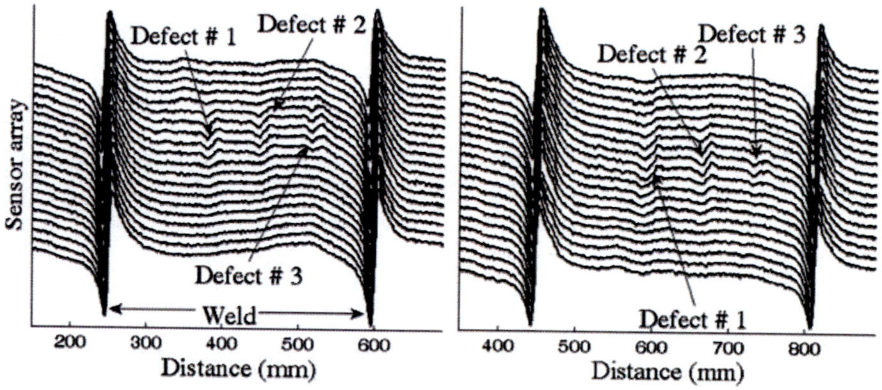

Fig. 8 **a** Radial MFL signal for external and **b** internal defects on test bench

4 Conclusion

The SEPIG-10 is a prototype tool for external inspection of CS pipeline. It is one of the promising methods of inspection of unpiggable pipelines using MFL technique. It provides accurate defect size estimation and does not require any sensor to surface coupling, surface preparation or removal of paint. The tool can detect defects on either surface of the pipe wall with all the advantages associated with MFL inspection. The tool can be used with minimum modifications to inspect CS pipelines of different sizes existing in various process plants and industries.

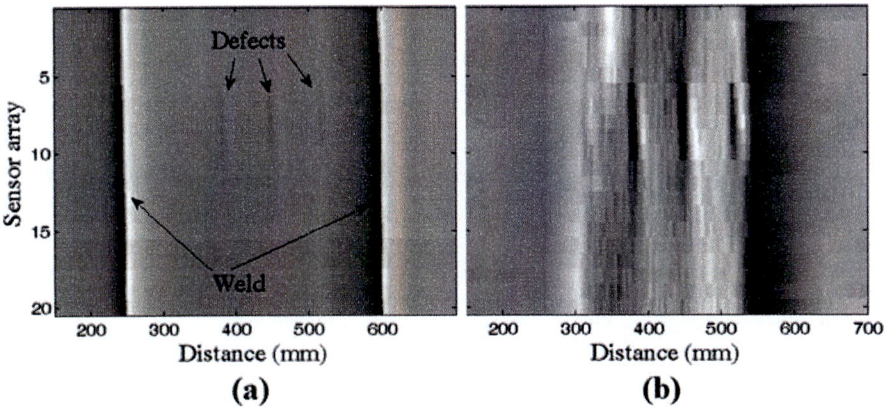

Fig. 9 **a** C-Scan image of raw MFL signal shown in Fig. 8a and **b** its defect signatures enhanced by DWT and weld suppression

References

1. Mukhopadhyay S, Lahiri SK, Saha S, Bahuguna SK, Dhage S, Ramrane SC, Mukherjee D, Taly YK (2012) Development of next generation in-line inspection technologies. In: PETROTECH-12, 10th international oil and gas conference and exhibition under the aegis of Ministry of Petroleum and Natural Gas, Government of India
2. Luciano N-B, Alberto S-CJ, Carlos P-OJ, Manuel R-AJ (2010) Development of an ultrasonic thickness measurement equipment prototype. In: 20th international conference on electronics communications and computers (CONIELECOMP), Cholula, 2010, pp 124–129
3. Mukherjee D, Manral SG, Ramrane SC, Chandra Y, Dhage S, Bahuguna SK, Saha S, Lahiri SK, Mukhopadhyay S, Kumar A (2013) Development of secondary sensor system based on Eddy current technology for in-line inspection tool. BARC Newsletter
4. Guidance on achieving successful In-Line-Inspection—October 2018. Pipeline Operators Forum. www.pipelineoperators.org
5. Mills GH, Jackson AE, Richardson RC (2017) Advances in the inspection of unpiggable pipelines. Robotics 6(4):36
6. Yukawa T, Suzuki M, Satoh Y, Okano H (2006) Design of magnetic wheels in pipe inspection robot. IEEE Int Conf Syst Man Cybern 1:235–240
7. Pipe crawler—Ultrasonic inspection robots. https://inspector-systems.com/pipe-crawler/ultrasonic-inspection-robots.html. Accessed on 1 Oct 2019
8. Crawler for thickness measurement and external corrosion mapping of pipelines. https://www.img-us.com/solutions-for-non-destructive-testing-ndt/crawler-for-thickness-measurement-and-external-corrosion-mapping-of-pipelines/. Accessed on 3 Oct 2019
9. Pople A (2003) Magnetic flux leakage pigs or ultrasonic pigs—The case for combined intelligent pig inspections. In: 6th International conference, pipeline rehabilitation and maintenance, 6–10 Oct 2003, Hilton, Berlin, Germany
10. Kim D-K, Yoo H-R, Cho S-H, Koo S-J, Kim D-K, Yoo J-S, Rho Y-W (2016) Inspection of unpiggable natural gas pipelines using in-pipe robot. In: Recent adnavces in electrical engineering and related sciences. AETA 2016 Lecture notes in electrical engineering, vol 415. Springer, Cham
11. Saha S, Ramrane SC, Mukherjee D, Chandra Y, Lahiri SK, Marathe PP, Bagchi AC (2017) Development of external pipeline inspection gauge for monitoring the health of industrial carbon steel pipelines. BARC Newsletter

12. Specifications and requirements for in-line inspection of pipelines—Version 2016, Pipeline Operators Forum. www.pipelineoperators.org
13. Mukherjee D, Saha S, Lahiri SK, Mukhopadhyay S (2014) Application of wavelets in non-stationary signal processing. In: Third national symposium on advances in control and instrumentation (SACI), Mumbai
14. Mukherjee D, Saha S, Mukhopadhyay S (2013) Inverse mapping of magnetic flux leakage signal for defect characterization. NDT & E Int 54:198–208
15. Donoho DL (1995) De-noising by soft-thresholding. IEEE Trans Inf Theory 41(3):613–627
16. Joshi AM (2006) Inverse problems in non-destructive evaluation of gas transmission pipelines using magnetic flux leakage. Ph.D. dissertation, Michigan State University
17. Mao B, Lu Y, Wu P (2014) Signal Processing and defect analysis of pipeline inspection applying magnetic flux leakage methods. Intel Serv Robot 7(4):203–209
18. Qiang S (2011) Data fusion for MFL signal characterization. Appl Mech Mater 44–47:3519–3523
19. Saha S, Mukhopadhyay S, Mahapatra U, Bhattacharya S, Srivastava GP (2010) Empirical structure for characterization of metal loss defects from radial magnetic flux leakage signal. NDT & E Int Elsevier Sci 43:507–512

Ultrasonic Waveguide Technique for Temperature Measurement Using $T(0,1)$ Wave Mode

Suresh Periyannan and Krishnan Balasubramaniam

1 Introduction

The temperature profile measurement is essential parameter in the industries such as glass and metal melting industries, thermal and nuclear power plants. Monitoring and recording the temperature at multiple locations are required in these industries' melter during heat treatment process. The radiation pyrometers and conventional thermocouples are the common sensors for measuring the temperature in these industries. Also, ensuring the temperature uniformity in the industries' melter is essential using these temperature sensors. The temperature profile of the molten glass conductivity, viscosity and density, glass melts, plenum off-gases are the important parameters in the nuclear power plants Woskov et al. [1]. However, these conventional sensors are having the accuracy issues due to the sensor's drift, while these sensors are involved into applications at prolong duration in operations [2]. Also, thermocouple has a reliability problem at the hot junction, when it is involved in the measurement at high-temperature environment. Now, another temperature measurement approach is more important for solving such a kind of these issues.

In this work, we would like to demonstrate the bent waveguide concept for measuring the temperature at different depths of the laboratory furnace, which may mimics a melter environment using ultrasonic waveguide technique. In general, the transducer must be positioned at the measurement region of interest while using bulk ultrasonic wave technique [3–6]. Titanate, bismuth, and lithium niobate crystals were used to develop the ultrasonic high-temperature transducers [5–8] that

S. Periyannan (✉)
Department of Mechanical Engineering, National Institute of Technology Warangal, Warangal, India
e-mail: sureshp@nitw.ac.in

K. Balasubramaniam
Department of Mechanical Engineering, Indian Institute of Technology Madras, Chennai, India

C. K. Mukhopadhyay and R. Mulaveesala (eds.), *Advances in Non-destructive Evaluation*, Lecture Notes in Mechanical Engineering,
https://doi.org/10.1007/978-981-16-0186-6_10

can be used in to high-temperature applications with limitations. Most piezoelectric materials (PZT) cannot be used above their Curie temperature, as they become depolarized beyond that point then lose their ability to send and receive the ultrasonic signals. Ultrasonic waveguide method is an alternate approach to overcome those issues. Hence, an ultrasonic transducer can be separated from the hostile environmental region for avoiding the transducer issues. Then ultrasonic wave can be transmitted from through the waveguide at which region of interest would like to measure the temperature. Here, an appropriate transducer holder is designed and fabricated for holding the transducer and waveguide securely. The ultrasonic wave is coupled from transducer to the waveguide without air gap using this holder with appropriate couplants (silicon grease). Hence, the wave transferred between the two locations, thereby facilitating the use of efficient, low cost and conventional PZT transducers. Ultrasonic waveguide-based measurement technique has been extensively used for the development of sensors for density, level, viscosity, temperature, and rheology measurement of the surrounding fluid among others [8–18]. The main drawback of the earlier reported approach is targeted in the single region of measurement. Here, we would like to develop the ultrasonic waveguide technique, which can help us to measure the temperature at multiple target location in a hot chamber.

In this paper, consider the pair waveguides and each waveguide is introduced with single bent for temperature measurement using a single shear wave transducer. This bent waveguide approach overcomes the key issues of the temperature gradients with the multiple straight waveguides method (as reported, for example, in [15]). Here, the flexible wires of circular cross section used as waveguide. In this work, carried out using the shear wave transducer $T(0,1)$ fundamental torsional ultrasonic guided wave mode in the circular cylindrical wires. The reflected signals are obtained from each bend of pair waveguides, and the time of flight (TOF) and amplitude measurements can be used for learning more about physical properties of the waveguide's surrounding medium. For this work, the material properties of the waveguide and δTOF must be known as function of temperature. If this information is not available, then the earlier reported technique [18–22] may be utilized to determine these temperature dependent moduli of the waveguide materials.

The difference in TOF due to a change in temperature, henceforth called δTOF, was used to monitor instantaneous temperatures around the horizontal region (bend portion) of an each waveguide. A peak-tracking concept (described in more detail elsewhere, [15, 21–29]) is used to continuously monitor or measure the δTOF from each bend "horizontal" region of the waveguide, while bend region is subjected into heating cycle. Each bend region of the waveguide is assumed to represent a uniform temperature distribution region of interest at multiple levels of furnace, due to the relatively short length of this region. Since the measured δTOF is an average value of local temperature changes over the bend (horizontal) region of the waveguides, the configuration of the waveguides must be carefully designed to avoid the signals' overlaps. This technique significantly reduces the cost of instruments, when compared to wave transmission/reception in single waveguide approach. The performance of optimal configuration with multiple waveguides connected to a single transducer was demonstrated in a high-temperature furnace for operation up to 1100 °C.

2 Ultrasonic Guided Waves in Cylindrical Wire

Guided waves [30] can be thought of as a superposition of partial plane waves that are reflected within waveguide boundaries. The ultrasonic wave propagation in waveguides is characterized by the phase velocity, waveguide material properties, frequency, and its dimensions. In a circular cross-sectional waveguide, there are three different types of wave modes: longitudinal (*L*), flexural (*F*), and torsional (*T*), all these wave modes are propagating in the axial direction (z) of cylindrical coordinate system (r, θ, and z). While different wave modes can be formed/generated in circular cross-sectional waveguides, we are focused on the fundamental torsional $T(0,1)$ mode. This mode has no dispersion over a wide range of frequencies and can be easily generated/transmitted in wire-like flexible waveguides made of high-temperature thermocouple material Chromel.

Phase velocity and group velocity dispersion curves obtained using DISPERSE [31], for Chromel waveguide is shown in Fig. 1, for the frequencies and diameters of interest to this work. The elastic moduli and the density of the waveguide material found using an approach explained previously [21, 22] are shown in Table 1. Chromel was used here, because of its ductility, high melting point, re-configurable and easy availability [23–28]. It is desirable that the waveguide must exhibit minimal dispersion in the chosen frequency range; this ensures that the pulse width of the signals is relatively narrow and improves the time of flight measurements. Hence,

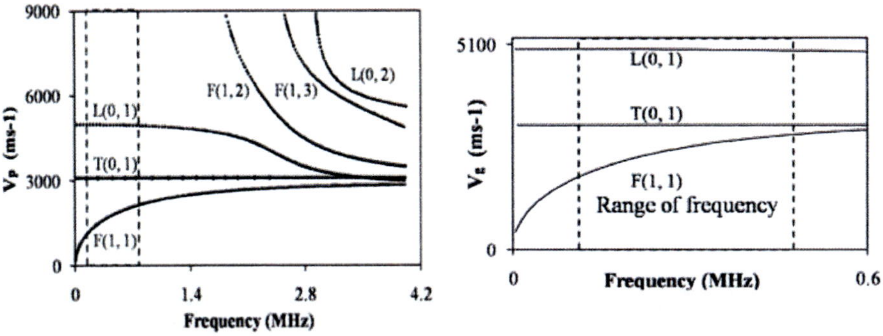

Fig. 1 Phase and group velocity dispersion curves of $L(0,1)$ and $T(0,1)$ modes for Chromel rod ($d = 1.2$ mm), respectively, and also frequency range of interest

Table 1 Chromel properties and dimensions of the waveguide used

Material	Density (ρ) Kg/m^3	Young's and shear modulus (GPa)	Velocities (*VL* and *VT*) m/s	Poisson's ratio (μ)	Bend lengths (c), wire dia (d) mm
Chromel	8740	196.4 and 80	4741 and 3025	0.23	29, 57, 130, 1.2

an operational frequency of 100–500 kHz range was chosen, using the dispersion curves. In this work, the range of working frequency is used with in the fundamental (low) frequency. Also, the temperature measurement may be improved using high frequency, but the wave may not travel long distance due to more attenuation at high frequency. Further, in order to maintain regions of low dispersion, appropriate diameter of the wire-like waveguide was chosen.

3 Temperature Measurement Using T(0,1) Mode

3.1 Single Bent Waveguide with a Single Transducer for Temperature Measurement

The $T(0,1)$ wave mode is less dispersive compared to $L(0,1)$ mode in the same material and thickness as shown in Fig. 1. The wavelength of the $T(0,1)$ mode is significantly smaller than that of $L(0,1)$ mode. The experimental setup and instruments used for this work from the earlier reported temperature measurement using $L(0,1)$ wave mode [25]. The piezoelectric ceramic-based conventional shear wave transducer was acoustically coupled to the bent waveguide using the transducer holder elsewhere [22–26]. The sensor holder setup ensured that the orientation between the transducer and the waveguide surface for transmission/reception of $T(0,1)$ mode from the shear wave transducer into the waveguide. Here, transducer orientation was followed perfectly to transmit and receive the $T(0, 1)$ wave modes in the waveguide and it was ensured that based on earlier studies [17, 18, 21]. Other than that, experimental setup and procedures are the same as described in the earlier reported works [21–26]. The sensor (transducer) holder is placed on the furnace outside (top side) for holding the one end of the waveguide with transducer. The other end of the waveguide sensor was kept inside the furnace. Hence, a single waveguide was connected initially to shear wave transducer and that other end (sensor region) was heated into different temperature while the furnace operated from low to high temperature.

The waveguide bent (sensor gage length) region was positioned in horizontal direction at high-temperature region of the furnace. The resistant heating type furnace used in the laboratory for studying this sensor. Furnace was operated from room temperature to 1100 °C. Figure 2a, b shows single bent waveguide and corresponding A-scan plots at room temperature and high temperatures. The δTOF's values were measured from bend length of waveguide sensor at different temperatures using Eq. (1) as shown in Fig. 2c. The bend region of waveguide was colocated with the K-type thermocouples for measuring the temperature at nearby waveguide. In the experiment, steady-state heating process (35–1100 °C) was conducted. The A-scan signals were collected simultaneously from sensor (bent) region at different temperature as shown in Fig. 2b.

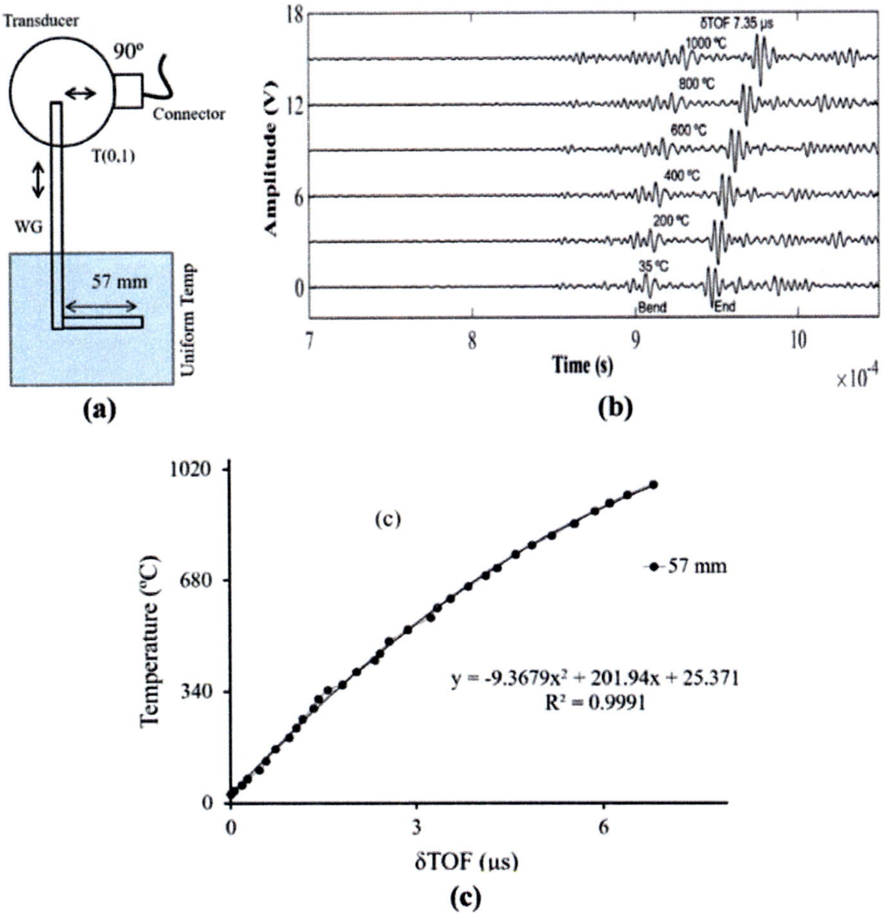

Fig. 2 **a** Schematic of experimental setup for a single waveguide connected to single transducer, **b** A-scans signals collected from bent waveguide at various temperatures and **c** shows the δTOF values of the bent region at various temperature

Reflected A-scan signals from bent portions were continuously monitored using a signal's peak-tracking technique method, which reported elsewhere [22–27]. Subsequently, the δTOF between bend and end (one sensor) of waveguide was measured using Eq. (1). The δTOF's of each sensor recorded at different temperatures inside the furnace.

Instantaneous time of flight difference (δTOF) of a waveguide is shown [25].

$$(\delta\mathrm{TOF}_{n+1})_i = \left[\mathrm{TOF}_{(n+1)i} - \mathrm{TOF}_{ni}\right] - \left[TOF_{(n+1)} - TOF_n\right] \quad (1)$$

where

TOF_{ni}, TOF_n Instantaneous (i) TOF at various temperature and (ii) TOF at room temperature, respectively.

$(\delta TOF_{n+1})_i$ Instantaneous change in TOF between the reflections from each sensor location-n.

If $n = 0$, the δTOF value is obtained from the straight region (vertical) of waveguide. If $n = 1$, the δTOF value is obtained from the 1st bend region of waveguide "sensor 1". $n = 0, 1, 2, 3$.For multiple bent waveguide sensors.

3.2 *Pair of Bent Waveguides with Single Transducer for Temperature Measurement*

A similar experimental setup was used to design the pair of bent waveguide sensors for sensing the temperature using a single shear wave transducer as shown in Fig. 2. Figure 3a shows a pair of bent waveguides connected to a single shear wave transducer (0.5 MHz). The same transducer was oriented 90° to the axes of the waveguides [21] for generating/receiving the $T(0,1)$ wave mode in the bent waveguides. All the waveguides were designed in order to avoid any overlap of $T(0,1)$ mode reflections from bend and end of the waveguides. Figure 3a–c shows a pair of waveguides (bend lengths: 29, 130 mm) and corresponding A-scan plot. The δTOF's values were measured from each bend length at different temperatures. The sensitivity of longer and shorter bend lengths "130, 57, and 29 mm" was found the δTOF's values 16.1 μs, 7.35 μs, 4.01 μs, respectively, and could be observed that in linear approximations as shown Fig. 3c. Each bent waveguide sensor's difference in time of flight was measured instantaneously using Eq. (1) at various temperature. The waveguide sensor to be calibrated based on the δTOF and co-located thermocouple's data at nearby sensor region during the calibration experiment. Then, emphirical equation to be formed based on the calibration (e.g., Figure 2c and more details [23, 25]) that can be used while the waveguide sensor is subjected or located into the unknown temperature region.

4 Conclusions

Ultrasonic guided $T(0,1)$ wave mode-based temperature sensor used on single bent waveguide and pair of bent waveguide using single shear wave transducer. The single bend waveguide and pair of waveguide with bend configurations were designed to demonstrate the measurement of temperature in a high-temperature furnace. A single waveguide and pair of waveguide sensor's signal were collected at different temperatures. Thermocouples were used simultaneously to measure the temperature near by the waveguide sensor for the initial calibration analysis. The resolution of ultrasonic temperature measurement using $L(0,1)$ mode with 57 and 29 mm bend waveguides was limited to 2.4 °C and 4.7 °C, respectively, from the early reported work [25]. The similar experimental setup and procedures were used the modes

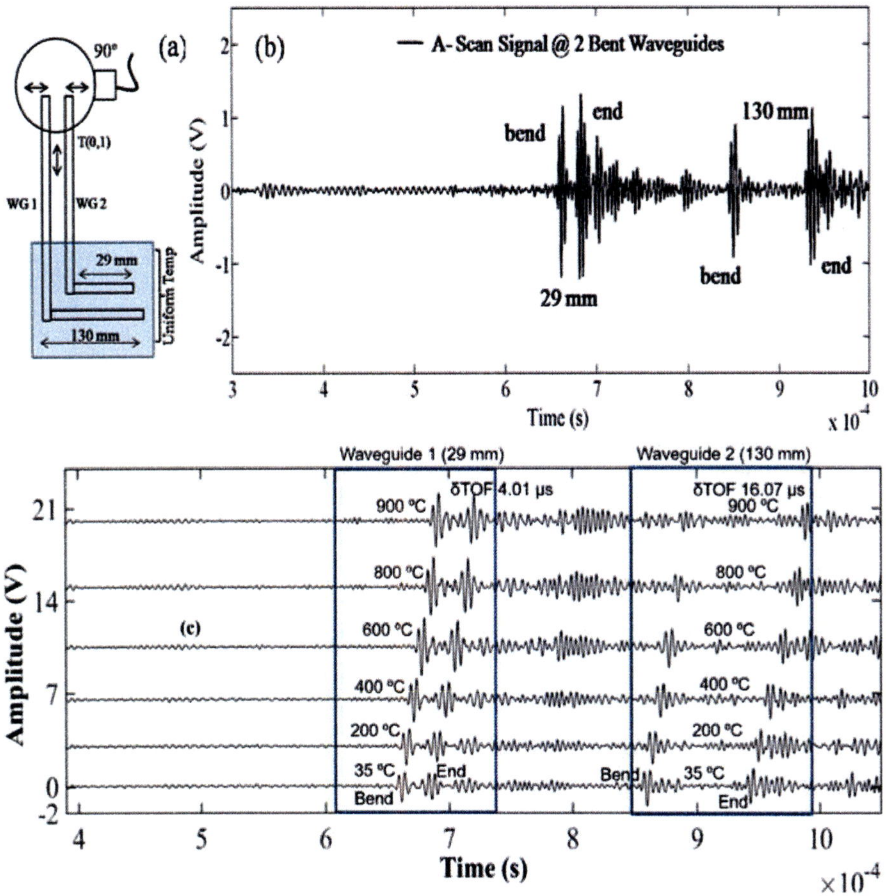

Fig. 3 **a** Schematic of experimental setup for a pair of waveguides connected to single transducer, **b** A-scans signals collected from a pair of bent waveguides, **c** The sensitivity results obtained from different bent lengths of waveguide sensors at different temperatures

with lower group velocities such as the $T(0,1)$ mode and improved the resolutions of temperature measurement 1.4 and 2.5 °C using the same bend lengths 57 and 29 mm, respectively. The maximum temperature difference between the ultrasonic waveguide reading and the thermocouple was observed that 5 °C in the range of measured temperatures from 30 to 900 °C. If a higher sampling rate is used, for instance a 1 GHz sampling rate "this work 100 MHz used," this can improve the resolution as well as reduce error between the thermocouple and waveguide sensors. Multiple ultrasonic bent sensors can be introduced in a single wire for the temperature measurement for measuring the temperatures at multiple (different depths) levels inside a furnace.

References

1. Woskov PP, Sundaram SK, Daniel WE (2008) Millimeter-wave measurement of high level and low level activity Glass melts, MIT, PSFC/RR-08-4, 1–28
2. Bentley RB (1990) Long-term drift in mineral-insulated Nicrosil-sheathed Type K thermocouples. Sens Actuators A Phys 24:21–26
3. Huang KN, Huang CF, Li YC, Young MS (2003) Temperature measurement system based on ultrasonic phase-shift method. IEEE, pp 293–295
4. Baba A, Searfass CT, Tittmann BR (2010) High temperature ultrasonic transducer up to 1000 °C using lithium niobate single crystal, App Phys Lett 97:232901
5. Parks DA, Zhang S, Tittmann BR (2013) High-Temperature (> 500 °C) ultra-sonic transducers: an experimental comparison among three candidate piezoelectric materials ultrasonics ferroelectrics and frequency control. IEEE Trans 60:1010–1015
6. Tsai WY, Chen HC, Liao TL (2005) An ultrasonic air temperature measurement system with self-correction function for humidity. Meas Sci Technol 16:548–555
7. Tsai WY, Chen HC, Liao TL (2006) High accuracy ultrasonic air temperature measurement using multi-frequency continuous wave. Sens Actuators A Phys 132:526–532
8. Lynnworth LC (1989) Ultrasonic measurements for process control: theory, techniques applications. Academic Press Inc. New York, U.S.A.
9. Kazys R, Sliteris R, Raisutis R, Zukauskas E, Vladisauskas A, Mazeika L (2013) Waveguide sensor for measurement of viscosity of highly viscous fluids. Appl Phys Lett 103:204102
10. Balasubramaniam K, Shah VV, Costley D, Bourdeaux G, Singh JP (1999) High temperature ultrasonic sensor for the simultaneous measurement of viscosity and temperature of melts. Rev Sci Instrum 70(12):1–6
11. Balasubramaniam K, Shah VV, Costley D, Singh JP (2001) Viscosity and temperature measurements at very high temperature by ultrasound reflection, U.S. Patent 6,296,385
12. Shah VV, Balasubramaniam K (2000) Measuring newtonian fluid impedance using the phase of a reflected ultrasonic shear wave. Ultrasonics 38:921–927
13. Prasad VSK, Balasubramaniam K, Kannan E, Geisinger KL (2008) Viscosity measurements of melts at high temperatures using ultrasonic guided waves. J Mater Process Technol 207:315–320
14. Pandey JC, Raj M, Lenka SN, Periyannan S, Balasubramaniam K (2011) Measurement of viscosity and melting characteristics of mould powder slags by ultrasonics. J Iron Making Steel Making 38:74–79
15. Periyannan S, Balasubramaniam K (2015) Multi-level temperature measurements using ultrasonic waveguides. Measurement 61:185–191
16. Nishanth R, Lingadurai K, Suresh P, Balasubramaniam K (2017) Ultrasonic waveguide-based distributed temperature measurement on a solid surface. Insight 7(59)
17. Nishanth R, Balasubramaniam K, Suresh P (2018) Interpretation of Axi and Non-axi symmetric reflectors in the ultrasonic sensor developments, IEEE Sens 18(14)
18. Nishanth R, Balasubramaniam K, Suresh P (2019) Ultrasonic waveguide based level measurement using flexural mode F(1, 1) in addition to the fundamental modes. Rev Sci Instrum 90(4):045108
19. Greer ZL (2001) Temperature frequency and young's modulus an aluminium tuning fork. ISB J Phys 5:1–4
20. Balasubramaniam K, Periyannan S (2020) Waveguide technique for the simultaneous measurement of temperatures dependent properties of materials. U.S patent no. US 10794870B2
21. Periyannan S, Balasubramaniam K (2015) Simultaneous moduli measurement of elastic materials at elevated temperatures using an ultrasonic waveguide method. Rev Sci Instrum 86:114903
22. Periyannan S, Balasubramaniam K (2016) Moduli determination at different temperatures by an ultrasonic waveguide method. Exp Mech 1–14. https://doi.org/10.1007/s11340-016-0157-y
23. Periyannan S, Rajagopal P, Balasubramaniam K (2016) Re-configurable multi-level temperature sensing by ultrasonic "spring-like" helical waveguide. J App Phys 119:144502

24. Periyannan S, Rajagopal P, Balasubramaniam K (2015) Robust ultrasonic waveguide based temperature sensing. Phys Procedia 70:514–518
25. Suresh P, Rajagopal P, Balasubramaniam K (2017) Ultrasonic bent waveguides approach for distributed temperature measurement. Ultrasonics 74:211–220
26. Balasubramaniam K, Suresh P (2019) A novel ultrasonic waveguide technique for distributed sensing and measurements of physical and chemical properties of surrounding media. Patent no: US 10520370B2
27. Suresh P, Rajagopal P, Balasubramaniam K (2017) Multiple temperature sensors embedded in an ultrasonic spiral-like waveguide. AIP Adv 7:035201
28. Periyannan S, Rajagopal P, Balasubramaniam K (2016) Torsional mode ultrasonic helical waveguide sensor for re-configurable temperature measurement. AIP Adv 6:065116
29. Suresh P, Balasubramaniam K (2020) Integrated thermocouple waveguide sensor system and method to measure physical properties of waveguide material and surroundings, Patent Application No. 201941032842. Patent No: US10794870B2
30. Rose JL (1999) Ultrasonic waves in solid media. Cambridge University Press, and Cambridge, UK, pp 143–152
31. Pavlakovic BN, Lowe MJS, Cawley P, Alleyne DN (1997) DISPERSE: a general purpose program for creating dispersion curves. Rev Prog Quant Non-Destruct Eval 16:185–192

Rapid Non-destructive Evaluation of Fireproof Coating Degradation Using Terahertz Line Scanner

A. Mercy Latha, A. S. Nirmala Devi, Harkrishnan Kaimal, and Bala Pesala

1 Introduction

Typically, aero engine parts, automobile exhausts, furnace parts, gas turbine parts are exposed to very high temperatures and rapid temperature gradients. Hence, fireproof or heat-resistant coatings are essential for protecting these parts from thermal and structural failures [1, 2]. In addition to the thermal protection, these coatings also form a physical barrier between the substrate and the environment thereby preventing environmental damages such as corrosion. [3]. However, defect-free and uniform coating during the manufacturing process is necessary for assuring the quality and functionality of these coatings in field conditions.

During the service life of these coated parts, unpredictable harsh environmental conditions may cause moisture or other hygroscopic species to diffuse into the coating resulting in deterioration/corrosion of the substrate/coating [4, 5]. Such regions of moisture ingression are not easily detectable until they cause structural failures. Hence, it is extremely important to identify regions of moisture ingression and any substrate material degradation induced by them, in advance [6].

Several NDE techniques have been explored for evaluating coating deteriorations such as ultrasound, eddy current, and thermography. Traditional ultrasound technique requires proximity and a coupling fluid for NDE which is cumbersome in field applications [7–9]. Eddy current method is very sensitive to magnetic permeability

A. Mercy Latha · A. S. Nirmala Devi · B. Pesala (✉)
CSIR-Central Electronics Engineering Research Institute (CEERI), CSIR Madras Complex, Chennai, India
e-mail: balapesala@gmail.com

A. S. Nirmala Devi · B. Pesala
Academy of Scientific and Innovative Research (AcSIR), Ghaziabad, India

H. Kaimal
Indian Institute of Technology Madras, Chennai, India

C. K. Mukhopadhyay and R. Mulaveesala (eds.), *Advances in Non-destructive Evaluation*, Lecture Notes in Mechanical Engineering,
https://doi.org/10.1007/978-981-16-0186-6_11

variations and is normally suitable for conductive coatings only [10–12]. Thermographic testing is limited to the evaluation of surface and sub-surface defects only as it is based on detecting the surface temperature [13–16]. Further, it is hard to delineate the effects of coating material degradation and moisture ingression using passive thermography technique.

Alternatively, terahertz (THz) imaging is emerging as an attractive non-contact technique for NDE [17, 18]. THz waves have the ability to penetrate most dielectric materials revealing details about small defects, coating non-uniformity with significant contrast. Further, water has strong absorption in the THz frequency range and hence moisture ingression can be easily detected. In conventional single emitter–detector THz imaging configuration, image acquisition speed is greatly limited by the speed of the mechanical stages required to perform the raster scan. This had been a major shortcoming preventing its usage in real-time NDE applications [19]. Recent advances in THz detector arrays and THz cameras have facilitated the realization of rapid THz imaging systems [20].

In this work, THz NDE using a line scanner has been done for identifying coating degradations in terms of moisture ingression, substrate defects, and coating impurities. The setup consists of a high-power 100 GHz IMPATT diode source and line scanner with 256 individual GaAs-based detectors [21]. This system can offer high imaging speed of 5000 fps suitable for real-time monitoring of coating process for quality control. The total image acquisition time for a sample of dimensions 140 $\times$ 100 mm^2 is ~2.33 min, which is >1000 times faster than the conventional single emitter-detector raster scan configuration.

2 Experimental Setup

The THz source is a 100 GHz IMPATT diode producing continuous wave THz signal of 80 mW power. The IMPATT diode is a silicon double drift diode with 0.6 μm transit region, mounted on a copper heat sink. The source is fitted with a cylindrical PTFE lens which ensures proper illumination of the entire line detector. The THz detector is a line scanner comprising of a 1D array of 256 individual GaAs detectors. The beam profile across the line scanner is Gaussian as shown in Fig. 1 and so the acquired images are normalized with respect to this reference profile. The scanner has a minimum detectable power of 14 nW at 100 fps and is capable of imaging at very high speeds of 5000 fps (5 kHz).

By using 1D array of detectors, with just 1D movement of the sample perpendicular to the direction of the array, 2D images of the sample can be generated. This eliminates the time-consuming 2D raster scans, thereby reducing the image acquisition time. In most of the industrial production units, the samples are already moving on a conveyor belt and hence additional translation stages for the sample movement are not required. In such applications, the THz line scanner can be conveniently placed below the conveyor belt.

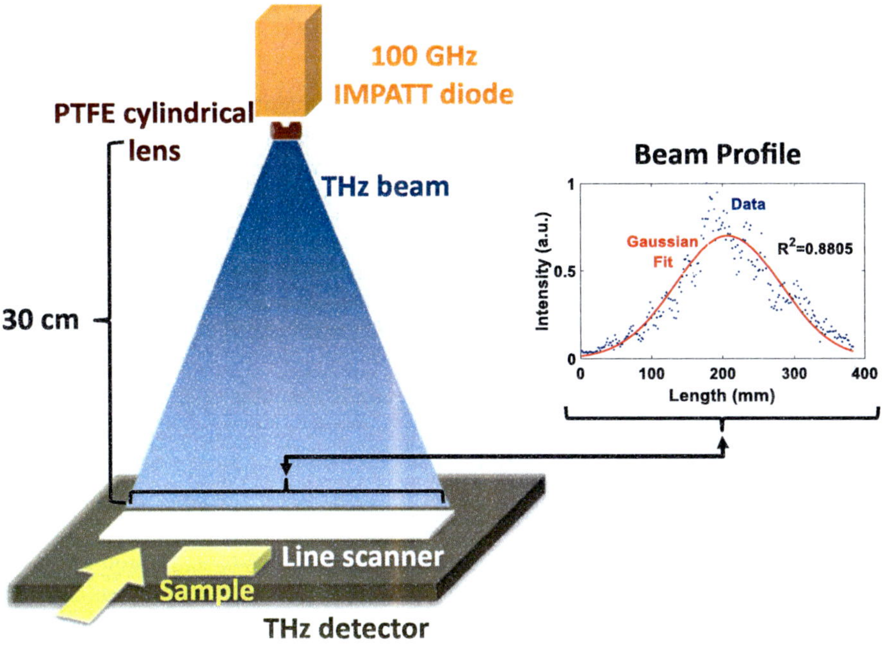

Fig. 1 Schematic of the THz imaging setup showing the THz source and detector modules

3 THz Images

Initial set of experiments have been performed on an acrylic substrate of dimensions 140×100 mm^2. The samples have been coated with Hacsol™ heat-resistant spray. The samples have been dried for 30 min after the spraying process. THz images are obtained by moving the sample using motorized 1D translational stage and averaging for ten times. The direction of sample movement is shown in Fig. 2.

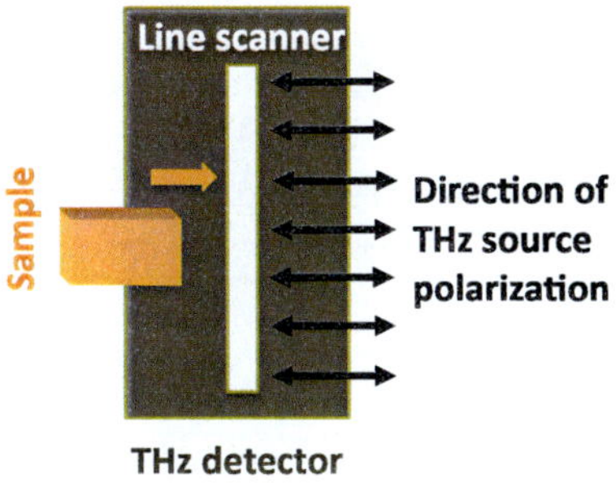

Fig. 2 Direction of sample movement

3.1 Moisture Ingression

The sample pretreatment process flow and their corresponding THz images are shown in Figs. 3 and 4, respectively. THz image of the uncoated sample, which is moved as in Fig. 2a, is shown in Fig. 4a. The THz image of the sample after spray coating at ambient conditions is shown in Fig. 4c which shows the non-uniformities in the coating. The sample has been placed in hot-air oven and heated at 70 °C for 24 h to obtain the THz image of the dry sample (Fig. 4b), which has been used as a reference for the moisture ingression studies. Thereafter, the sample has been placed in humidity chamber at 95% RH with a stable temperature of 40 °C. THz images of the sample after 6 and 24 h are shown in Fig. 4c, d, respectively. It can be clearly seen from these images that the transmitted THz intensity has reduced at several regions where moisture ingression has taken place as water in these regions has absorbed most of the THz radiations, thereby reducing the transmitted power. Regions of moisture ingression can be estimated directly by normalizing the corresponding THz images ($\mathrm{In}(x,y)$) with the image of the dry sample and finally plotting $1 - \mathrm{In}(x,y)$ to highlight these regions. Figure 4f, g shows regions of moisture ingression in 6 and 24 h, respectively. It is obviously noticeable that in 24 h, the moisture ingress regions (as seen in bright red regions) have increased in comparison with 6 h.

Table 1 summarizes the mean value of the normalized intensity for the obtained THz images under different sample pretreatments. It can be observed that the dry sample (Fig. 4b) which has been heated at 70 °C for 24 h) has the highest mean normalized intensity value. As the moisture ingression increases, the mean value reduces indicating the absorption of the THz radiation by water with the lowest value obtained for the sample placed at 95% RH for 24 h.

From Fig. 4g, it is obvious that even after 24 h, moisture has not diffused throughout the sample. Certain regions in the sample have shown no trace of moisture ingression even after 24 h, whereas at few other regions, increased moisture ingression has been observed from 6 to 24 h. This clearly depicts existence of two different regions in the sample. Hence, these distinct regions in the samples have been imaged using an optical microscope with 45× magnification. The optical images of the coated sample have revealed the presence of two distinct regions. The first region as shown in Fig. 5a is highly porous with pores of different sizes ranging from 10 to ~140 μm. This region corresponds to the region where the moisture ingression has been appreciably noticed. The second region as shown in Fig. 5b appears to be less porous with very few pores of sizes ~40 μm, which corresponds to regions with no significant moisture ingression.

3.2 Substrate Defects

THz imaging has the capability to identify substrate defects that are underlying beneath the coating which cannot be identified by visual inspection. In order to

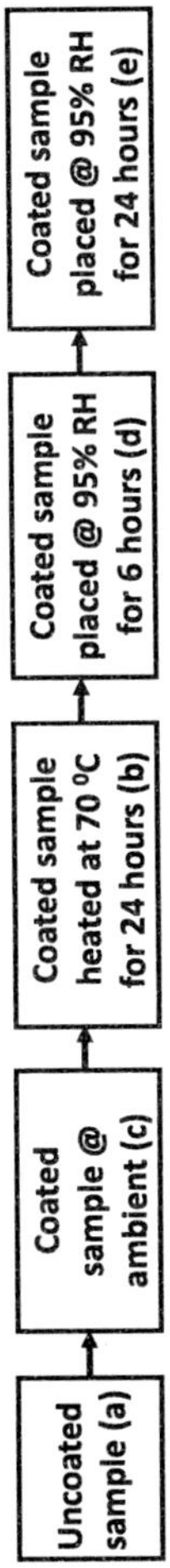

Fig. 3 Sample pretreatment flow before the THz imaging (respective alphabets indicated are mapped to the captions of THz images in Fig. 4)

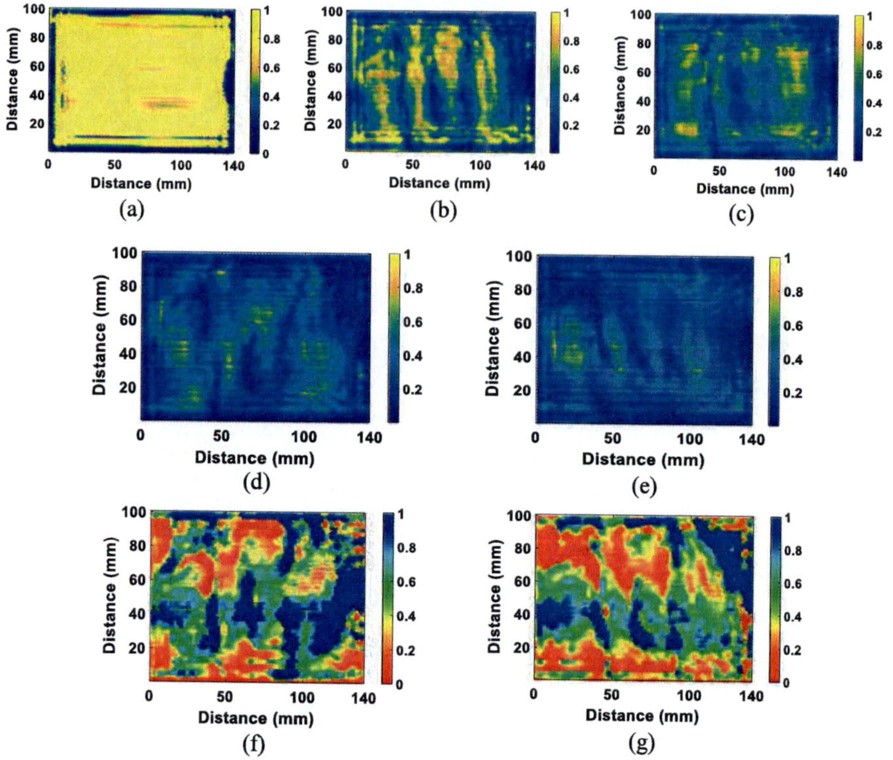

Fig. 4 THz images acquired with sample movement as in configuration indicated in Fig. 2. THz image of **a** uncoated acrylic substrate, **b** coated sample heated at 70 °C for 24 h in hot-air oven, **c** coated sample in ambient temperature (~27 °C) and humidity (RH ~ 45%), **d** coated sample placed in humidity chamber with RH of 95% for 6 h, **e** coated sample placed in humidity chamber with RH of 95% for 24 h, **f** region of moisture ingression after 6 h, and **g** regions of moisture ingression after 24 h

Table 1 Mean value of normalized THz intensity at various sample pretreatments

S. No.	Sample	Mean Value of Normalized Intensity
1	Heated at 70 °C for 24 h	0.4454
2	Ambient temperature (~27 °C) and humidity (RH ~ 45%)	0.3673
3	Placed in humidity chamber with RH of 95% for 6 h	0.2800
4	Placed in humidity chamber with RH of 95% for 24 h	0.2331

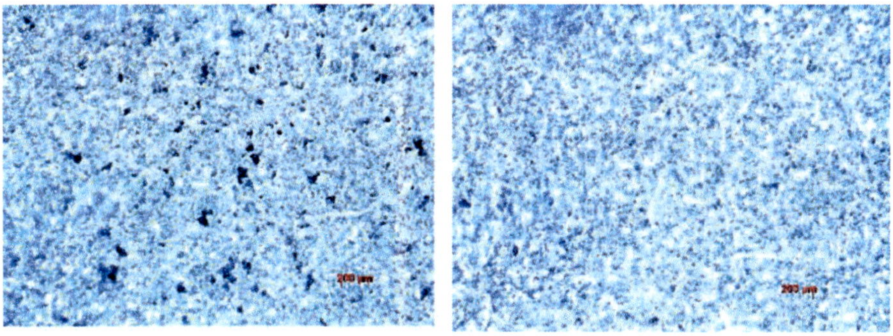

(a) (b)

Fig. 5 Optical microscope images of spray coated sample showing **a** highly porous regions corresponding to regions of high moisture ingression and **b** less porous regions corresponding to regions with less moisture ingression

identify such substrate defects using THz line scanner, artificially induced defects in terms of three circular indentations of diameter 3 mm each has been made on the acrylic samples of dimension $90 \times 60 \times 3.5$ mm^3. The indentations are of different depths of 1, 2, and 2.5 mm from left to right in Fig. 6a. Figure 6b, c shows the coated sample and its THz image where circular indentations are clearly visible. However, the diameter of the observed circular indentations in the THz image are increased by a factor of 3.8 and 3.5 in the *X*- and *Y*-directions, respectively, which is due to the diverging THz beam with divergence angle of 65.2°. The imaging resolution achieved is 3 mm. Further, in order to compare the THz images with other industrial imaging technique, passive IR thermography has been carried out as it is also a non-contact technique similar to THz imaging. However, the passive thermal image which is shown in Fig. 6d does not reveal any information about the substrate defects due to the presence of circular indentations at higher depths (>1 mm from the surface) and poor thermal conductivity of the acrylic substrate (0.2 W/m K).

3.3 Foreign Particles

THz imaging can enable identification of coating impurities in terms of any metal particles, etc., which get added during the coating process. Metal particles may contaminate the coating when metallic residues are not properly cleaned post-machining. In such cases, THz imaging can identify the metal particles as metals are not transparent to the THz radiations. Here, metal particles of sizes ranging between 1 and 2 mm have been sprinkled before the spray coating. Figure 7a–c shows the coated sample and its THz and passive thermal images, respectively. It could be observed that the metal particles can be identified in the THz image but not in passive IR thermography images.

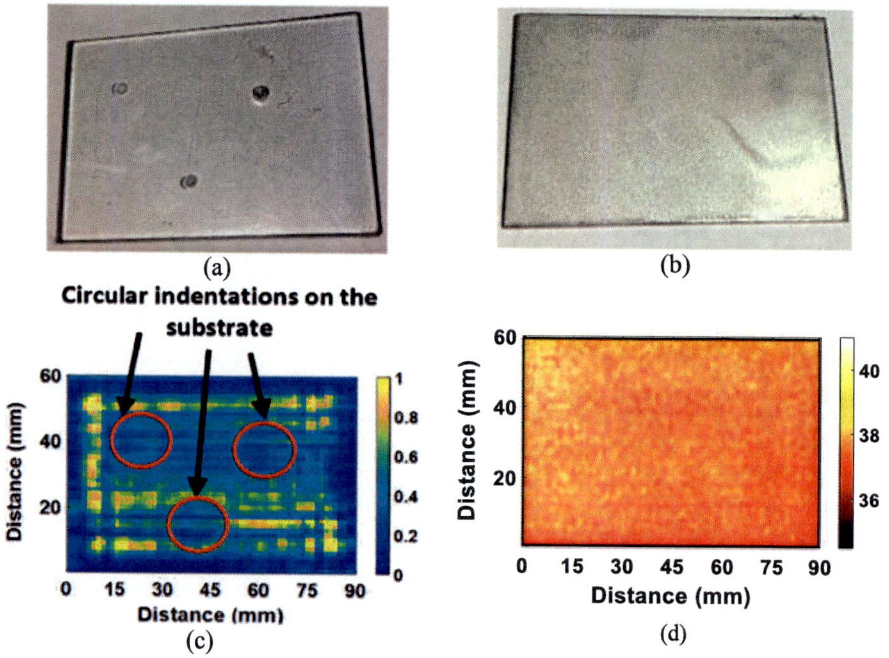

Fig. 6 Sample **a** before coating showing circular indentation on the surface, **b** after coating, **c** THz image of the corresponding coated sample depicting the circular indentations on the substrate (marked in red), **d** passive thermal image of the sample

4 Conclusions

The capabilities of THz line scanner in non-destructive evaluation of the coating non-uniformity, coating purity, substrate defects, moisture ingression have been successfully demonstrated on an acrylic substrate. Moreover, it has been shown that the THz images are superior to the corresponding passive IR thermography images in identifying substrate defects and foreign particles. Further, dependence of imaging intensity on the alignment of the defect with respect to the THz beam polarization has been observed which will be extensively investigated in future. In addition, studies on the practical metal substrate will be carried out by aligning the line scanner in reflection mode.

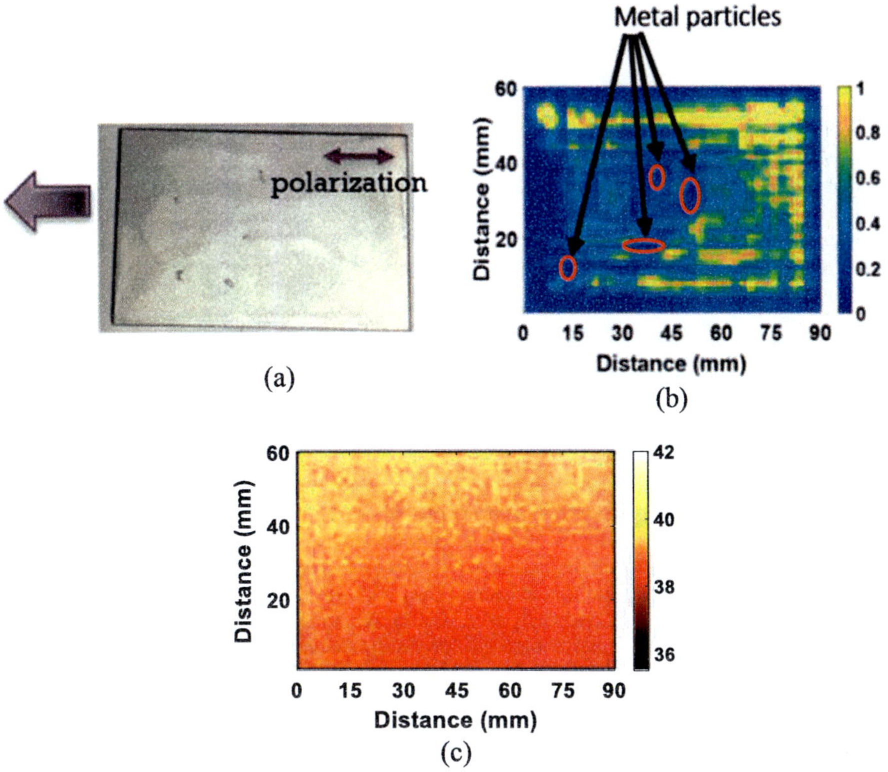

Fig. 7 **a** Coated sample with foreign (metal) particles, **b** the corresponding THz image depicting the foreign particles (marked in red), **c** passive thermal image of the sample

Acknowledgements The authors would like to acknowledge the funding of CSIR through the project MLP-0114 for this research activity. The authors would also like to thank Dr. S. Md. Iqbal, Chief Scientist and his workshop team for their help in fabrication of the samples.

References

1. Urbahs A, Savkovs K (2019) Heat resistant conglomerate Ti-Al-Si-Cr-N coatings for Titan alloy-based parts of gas turbine engines. Eng Rural Dev 18:1042–1047
2. Blomqvist C (2014) Thermal barrier coatings for diesel engine exhaust application, 1–67
3. Nazir MH, Khan Z, Stokes K (2015) Modelling of metal-coating delamination incorporating variable environmental parameters. J Adhes Sci Technol 29:392–423
4. Cheruvu NS, Chan KS, Viswanathan R (2006) Evaluation, degradation and life assessment of coatings for land based combustion turbines. Energy Mater Mater Sci Eng Energy Syst 1:33–47
5. Coatings for high-temperature structural materials. The National Academies Press, Washington, D.C. (1996)
6. Devasahayam S (2006) Effect of moisture-ingress on adhesion energy in a metal oxide-polymer system. J Appl Polym Sci 99:2052–2061

7. Steffens H-D, Crostack H-A (1981) Methods based on ultrasound and optics for the non-destructive inspection of thermally sprayed coatings. Thin Solid Films 83:325–342
8. Leckey CAC, Wheeler KR, Hafiychuk VN et al (2018) Simulation of guided-wave ultrasound propagation in composite laminates: benchmark comparisons of numerical codes and experiment. Ultrasonics 84:187–200
9. Roy S, Bose T, Debnath K (2020) Detection of local defect resonance frequencies for defect imaging: a nonlinear ultrasound-based approach 1163–1172
10. Grosso M, Pacheco CJ, Arenas MP et al (2018) Eddy current and inspection of coatings for storage tanks. J Mater Res Technol 7:356–360
11. Huang R, Lu M, Peyton A et al (2020) A novel perturbed matrix inversion based method for the acceleration of finite element analysis in crack-scanning eddy current NDT. IEEE Access 8:12438–12444
12. Yin L, Ye B, Zhang Z et al (2019) A novel feature extraction method of eddy current testing for defect detection based on machine learning. NDT E Int 107:102108
13. Shepard SM (1997) Introduction to active thermography for non-destructive evaluation. Anti-Corros Methods Mater 44:236–239
14. Maier F, Zagar BG (2009) Measurement of paint coating thickness by thermal transient method. IEEE Trans Instrum Meas 58:1958–1966
15. Milovanović B, Pečur IB (2016) Review of active IR thermography for detection and characterization of defects in reinforced concrete. J Imaging 2:1–27
16. Balageas D, Maldague X, Burleigh D et al (2016) Thermal (IR) and other NDT techniques for improved material inspection. J Nondestruct Eval 35:1–17
17. Anastasi RF, Madaras EI, Seebo JP et al (2007) Terahertz NDE application for corrosion detection and evaluation under Shuttle tiles. Nondestruct Charact Compos Mater Aerosp Eng Civ Infrastructure, Homel Secur
18. Zhang J, Chen J, Wang J et al (2019) Nondestructive evaluation of glass fiber honeycomb sandwich panels using reflective terahertz imaging. J Sandw Struct Mater 21:1211–1223
19. Guerboukha H, Nallappan K, Skorobogatiy M (2018) Toward real-time terahertz imaging. Adv Opt Photonics 10:843
20. Han S-P, Ko H, Kim N et al (2014) Real-time continuous-wave terahertz line scanner based on a compact 1 × 240 InGaAs Schottky barrier diode array detector. Opt Exp 22:28977
21. THz scanner [Internet]. Available from: https://terasense.com/products/thz-scanner/

A Non-destructive Approach of Dynamic Characterization of CFRP Composite Subjected to Hygrothermal Ageing

Nilesh Bagale, M. R. Bhat, and S. B. Kandagal

1 Introduction

Environmental factors which the structural materials are being exposed to, deteriorate the material properties over a period of time which in turn affects the durability and integrity of structures. Such factors, depending upon the environmental conditions include heat, moisture, ultraviolet rays, radiations, high energy electrons, etc. Degradation in material properties due to these factors is a gradual and permanent phenomenon and is known as *Ageing*. The aging happening specifically due to the combined effect of heat and moisture is known as *Hygrothermal ageing*. *Hygro* refers to water/humidity and *thermal* to heat. Hygrothermal aging can lead to significant degradation in physical and mechanical properties of a material.

In recent years, fibre-reinforced polymer composites are being extensively used in the aerospace industry due to their specific-strength and stiffness as well as excellent thermal properties [1]. Despite possessing excellent material properties, they are susceptible to heat and moisture affecting their performance in the long term. Thus, hygrothermal ageing poses a challenge questioning the durability of such materials. Many research studies show that hygrothermal ageing in polymer composites is known to cause degradation in strength and stiffness, plasticization of matrix, reduction in glass-transition temperature and deterioration of fibre-matrix bond [2–5].

Estimation of loss in strength and stiffness of composite material along with the prediction of residual life of a structure becomes vital in order to avoid unwarranted failures. This requires a thorough investigation of material behaviour undergone similar ageing as the environmental conditions of a structure. In this scenario, non-destructive evaluation techniques can prove to be an effective tool in detecting and

Supported by Aerospace Engineering, Indian Institute of Science.

N. Bagale (✉) · M. R. Bhat · S. B. Kandagal
Department of Aerospace Engineering, Indian Institute of Science, Bengaluru 560012, India
e-mail: bagale.nilesh0@gmail.com

C. K. Mukhopadhyay and R. Mulaveesala (eds.), *Advances in Non-destructive Evaluation*, Lecture Notes in Mechanical Engineering,
https://doi.org/10.1007/978-981-16-0186-6_12

evaluating the material degradation due to ageing where a correlation between NDE parameters and the extent of ageing can be achieved [6, 7]. This can further facilitate prediction of balance life of a structure.

In the present study, the damping property of CFRP and its variation due to hygrothermal ageing have been evaluated using free vibration analysis and have been further correlated with the attenuation coefficient obtained through ultrasound NDE. The study was performed on CFRP specimens that have undergone up to 500 h of ageing. Good correlation has been observed between logarithmic decrement of vibration response signal amplitude and ultrasound attenuation coefficient.

2 Experimental

2.1 Fabrication of CFRP Composite

CFRP composite laminate was fabricated using 12 layers of 250 GSM woven bidirectional carbon-fabric with epoxy (Araldite as resin and Aradur as hardener) as the matrix. The fabrication was carried out by hand lay-up technique using vacuum bagging. Laminate fabricated was quasi-isotropic. The thickness obtained was 2.67 mm with a standard deviation of 0.05 mm. A total of 10 specimens were obtained with dimension of 25.4 mm × 152.4 mm from the CFRP laminate.

2.2 Accelerated Hygrothermal Ageing

To accelerate the hygrothermal ageing in CFRP, specimens were immersed in a hot water bath at 80 °C for up to 500 h. Moisture uptake, vibration characteristic and NDE parameter were evaluated at 70, 110, 177, 360 and finally 500 hours of ageing treatment. The moisture absorption in specimens as the ageing was prolonged is shown in Fig. 1. On an average up to 3 per cent moisture gain was observed. The combined effect of moisture gain and heat on CFRP is expected to alter its properties and further reflect corresponding changes in vibration damping and ultrasound energy absorption characteristics.

2.3 Free Vibration Analysis

To evaluate the effect of the extent of hygrothermal ageing on test specimens, vibration tests were performed at different intervals of treatment. The schematic for setup is as shown in Fig. 2. The study was performed with support conditions as a cantilever beam, where the excitation was given at the free end. A 2.5 GS/s 200 MHz

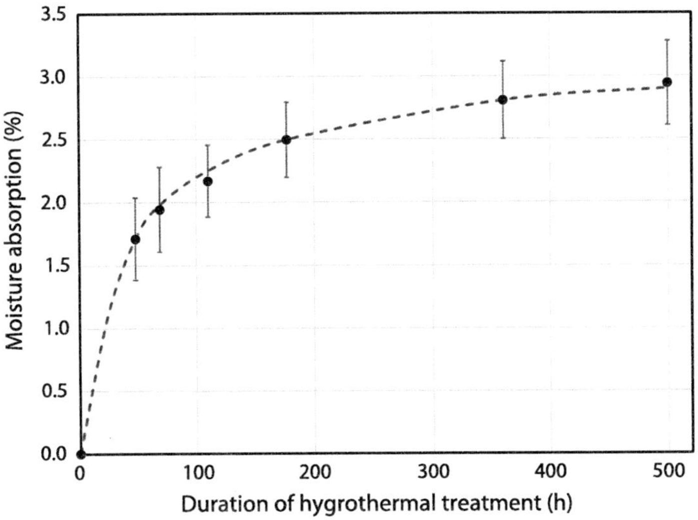

Fig. 1 Moisture absorption in CFRP specimens subjected to hygrothermal treatment

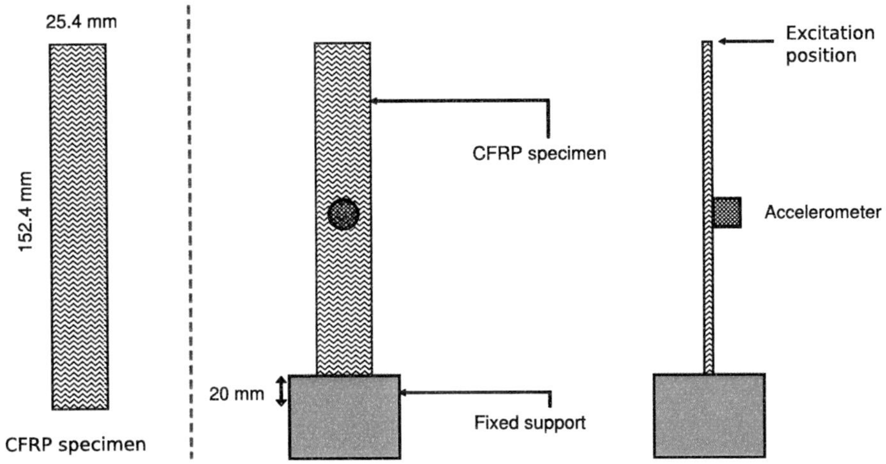

Fig. 2 A schematic of free vibration setup

digital oscilloscope was used for output signal analysis. A piezoelectric PCB 336A SN786 accelerometer was used to capture the dynamic response of test specimens. A sampling interval of 16 μs was chosen for each test. Further, the signals were analysed using MATLAB software. The vibration damping in terms of logarithmic decrement (δ) of the peak amplitude of response signal was evaluated. In Fig. 3, the effect of hygrothermal ageing on the damping parameter 'δ' has been presented. About 65% increase in δ for about 500 h of ageing was observed.

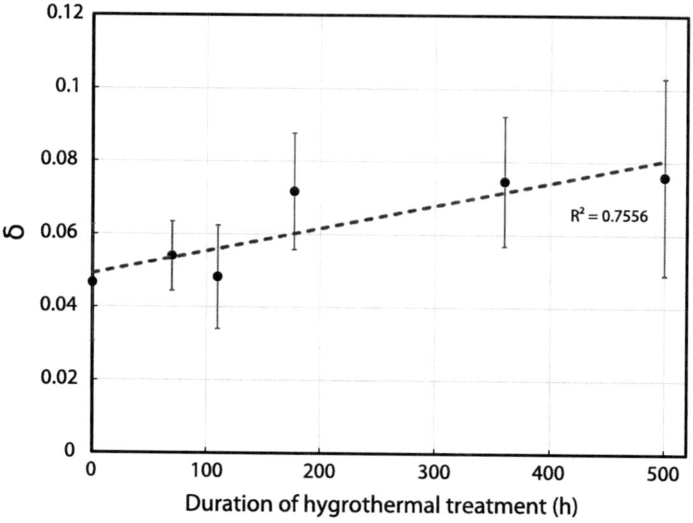

Fig. 3 A variation in the vibration damping parameter due to hygrothermal treatment

2.4 NDE Using Ultrasound

At different intervals of ageing, to evaluate ultrasound attenuation in CFRP, ultrasound scans were performed on specimens in the pulse-echo mode. A focused-beam transducer with 50 mm focal length having 2.25 MHz frequency response was used for the evaluation. Ultrasound scans were analysed using data acquisition software and post-processor. Ultrasound pulses were delivered normal to the specimens. The resolution for scanning was kept as 1 mm. Through the scans, front-wall and back-wall echo amplitude were obtained over the specimens for each interval of hygrothermal ageing. A typical A-scan image is as shown in Fig. 4 indicating front-wall and back-wall echo.

Using ultrasound principles, the echoes are related as follows:

$$\frac{\mathrm{BW}}{\mathrm{FW}} = (1 - \gamma)^2 \mathrm{e}^{-2\alpha z} \tag{1}$$

where FW denotes the front-wall amplitude; BW, the back-wall amplitude; γ, the reflection coefficient for CFRP-water pair; α, the ultrasound attenuation coefficient of CFRP; z, the thickness of CFRP specimens. Using the Eq. 1, α was evaluated for each test specimen at each interval of hygrothermal ageing. The trend of this coefficient with respect to the duration of hygrothermal treatment is as shown in Fig. 5. A consistent increase in the coefficient was observed indicating an increase in the material damping as a result of ageing.

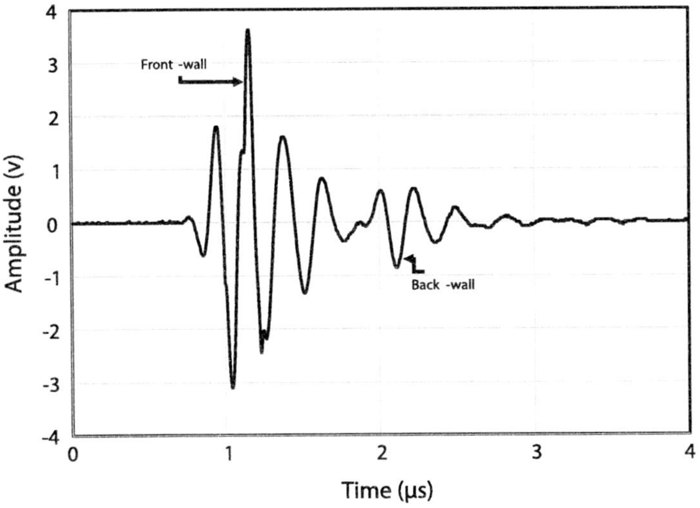

Fig. 4 A typical ultrasound A-scan image

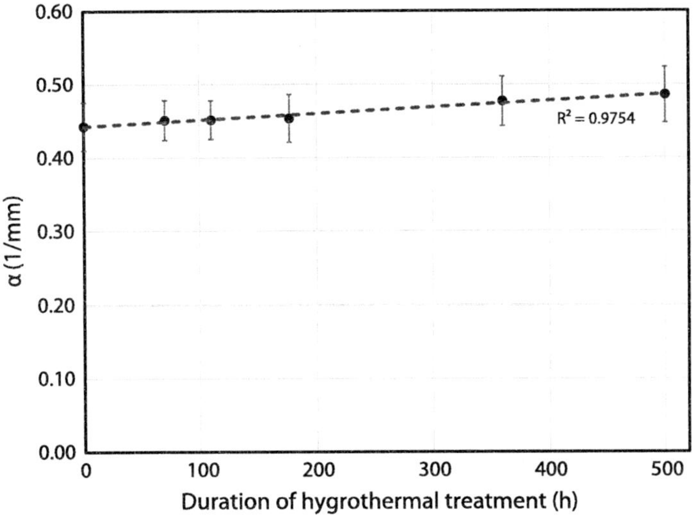

Fig. 5 A variation in ultrasound attenuation coefficient due to hygrothermal treatment

2.5 Correlation Between Damping and Ultrasound Attenuation

In order to determine the vibration damping in CFRP using ultrasound technique as NDE tool, a correlation between the damping parameter (δ) and ultrasound attenuation coefficient (α) was obtained. The trend of this correlation is as shown in Fig. 6.

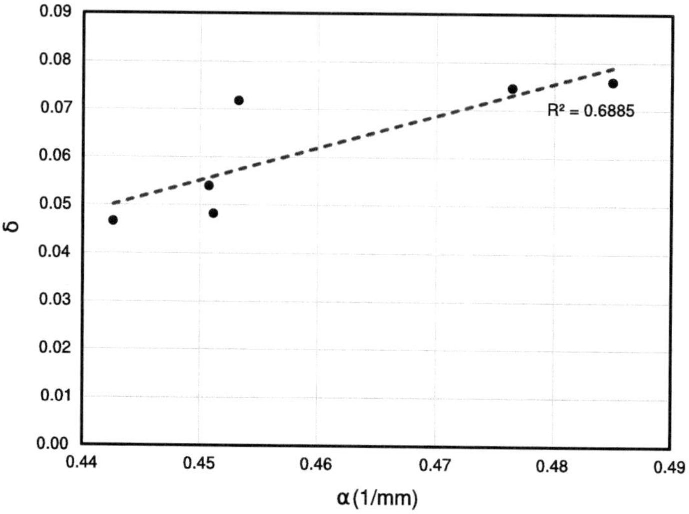

Fig. 6 A correlation between logarithmic decrement of the response signal and ultrasound attenuation coefficient of CFRP subjected to hygrothermal treatment

Since both the experimental results have revealed an increase in energy absorption with HT ageing, this correlation has come out with a positive slope. Further increasing the number of dataset, a fit with a better correlation can be obtained.

3 Results and Conclusion

Through this experimental investigation, it was observed that the hygrothermal ageing has resulted in moisture uptake and altering material damping as well as ultrasound attenuation in CFRP composite. The moisture uptake followed Fickian model eventually reaching saturation. Free vibration analysis, as well as ultrasound NDE, has shown an increase in the energy absorption in CFRP composite with HT ageing. Both the results show a good agreement which resulted in establishing a correlation between them. This has shown the effectiveness of ultrasound as NDE tool to evaluate the effect of ageing in CFRP. An unconventional approach has been presented in this study. This approach can facilitate economic inspection of the aged health of a composite material.

References

1. Quilter BA, Analysis S, Esdu IHS. Composites in aerospace applications
2. Halpin J (1969) Effects of environmental factors on composite materials. no AFML-TR-67-423, p 62

3. Moy P, Karasz FE (1980) Epoxy-water interactions. Polym Eng Sci 20(4):315–319
4. Cysne AP et al (2017) Accelerated aging effects on carbon fiber/epoxy composites, vol 110, 298–306
5. Botelho EC, Rezende MC (2010) Evaluation by free vibration method of moisture absorption effects in polyamide/carbon fiber laminates. J. Thermoplastic Compos Mater 23(2):207–225
6. Bagale NC, Dr. Bhat MR (2017) Characterization of porosity and hygrothermal conditioning of carbon/epoxy composite material. In: International conference on composite materials and structures (ICCMS)
7. Bagale NC, Bhat MR (2019) Effects of hygro-thermal ageing on polymer composites using acoustic wave propagation techniques. In: International conference on sound and vibrations (ICSV-26), Montreal-Canada

Pore Evaluation and Distribution in Cement Mortar Using Digital Image Processing

Sreelekha Gargepuram, Smrithy Subash, and Sumedha Moharana

1 Introduction

Concrete is the most economic and easily available construction material in worldwide. However, inherent pores, present in this material with varying sizes and shapes, make less durable after certain period. Porosity has ability to alter the strength and functionality of concrete [1–4]. Fresh cement mortar contains sand and paste, having good workability in a plastic phase, as time progress, the setting occurs along hydration, which alter most microstructural property of cement mortar and creation of variety of pores. The hardened cement paste during different hydration stages consists of very poorly crystallized hydrates of the various compounds, referred to collectively as gel, of crystals of $Ca(OH)_2$, some minor components, un-hydrated cement and the residue of the water-filled spaces in the freshly prepared cement mortar [5, 6]. Voids which are filled with water are called capillary pores but, within the gel itself, there exist interstitial voids, called gel pores. The nominal diameter of gel pores is about 3 nm, while capillary pores are one or two orders of magnitude larger, i.e. greater than 10 nm [7, 8]. The pore types with the greatest impact on concrete shrinkage are capillary and gel pores, especially for pores with sizes ranging from 2.5 to 50 nm and from 50 to 100 nm [5].

The volume of capillary pores formed due to absence of hydration products gets reduced with progress of hydration. The water–cement ratio and degree of hydration mainly contribute to capillary porosity [6, 9]. The interconnected capillary pores are mainly responsible for the permeability of the hardened cement paste and for its vulnerability to cycles of freezing and thawing and acid attacks which further deteriorates the durability. The gel pores are much smaller compared to the capillary pores that is less than 2 or 3 nm in nominal diameter. The total volume of gel

S. Gargepuram · S. Subash · S. Moharana (✉)
Department of Civil Engineering, Shiv Nadar University, Dadri 201314, India
e-mail: sumedha.maharana@snu.edu.in

C. K. Mukhopadhyay and R. Mulaveesala (eds.), *Advances in Non-destructive Evaluation*, Lecture Notes in Mechanical Engineering,
https://doi.org/10.1007/978-981-16-0186-6_13

increases with the decomposition of hydration products, and therefore, the total volume of gel pores also increases [10, 11]. On the other hand, the volume of capillary pores decreases with the progress of hydration [3, 9, 11]. The pore structures of cement pastes can be described by several aspects: total porosity, connectivity of pore network and pore size distribution, etc. Total porosity and pore size distribution can be determined by scanning electron microscopy (SEM), mercury intrusion porosimetry (MIP), nuclear magnetic resonance (NMR), etc. [1, 4, 6, 7, 12].

This paper comprises the study of pore evaluation through digital image processing using ImageJ. The mortar samples are prepared not only by varying the batching process and water–cement ratio but also by adding an admixture (fly ash) as partial replacement of cement. Samples are cured for 28 days by pond curing, and then, the optical image was acquired for all mortar samples casted for this study. All those images were further analysed for different types of pores and pore sizes present in the sample.

2 Materials and Methodology

The porosity of the mortar cylindrical samples was prepared in lab, varying three parameters—w/c ratio, the batching styles and also by partially replacing cement (*C*) by other pozzolanic material (fly ash). An OPC of 43 grade along with standard sand (*S*) and water (*W*) is used as the raw materials. The cement is partially replaced by fly ash (*F*) and is taken as 50% replacement for cement. A cement to sand ratio of 1:3 was kept constant throughout this investigation. The ratios of samples used in this study are as given in Table 1.

Required proportions of cement and sand were taken first for dry mix. Then water was added into it and mixed well without lumps for 3–5 min using trowels. Then the mix was filled into the moulds (petri dishes) and properly tamped to remove entrapped air. While batching and mixing, three different person did mix in different style (i.e.

Table 1 Mix Ingredients for Cement Mortar Samples

Sample No	Water (g)	Cement (g)	Sand (g)	Fly ash (g)	W/C ratio	Varying factor	W:C:F:S
1	18.9	45	135	–	0.42	Mixing 1	0.42:1:0:3
2	18.9	45	135	–	0.42	Mixing 2	0.42:1:0:3
3	18.9	45	135	–	0.42	Mixing 3	0.42:1:0:3
4	13.5	45	135	–	0.3	W/C ratio lowered	0.30:1:0:3
5	22.5	45	135	–	0.5	W/C ratio increased	0.50:1:0:3
6	18.9	22.5	135	22.5	0.42	Flyash substitution	0.42:0.5:0.5:3

harshly mixing (*sample 2*), mixing for longer time (*sample 3*) and proper batching and mixing (*sample 1*) (prescribed by IS codal guidelines). After unmoulding the samples they are kept for 28 days for curing. Once cured in water, the samples were kept to dry for 4 days and the microscopic images were taken. Images were taken at 5× magnification using an Optical Microscope. The microscopic images were taken for top surface only. The images obtained from microscope were further processed using ImageJ, from which the intensity and distribution of pores were obtained.

3 Results and Discussion

The images obtained from optical microscope were analysed using ImageJ, and the images were as given below. The total area of images analysed was 56,77,600 μm^2. Total pore area and number of pores were calculated from the images processed from imageJ.

It is very clear pertinent from the images given in Table 2 that the least capillary pores are shown in sample 1 with 0.42 w/c compared to sample 2 and sample 3. Maximum pores were observed in sample 4 with 0.3 w/c because of improper hydration cement compounds. The fly ash substituted sample 6 also has moderate number of pores with a w/c of 0.42. Sample 5 with a w/c of 0.5 has large number of pores when compared to other water–cement ratio. The sample 3 has more irregular surface texture, and hence, optical images have lots of variation.

The pore size and number of pores present in mortar samples are shown in Table 3. Interlayer spaces and gel pores mainly contribute forms shrinkage and creep in later stage yields the variation concrete strength, whereas capillary pores and micro-cracks formed in interfacial transition zones, significantly affect the permeability and durability of cement mortar [13]. Most of the samples have large capillary pores of smaller pore diameter, but sample 4 and sample 5 have maximum number of capillary pores of smaller and larger pore diameter. It may be due to decreased w/c which further leads to reduced workability and improper hydration. For sample 5, increasing water–cement ratio also increases the more shrinkage cracks because of higher rate of water evaporation from mortar sample and yield more number of large capillary (Abraham's Law).

On the other hand, minimum capillary pores are seen in sample 1, which is mainly due to well batching and adequate w/c. Other than large capillary pores, interfacial transition zones and micro cracks can be seen in samples 4, 5 and 6. Hence, it can be clearly inferred that the samples 4, 5 show more permeable pores because of improper hydration and initial curing. Sample 4 has shown maximum pore area ratio as it consists of maximum number of pores with maximum pore diameter, and the least is shown by sample 1 as the latter is well batched and mixed compared to other cement mortar samples. Samples with extremely low water/cement (w/c) ratios undergo significant self-desiccation if no external water is supplied during the initial hydration process. This self-desiccation may cause cracks in the premature concretes [1]. This caused more pores in sample 4 compared to other samples. The additional

Table 2 Images of samples

Sample No	Sample type	Surface Image	Microscopic Image	Threshold Image
1	0.42T			
2	0.42S			
3	0.42R			
4	0.3C			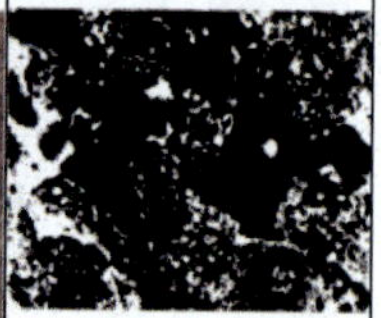
5	0.5C			
6	0.42FC			

Table 3 Pore size and classification of different samples

Pore size (microns)	Pore classification	Sample 1	Sample 2	Sample 3	Sample 4	Sample 5	Sample 6
0–0.0005	Interlayer spaces	0	0	0	0	0	0
0.0005–0.0025	Gel micro pores	0	0	0	0	0	0
0.0025–0.01	Gel small pores	0	0	0	0	0	0
0.01–0.05	Capillary median pores	0	0	0	0	0	0
0.05–10	Capillary large pores	29	75	71	1370	633	284
10–20	Capillary large pores	0	2	2	80	73	13
20–50	Cracks in interfacial transition zones	0	0	0	29	12	1
50–200	Micro cracks	0	0	0	8	2	0
Total no of pores		**29**	**77**	**73**	**1487**	**720**	**298**

mixing water can help ease mixing and workability but causes increased porosity, which yields degradation of durability and structural performances. The larger water content in cement mortar leads to coarse pore distribution and has contributed to pores in sample 5 [2]. For sample 6, using fly ash only aids to reduce the formation of large diameter capillary pores but has percentage of pore distribution lesser than sample 4 and sample 5 because the fly ash substituted for cement fills the voids, thereby reducing the percentage of pores (Table 4).

Table 4 Pore area in %

Samples	Sample 1	Sample 2	Sample 3	Sample 4	Sample 5	Sample 6
Min dia (μ)	3.34	3.34	3.34	3.34	3.34	3.34
Max dia (μ)	6.68	12.5	12.5	143.12	63.62	20.58
Pore total area (A) (μm^2)	341.3697	1619.319	1190.418	99522.4	37200.55	6486.025
Image total area (A) (μm^2)	5,677,600	5,677,600	5,677,600	5,677,600	5,677,600	5,677,600
a/A (%)	0.006	0.0176	0.013	1.7529	0.6552	0.11424

4 Conclusions

This paper comprises the pores and pore size distribution in the cement mortar sample using image processing tool—ImageJ. The cement mortars are prepared by varying many factors such as w/c, batching and mixing process and addition of pozzolanic material to the mix. After curing, the optical microscopic images of cylindrical cement mortar sample were obtained and analysed through digital image processing tool software. The number of pores, size of pores and their distribution across the surface scan area were obtained through ImageJ software. A detailed analysis has made to find exact pore sizes and their percentage distribution for cement mortar sample for different style of batching and water–cement ratio and fly ash. From this study, it is found that, the pore formation and its distribution measures through digital image processing tool have similarity with established theory of cement hydration and strength development process. From the experimental results and pore analysis, this method can be utilized for non-destructive evaluation of cement composite along with post-processing computational approach.

References

1. Igarashi SI, Watanabe A, Kawamura M (2005) Evaluation of capillary pore size characteristics in high-strength concrete at early ages. Cem Concr Res. https://doi.org/10.1016/j.cemconres.2004.06.036
2. Kim YY, Lee KM, Bang JW, Kwon SJ (2014) Effect of W/C ratio on durability and porosity in cement mortar with constant cement amount. Adv Mater Sci Eng. https://doi.org/10.1155/2014/273460
3. Kumar R, Bhattacharjee B (2003) Porosity, pore size distribution and in situ strength of concrete. Cem Concr Res. https://doi.org/10.1016/S0008-8846(02)00942-0
4. Winslow D, Liu D (1990) The pore structure of paste in concrete. Cem Concr Res. https://doi.org/10.1016/0008-8846(90)90075-9
5. Guo Y, Qian J, Wang X (2013) Pore structure and influence of recycled aggregate concrete on drying shrinkage. Math Probl Eng. https://doi.org/10.1155/2013/912412
6. Meddah MS, Tagnit-Hamou A (2009) Pore structure of concrete with mineral admixtures and its effect on self-desiccation shrinkage. ACI Mater J. https://doi.org/10.14359/56548
7. Joshi SK (2014) On application of image processing: study of digital image processing techniques for concrete mixture images and its composition. Int J Eng Res Technol 3(3). ISSN 2278-0181
8. Zhao H, Xiao Q, Huang D, Zhang S (2014) Influence of pore structure on compressive strength of cement mortar. Sci World J. https://doi.org/10.1155/2014/247058
9. Kondraivendhan B, Bhattacharjee B (2010) Effect of age and water-cement ratio on size and dispersion of pores in ordinary portland cement paste. ACI Mater J. https://doi.org/10.14359/51663578
10. Odler I, Rößler M (1985) Investigations on the relationship between porosity, structure and strength of hydrated Portland cement pastes. II. Effect of pore structure and of degree of hydration. Cem Concr Res https://doi.org/10.1016/0008-8846(85)90113-9
11. Chen X, Wu S (2013) Influence of water-to-cement ratio and curing period on pore structure of cement mortar. Constr Build Mater. https://doi.org/10.1016/j.conbuildmat.2012.09.058
12. Gal E, Kryvoruk R (2010) Properties of concrete. In: Computational modelling of concrete structures

13. Mehta PK, Monteiro PJM (2014) Concrete: microstructure, properties, and materials. McGraw-Hill, fourth edition

Thickness Estimation of Marine Structures Using an ROV-Based Pulsed Eddy Current Technique

Ashish Antony Jacob, Santhosh Ravichandran, Vineet Upadhyay, Prabhu Rajagopal, and Krishnan Balasubramaniam

1 Introduction

Jacket-type steel members are the most common type of near and offshore structures where tubular members are welded together to either form or protect the load-carrying member. Tubular joints are subject to damage as a result of fatigue, marine growth and severe corrosion from the environment. These structures are inspected for loss of wall thickness periodically to prevent catastrophic damage and loss of lives. Conventional ultrasonic testing (UT) is widely used for the estimation of thickness of metallic structures; however, especially in the case of marine structures, direct access to the structure is hindered by marine growth. Marine growth is removed conventionally using powered brushes, high-pressure water jets or in some cases, manually using chisels and hammers; however, they are time consuming and expensive. One of the alternative technologies which can be used for wall thickness estimation over marine growth of thickness > 10 mm is pulsed eddy current (PEC) which uses a stepped input signal to detect wall-thinning areas without removing the protection coatings or biomass [1]. PEC has gained wide popularity in various domains including multi-layered carbon fibre in the aircraft industry [2], defect detection and characterization in the inner surface of pipes under insulation in the oil and gas industry [3] and estimation of thickness under insulation [4].

The principle of PEC technique is shown in Fig. 1 and is explained briefly as follows. An alternating current-driven coil induces eddy currents in the specimen

A. Antony Jacob (✉) · S. Ravichandran · V. Upadhyay
Planys Technologies Pvt. Ltd, No. 5, Balaji Nagar Main Road, Puzhuthivakam, Chennai, Tamil Nadu 600091, India
e-mail: ashish.aj@planystech.com

P. Rajagopal · K. Balasubramaniam
Centre for Nondestructive Evaluation, Machine Design Section, Indian Institute of Technology (IIT), Madras, Chennai, Tamil Nadu 600036, India

C. K. Mukhopadhyay and R. Mulaveesala (eds.), *Advances in Non-destructive Evaluation*, Lecture Notes in Mechanical Engineering,
https://doi.org/10.1007/978-981-16-0186-6_14

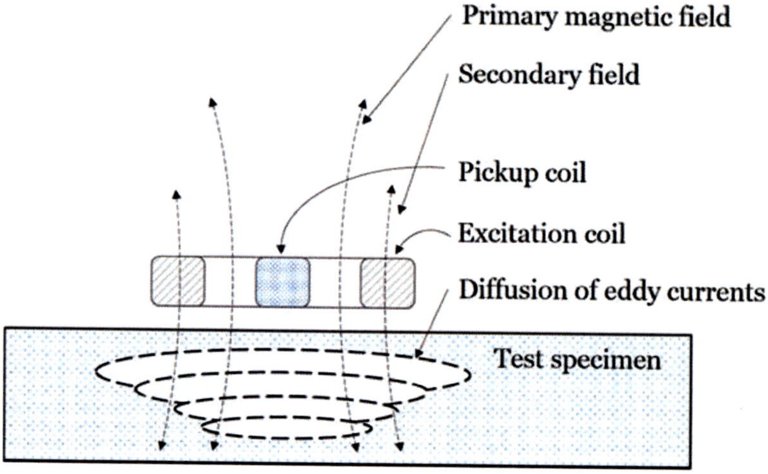

Fig. 1 An illustration of the working principle of eddy current testing

of interest as a result of electromagnetic coupling. The circulation of eddy currents in turn induce a secondary magnetic field which is distorted in the case of wall thickness loss. This distortion is sensed by another receiver coil from which the change in thickness is estimated.

The PEC technique can quickly estimate thickness of the metallic structures with insulations or marine growth up to a thickness of 150 mm. In the present pulsed eddy current system, the size of the area inspected below the probe can be obtained by estimating the footprint (FP) as 0.65 x (lift off, i.e. perpendicular distance of the probe from the surface) + FP_0 where FP_0 is the footprint at zero lift off. FP_0 is obtained from the datasheet of the chosen probe for the application at hand. The size of the area inspected using the pulsed eddy current probe can be visualized as a cone projected on the surface of interest. Higher the lift off of the probe from the surface, the bigger the area covered and vice versa. The footprint also determines the numbers of measurements taken both axially and radially on the surface of the structure.

This document is organized as follows. Following a brief introduction to the pulsed eddy current and its working principles. Then, the methodology of deployment using a remotely operated vehicle (ROV) is provided. Further, results are discussed from case studies performed using ROV Beluga wherein thickness of structures is compared using ROV-based conv. UT and PEC are benchmarked against UT measurements taken by divers. Key advantages and limitations to the conventional methodologies of inspection are discussed before which the report concludes with a summary.

2 Methodology

The PEC probes are mount on Beluga, shown in Fig. 2, which is a compact, modular and portable ROV that can sustain 2–4 m of wave height, moderately rough swell, up to 1 m/s of sea current and can dive up to 200 m of depth. Beluga possesses a state-of-the-art dual hull heavy bottom hydrodynamic design, providing it with static and dynamic stability in rough environmental conditions. Beluga is equipped with eight thrusters providing control in 5 degrees of freedom along with forward thrust of 17.5 kg-f and an operating speed of 3–4 knots. Beluga also carries an inertial measurement unit for orientation feedback, a pressure sensor for depth feedback and a global positioning system for position feedback on the water surface. Beluga's modular architecture allows it to carry a variety of different payloads such as high-definition cameras, high-intensity lights, laser for scaling and crack measurement, altimeter, side-scan sonar, an ultrasonic thickness measurement unit, cathodic potential measurement unit, bio-fouling cleaning equipment, 2D and 3D imaging/ scanning SONARs [5].

PEC transducers were mounted on ROV Beluga to estimate thickness of steel layers covered by marine growth. In order to mount the probe on the ROV, customized probe mounts are used; they are vibration resistant and have the ability to minimize dependence of operator skill to estimate scans of satisfactory signal quality. The probe mount is also designed to minimize any damage to the probe resulting from impacts against the specimen of interest.

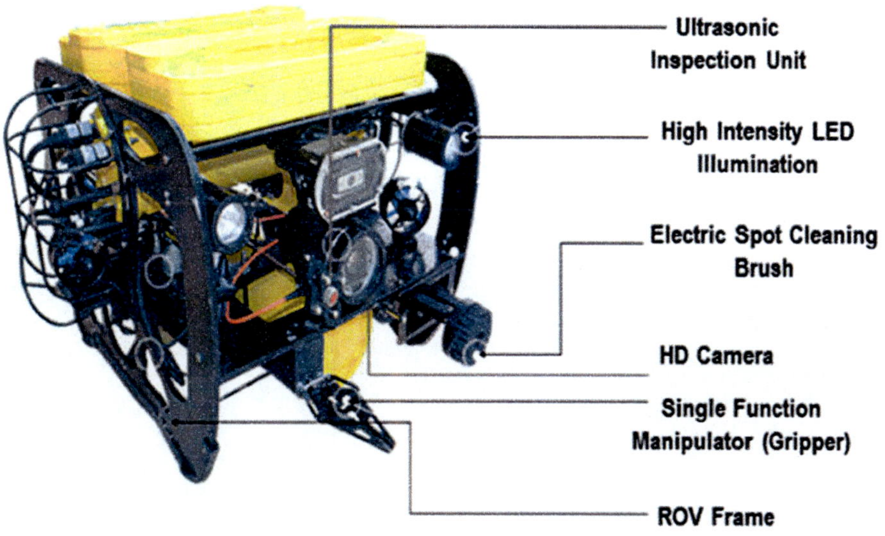

Fig. 2 Photograph of ROV Beluga highlighting its key features

3 Results and Discussion

3.1 Case Study: Thickness Estimation of Steel Jackets Covered with Marine Growth at a Port in TN, India

Pulsed eddy current technique was used to estimate thickness of the steel jacket without removing marine growth of thickness ranging from a few centimetres at 8 m depth to 10 cm at the water. A photograph of arrangement of pipes in the jetty is shown in Fig. 3 below, from which we can infer that removal of marine growth from the surface and performing surface grinding to enable use of conventional ultrasonic testing will be challenging and cumbersome. The thickness of the steel jackets is of interest since any major loss in wall thickness can allow water to seep into the concrete piles and cause corrosion, ultimately threatening structural integrity of the jetty. The outer diameter of the piles is 1.2 m wide and the design thickness of the steel jackets is 6 mm.

Splash-zone-rated (15 m depth) PEC probes, shown in Fig. 4, are used for the estimation of thickness in the present case. Each pile of interest is divided along the circumference and vertically across depth from the water surface to the seabed, following which the PEC probe mounted on ROV Beluga is used to navigate to each of these sections and record measurements of thickness as shown in Fig. 5 a, b.

The inspection procedure consisted of the following steps,

(1) Calibration mode where the probe is placed a region of nominal or known thickness and calibrate for optimum measurement parameters

Fig. 3 Photograph of the arrangement of piles in the jetty

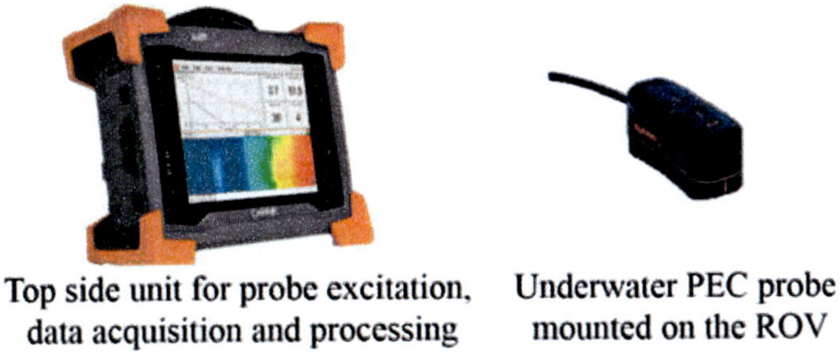

Fig. 4 Illustration of the pulsed eddy current probe used for underwater thickness estimation of metallic structures using an ROV

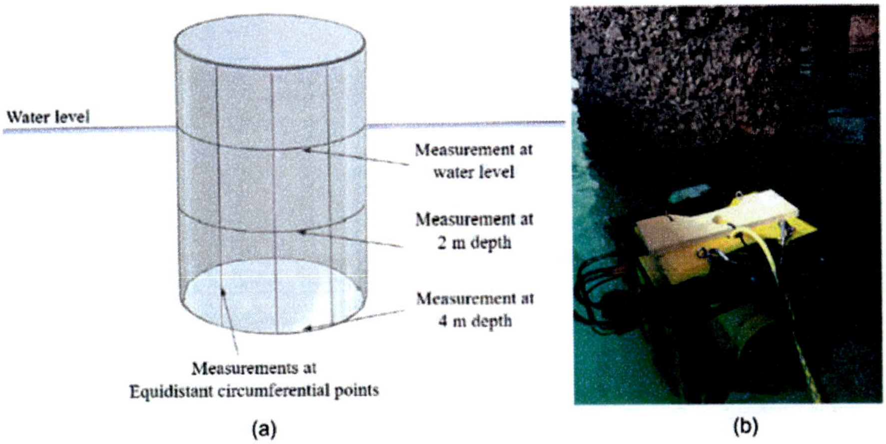

Fig. 5 Illustration of a sample division of a pile into different sections, and **b** the PEC probe mounted on ROV Beluga to navigate to the surface of interest and record measurements

(2) Survey mode where on the basis of the calibration, sample measurements are taken at different regions of the pile to check the validity of step (I)
(3) Acquisition mode is the actual data recording phase of the testing where the entire divided sections of the pipe, as shown in Fig. 6a are covered.

Multiple trials were conducted using the ROV mounted with the probe. Circumferential and vertical measurement trials were conducted on different piles. A sample set from the measurements during the inspection is presented in Figs. 6a, b. Figure 6a shows a snapshot from the PEC probe software indicating the time–voltage plot comparing the current scan and the calibrated scan.

From the colour chart showing the consolidated thicknesses estimated in Fig. 6b, we observed the estimated thickness decreased as depth increased. The maximum thickness loss from the design thickness is observed to be 1.2 mm which falls within

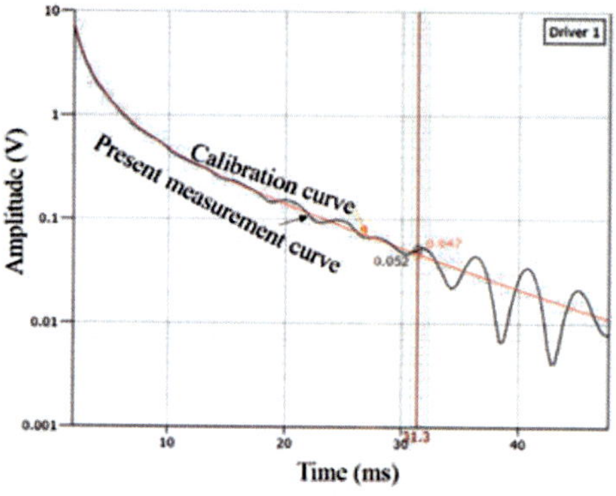

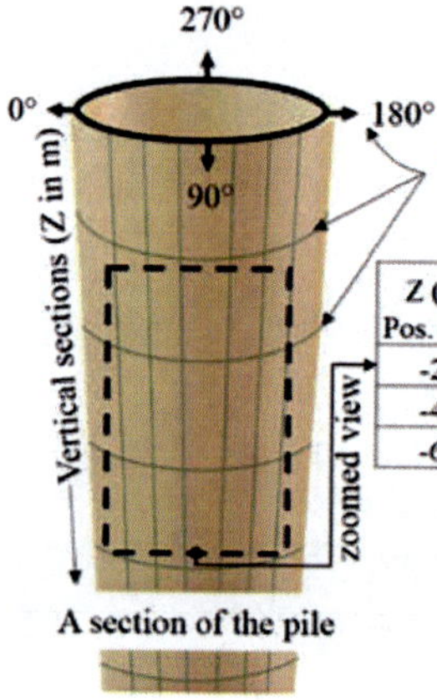

Z (m)/ Pos. (deg)	Thickness (mm)											
	0.0	30.0	60.0	90.0	120.0	150.0	180.0	210.0	240.0	270.0	300.0	330.0
-2.0	5.8	6.0	5.6	5.4	5.2	4.9	5.5	5.4	5.5	5.5	5.5	5.7
-4.0	5.5	5.4	5.5	5.5	5.5	5.6	5.1	5.5	5.3	5.4	5.3	4.8
-6.0	4.9	4.9	4.9	4.9	4.9	4.8	4.9	4.9	4.9	5.0	5.0	5.1

(b)

Fig. 6 Thickness estimation illustrated **a** in a snapshot taken from the PEC probe software, **b** a sample consolidated colour chart view of a section of a pile scan-zone showing estimated thicknesses from each measurement

the critical threshold of thickness loss. Further, following the examination of thickness measurement values of the other piles in the inspection study, it was concluded that the overall conditions was healthy and operable.

3.2 *Case Study: Thickness Estimation of Steel Piles Covered with Marine Growth at a port in Rotterdam, The Netherlands*

The case study presents the methodology used to inspect steel piles in a jetty, a photograph of which is shown in Fig. 7; further, the comparison of the estimated

Fig. 7 Photograph of the part of the jetty showing the steel piles to be inspected

thickness results from conventional ultrasonic testing methodology and from the pulsed eddy current methodology is discussed, and key learnings from the study are presented.

Calibration is an important step in both the methodologies, Conv. UT and PEC, an illustration of which is shown in Fig. 8a, b, respectively. In the case of conv. UT, the material velocity is calibrated against a pile of known thickness. For the PEC probes, calibration is performed on a pile of known thickness.

One of the key prerequisites of the conv. UT methodology of inspection is surface preparation which includes removal of marine growth. High-pressure cleaning jets

(a) (b)

Fig. 8 Photographs captured during the calibration phase of the inspection for the following techniques **a** conv. UT and **b** pulsed eddy current with 20 L/ min at 150 bar were used to remove layers of marine growth. A snapshot during the cleaning operation to remove marine growth and subsequently perform UT is shown in Fig. 9. Conventional ultrasonic testing was performed using a commercially available ultrasonic mini-ROV gauge as shown in Fig. 10a, b. Thickness was estimated using a 2.25 MHz–13 mm crystal diameter piezoelectric transducer in pulse-echo mode wherein subsequent echoes from the metallic specimen are used to estimated thickness.

Piles of interest are divided into sections both radially, along the circumference and axially, along its depth. The ROV is navigated to each of these sections and subsequently, the probe is placed on the surface of interest and the measurement is recorded. Thickness is estimated using the ROV from both methodologies, UT and PEC, and are compared. These results are further compared to the benchmark thickness results determined by the professional divers. Thickness estimation using conv. UT technique performed by professional divers is widely accepted to be the benchmark against which other techniques may be compared against.

In the Fig. 11, colours red, yellow and green are used to represent low, medium and high values of estimated thickness, respectively. Tables are populated with thickness in millimetres varying axially with depth in metres and radially along the circumference of the pile in degrees. Sample results indicate that the estimated thickness using ROV-pulsed eddy current technique shown in Fig. 11a fall within the expected deviations of the thickness estimated using the conv. UT technique used as shown in Fig. 11a, b. The maximum thickness loss is estimated to be 0.9 mm using the ROV-PEC methodology which is well above the critical threshold of thickness. On the basis of measurements from other piles, it was concluded that the overall condition and operability of the piles were healthy and safe.

Fig. 9 Snapshot taken during the removal of marine growth from the piles using high-pressure water jets; the cleaned area appears brighter in the photograph as a result of marine growth removal

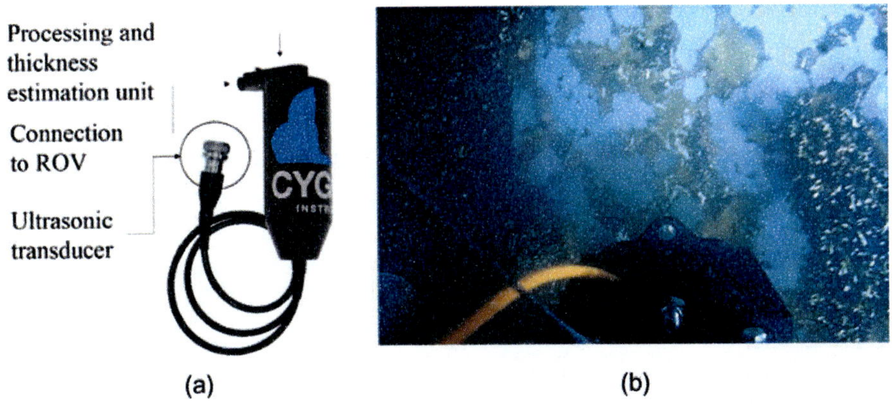

Fig. 10 **a** Illustration of a mini-ROV thickness gauge and its key parts **b** A snapshot taken during the UT measurement operation using ROV Beluga post the cleaning of the pile at the region of interest.

4 PEC: Key Application Areas, Advantages and Limitations

4.1 Key Applications Areas Include Inspection of Underlying Steel Structures

- Over marine growth without surface cleaning
- Over a galvanized steel weather jacket or fireproof jackets
- Over concrete or polymer coatings
- Over corrosion by-products including scabs or blisters without surface preparation

4.2 Advantages

- No risk to human life as in the case of a diver-based operation
- Data capture results and positions recorded are more repeatable
- Live remote monitoring of the inspection process
- Less dependency on operator skill
- Non-invasive methodology of inspection – no need to remove insulation or coatings; the pre-requisitery surface preparation step including cleaning and surface grinding in many cases can be eliminated
- Eliminates the need for cleaning of marine growth which conventionally requires the use of high-power equipment and related zone certifications in many cases
- Up to 3-times faster results than conv. UT methodology of inspection

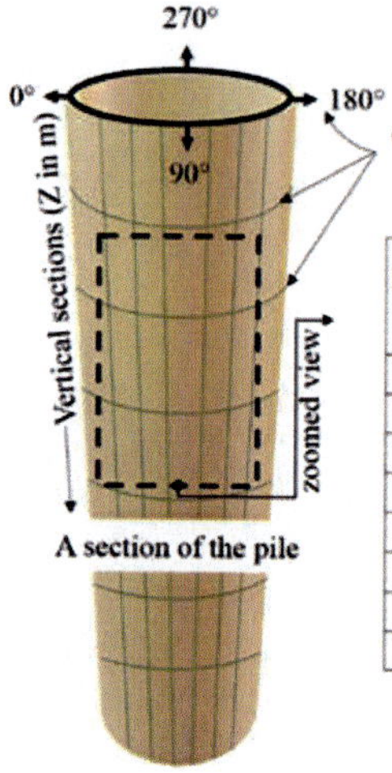

Pos. (deg)/ Z (m)	Thickness (mm)	
	0-180 deg.	180-360 deg.
0.00	13.10	13.7
-1.00	13.40	13.1
-2.00	13.90	13.7
-3.00	14.00	13.4
-4.00	13.60	13.8
-5.00	13.80	13.1
-6.00	13.30	NA
-7.00	13.50	NA

Thickness estimated by ROV PEC

Pos. (deg)/ Z (m)	Thickness (mm)	
	0-180 deg.	180-360 deg.
0.00	12.80	NA
-1.00	12.95	12.98
-3.00	NA	12.75
-4.00	13.66	NA
-5.00	NA	12.35
-6.00	12.89	NA

Thickness estimated by ROV UT

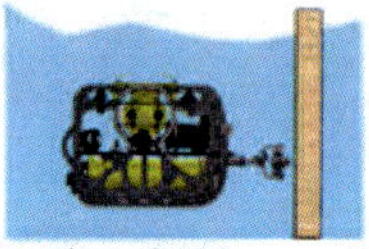

An illustration of thickness estimation using ROV

Design thickness: 13.8 mm

Critical threshold of thickness: 10.0 mm

(a)

Pos. (deg)/ Z (m)	Thickness (mm)					
	0-60	60-120	120-180	180-240	240-300	300-360
1.00	11.50	12.2	11.9	12.8	12.7	12.5
-0.50	13.00	13.1	13.2	11.6	13.1	12.2
-4.00	13.30	13.3	13.2	13.2	13.1	13
-7.50	13.20	13.3	13	13	13.2	13.2

Benchmark: Thickness estimated using Conv. UT by divers

(b)

Fig. 11 Illustration of the **a** division of the pile axially and radially; thickness estimated using ROV-based UT and PEC are given, **b** benchmark of thickness values estimated by professional divers

4.3 Limitations

- Results obtained have a 5–10% variability when compared to the conv. UT technique
- The smallest area inspected is limited to minimum footprint of the probe unlike the conv. UT technique wherein tiny spot measurements can be performed
- PEC is not designed to detect small-sized pitting—an averaged value of thickness of the area inspected is obtained
- The ROV-PEC system yields poor results in regions with significant wave or similar conditions

5 Summary and Conclusion

The pulse eddy current technique applied to estimation of thickness in steel structures with marine fouling is introduced. Key features of ROV Beluga are then presented. Specialized mounts, to compensate for rough sea currents and uneven marine growth which may otherwise distort the measurements, are used to mount the PEC probe on ROV Beluga. Further, case studies of using the PEC system mounted on the ROV at a port in TN, India, and in Rotterdam, the Netherlands, are presented. Further, key applications, advantages and limitations of the PEC technique compared to a diver-based modality of inspection using the conv. UT technique are discussed.

References

1. Rizzo P (2012) NDE/SHM of Underwater Structures: A Review. Adv. Sci. Technol. 83:208–216
2. Horan P, Underhil R, Krause TW (2014) Pulsed eddy current detection of cracks in F/A-18 inner wing spar at large lift-off using modified principal component analysis. Int. J. Appl. Electromagn. Mech. 45(1–4):287–292
3. Lai S, Chen DY, Chen H, Fu YW (2015) Pulsed Eddy Current Testing of Inner Wall Flaws in Pipe under Insulation. Procedia Eng. 130:1658–1664
4. Chen X, Lei Y (2014) Time-domain analytical solutions to pulsed eddy current field excited by a probe coil outside a conducting ferromagnetic pipe. NDT E Int. 68:22–27
5. Ashish AJ, Galipali M, Upadhyay V, Balasubramaniam K (2017) A Novel Ultrasonic Inspection Methodology for Submerged Metallic Structures using a Remotely Operated Vehicle (ROV). ISNT JNDE 1:14–16

Prediction of Failure Process in Geo-Material Using Finite Element Based Digital Image Correlation (DIC) Method

Chamanth Sai Reddy Vemulapati and Debasis Deb

1 Introduction

Getting deformation information through image analysis is one of the most active research areas in non-contact measurement methods. A very recent technique in this field is digital image correlation (DIC). The initial concepts of DIC are introduced by Peters and Ranson [1] from the University of South Carolina in the paper published in 1982, but gained popularity from the last decade with the advent of low-cost imaging systems (camera hardware) and ever-increasing computational capability of modern computers [2].

DIC is considered as the most advanced technique after strain gauge in the experimental mechanics, which allows visualization of surface deformation (strain) field of the object [3]. DIC method can be divided into two parts (i) collection of deformation data in a series of digital images (ii) processing of series of digital images using correlation algorithms for deformation and strain fields, as shown in Fig. 1.

Data in the form of images are collected by the image acquisition system (consists of camera and illumination system), and then processing is done on a computer to access the deformation behavior. The test sample surface should be randomly dotted (speckle pattern) to have unique correspondence between the consecutive images to get the deformation fields that take place on the measured surface. Artificial speckle patterns either (by paper or by paint, or by stamp) one of them be used, if the natural surface of the sample is not randomly speckled.

C. S. R. Vemulapati (✉) · D. Deb
Department of Mining Engineering, Indian Institute of Technology Kharagpur, Kharagpur, West Bengal, India
e-mail: chamanth@iitkgp.ac.in

D. Deb
e-mail: deb@iitkgp.ac.in

C. K. Mukhopadhyay and R. Mulaveesala (eds.), *Advances in Non-destructive Evaluation*, Lecture Notes in Mechanical Engineering,
https://doi.org/10.1007/978-981-16-0186-6_15

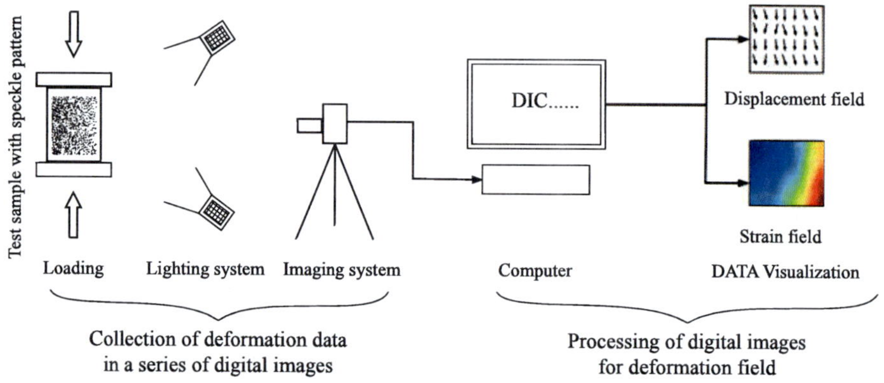

Fig. 1 Digital image correlation process

This paper describes about the finite element-based digital image correlation algorithm developed by the authors and examines its capability in prediction of failure process in geo-materials.

2 Finite Element-Based Digital Image Correlation Formulation

Finite element based digital image correlation is an alternative algorithm to commonly used subset-based local DIC algorithm because of its capability to enforce displacement continuity and displacement gradient continuity in the calculation of displacement field [4]. Moreover, FE-DIC offers the scope of incorporating the concept of extended finite element method (X-FEM), thereby improving the accuracy in the specimens having physical discontinuities (which appear on the surface during the loading or pre-existing on the surface) [5–7].

Let $f(\boldsymbol{x})$ (reference image) and $g(\boldsymbol{x})$ (deformed image) are the pixel values obtained from two consecutive images with a time gap of Δt. Assume $\boldsymbol{u}(\boldsymbol{x})$ as the very small displacement occurred within time Δt, as shown in Fig. 2. Hild and Roux [8, 9] proposed this method of introducing finite element framework into the image correlation process. Based on the law of conservation of optical flow, $f(\boldsymbol{x}) and\ g(\boldsymbol{x})$

Fig. 2 Deformation process

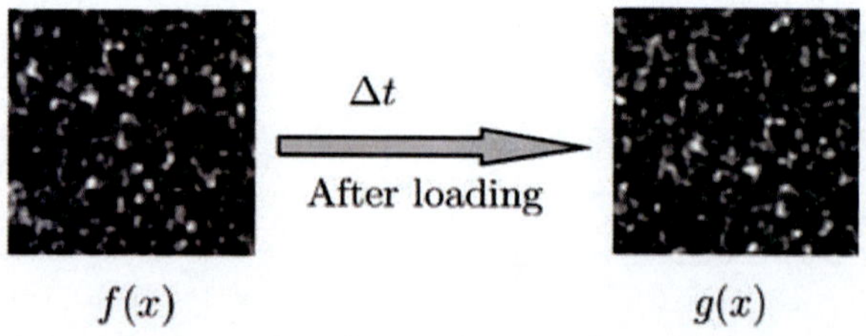

are related as shown in Eq. (1).

$$g(\boldsymbol{x}) = f(\boldsymbol{x} + \boldsymbol{u}(\boldsymbol{x})) \tag{1}$$

Assuming $f(\boldsymbol{x})$ and $g(\boldsymbol{x})$ are spatially differentiable, by applying first-order Taylor expansion to Eq. (1) results in

$$g(\boldsymbol{x}) \approx f(\boldsymbol{x}) + u(\boldsymbol{x}).\nabla f(\boldsymbol{x}) \tag{2}$$

Quadratic residual $\Psi(x)$ is calculated as

$$\Psi(\boldsymbol{x}) \approx (f(\boldsymbol{x}) + \boldsymbol{u}(\boldsymbol{x}).\nabla f(\boldsymbol{x}) - g(\boldsymbol{x}))^2 \tag{3}$$

The quadratic residual is integrated over the entire domain Ω for estimation of $\boldsymbol{u}(\boldsymbol{x})$ as given in Eq. (4). This equation resembles a least-square finite element formulation.

$$\psi = \int_{\Omega} (f(\boldsymbol{x}) + \boldsymbol{u}(\boldsymbol{x})\nabla f(\boldsymbol{x}) - g(\boldsymbol{x}))^2 d\boldsymbol{x} \tag{4}$$

Assuming that the domain Ω is divided into E number of noded elements with an area of Ω_e. The displacement filed $\boldsymbol{u}(\boldsymbol{x})$ is approximated with unknown nodal displacements $u_{\alpha n}$ (displacement at node n in dimension α) and shape functions $N_n(\boldsymbol{x})$ as:

$$u_e(x) = \sum_{n=1}^{n_e} \sum_{\alpha=1}^{2} u_{\alpha n} N_n(x) e_\alpha \tag{5}$$

where e_α is the unit vector of dimension α. After rewriting, Eq. (4) becomes as Eq. (6).

$$\psi = \sum_{i=1}^{E} \int_{\Omega_e} \left(\sum_{n=1}^{n_i} \sum_{\alpha=1}^{2} u_{\alpha n} N_n(x).\nabla f(x).e_\alpha + f(x) - g(x) \right) \Bigg)^2 \tag{6}$$

Minimizing Eq. (6) with respect to nodal displacements $u_{\alpha n}$ results in a linear system of equations represented in matrix form in Eq. (7).

$$\boldsymbol{M_e u_e} = \boldsymbol{b_e} \tag{7}$$

matrix $\boldsymbol{M_e}$ and vector $\boldsymbol{b_e}$ are obtained by assembling matrix elements $M^e_{\alpha n \beta m}$ (Eq. 8) and $b^e_{\alpha n}$ (Eq. 9).

$$M^e_{\alpha n \beta m} = \int_{\Omega_e} N_m(x) N_n(x) \frac{\partial f}{\partial x_\alpha} \frac{\partial f}{\partial x_\beta} dx \tag{8}$$

$$b^e_{\alpha n} = \int_{\Omega_e} [g(x) - f(x)] N_n(x) \frac{\partial f}{\partial x_\alpha} dx \tag{9}$$

To reduce the effect of variation in surface texture during the deformation, the derivatives in Eqs. (8) and (9) are substituted by averaging the contributions of both reference image and deformed image as $\frac{\partial f}{\partial x_\alpha}$ =0.5 $[\frac{\partial f}{\partial x_\alpha} + \frac{\partial g}{\partial x_\alpha}]$. Images $f(\boldsymbol{x})$ and $g(\boldsymbol{x})$ are raster greyscale images (rectangular grid of numbers), which are not differentiable directly. For the calculation of derivates B-splines image interpolation, a method given by Hou and Andrews [10] is used. Multilevel approach proposed by Besnard et al. [9] is implemented into the framework of FE-DIC for finding sub-pixel displacements. The algorithm is coded in C + + language.

3 Numerical Validation:

Numerically simulated speckle images were used to evaluate the performance of the FE-DIC algorithm. Four types of generalized deformation fields (translation, rotation, pure shear, and uni-axial tensile) were generated between a pair of images based on the procedure given by Pan et al. [11]. Reference image $f(x, y)$ and deformed image $g(x, y)$ are calculated as in Eq. (10 and 11).

$$f(x, y) = \sum_{k=1}^{s} I_k^0 exp[-\frac{[(x - x_k)^2 + (y - y_k)^2]}{R^2}] \tag{10}$$

$$g(x, y) = \sum_{k=1}^{s} I_k^0 \exp\left[-\frac{\left[(x - x_k - u_0 - \epsilon_{xx}x - \epsilon_{xy}y)^2 + (y - y_k - v_0 - \epsilon_{yx}x - \epsilon_{yy}y)^2\right]}{R^2}\right] \tag{11}$$

where s is the total number of speckles in the image, I_k^0 is the random peak intensity of each speckle at a random position (x_k, y_k), and R denotes the size of the speckle. The parameters u_0 and v_0 represent translation in x-and y-directions respectively and $(\epsilon_{xx}, \epsilon_{yy}, \epsilon_{xy}, \epsilon_{yx})$ denote strain tensors in two dimensions. Equations (10 and 11) are in image coordinates (unit of displacement is pixel).

In this study, the images were generated with $s = 7829$ and $R = 2.5$ and the other variables are varied according to loading conditions, as mentioned in Fig. 3.

$$E_{RMS} = \sqrt{\frac{1}{N} \sum_{m=1}^{N} (a_m - \hat{a}_m)^2} \tag{12}$$

Root mean square error is calculated based on Eq. (12) where N denotes the total number of nodal points, a_m is the displacement of mth node obtained from FE-DIC algorithm, and $\hat{a}_m$ is applied displacement of mth node based on Eq. (10 and 11).

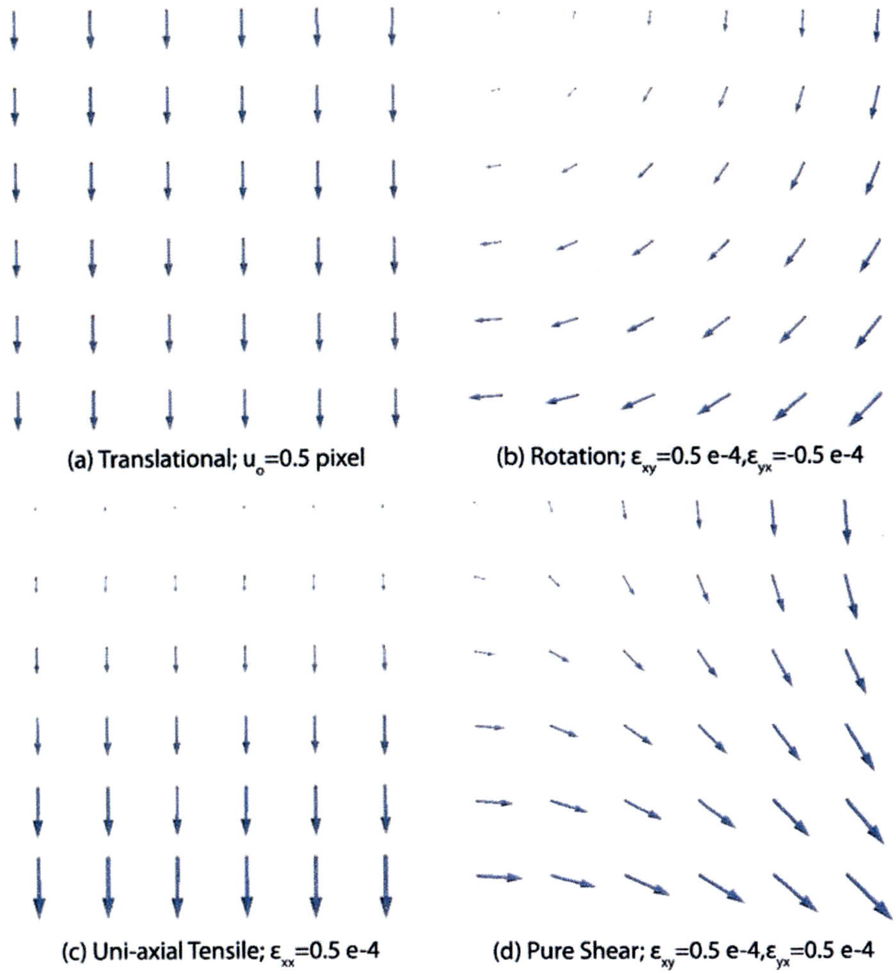

Fig. 3 Generalised loading conditions

The lower values of root mean square error indicate the accuracy of the FE-DIC algorithm. Similar results were observed for various values of deformation fields.

4 Laboratory Experimentation

150-mm-sized cube specimens made with cement concrete of M30 grade are used for experiments. One surface of the cube is pasted with a speckle sticker developed at IIT Kharagpur and patented in India [12]. A universal testing machine capacity of 3500 KN is used for uni-axial compressive tests with a loading rate of 0.45 mm/min.

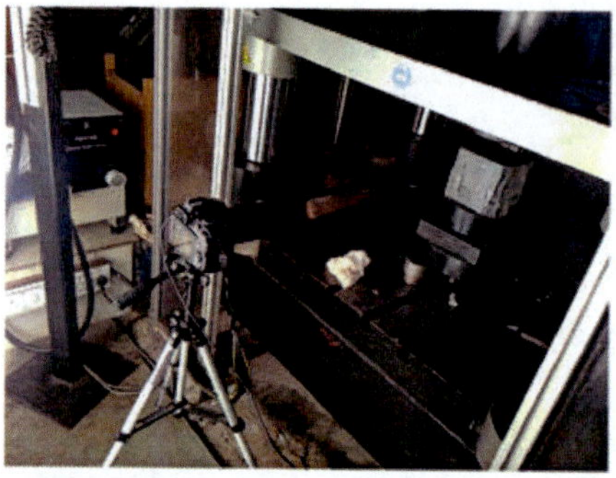

(a) Camera capturing the surface of the cube

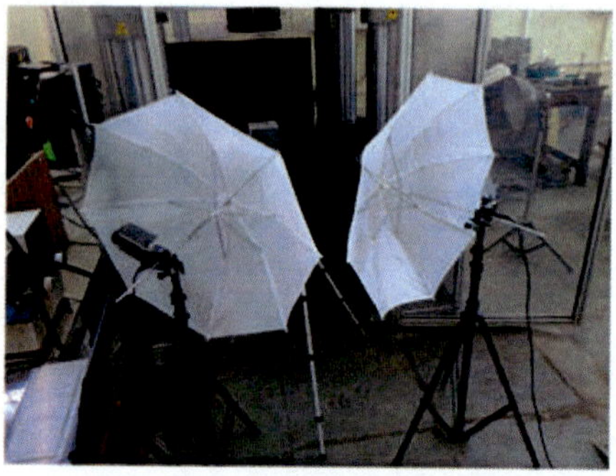

(b) Lighting System

(c) Camera controlled by laptop

(d) Specimen after the failure

Fig. 4 DIC Experimental Setup

A Nikon D3400 camera is used to capture the images of the speckled surface with a frequency of 12 images/min, and Simpex setup is used for lighting, as shown in Fig. (4). Captured images were analyzed for displacement using FE-DIC algorithm.

The results of one experiment are shown in Fig. 5; the left-side are the displacement plots (output by FE-DIC algorithm) labeled with time, measured from the starting of load, and right-side are the corresponding images of the surface of the specimen. The displacement pattern at 185 s (Fig. 5.e) spots a zone of crack propagation, well ahead of the crack appears on the surface (Fig. 5.l). Displacements in that zone are oriented in two opposite directions indicating the opening of a crack. Results show that DIC can predict the zones of failure in the specimen. Similar observations have been made in several other specimens tested under uni-axial loading conditions.

5 Conclusion

This research demonstrates that DIC is a useful technique to predict the zone of failure well ahead of cracks appear on the surface and to understand the internal

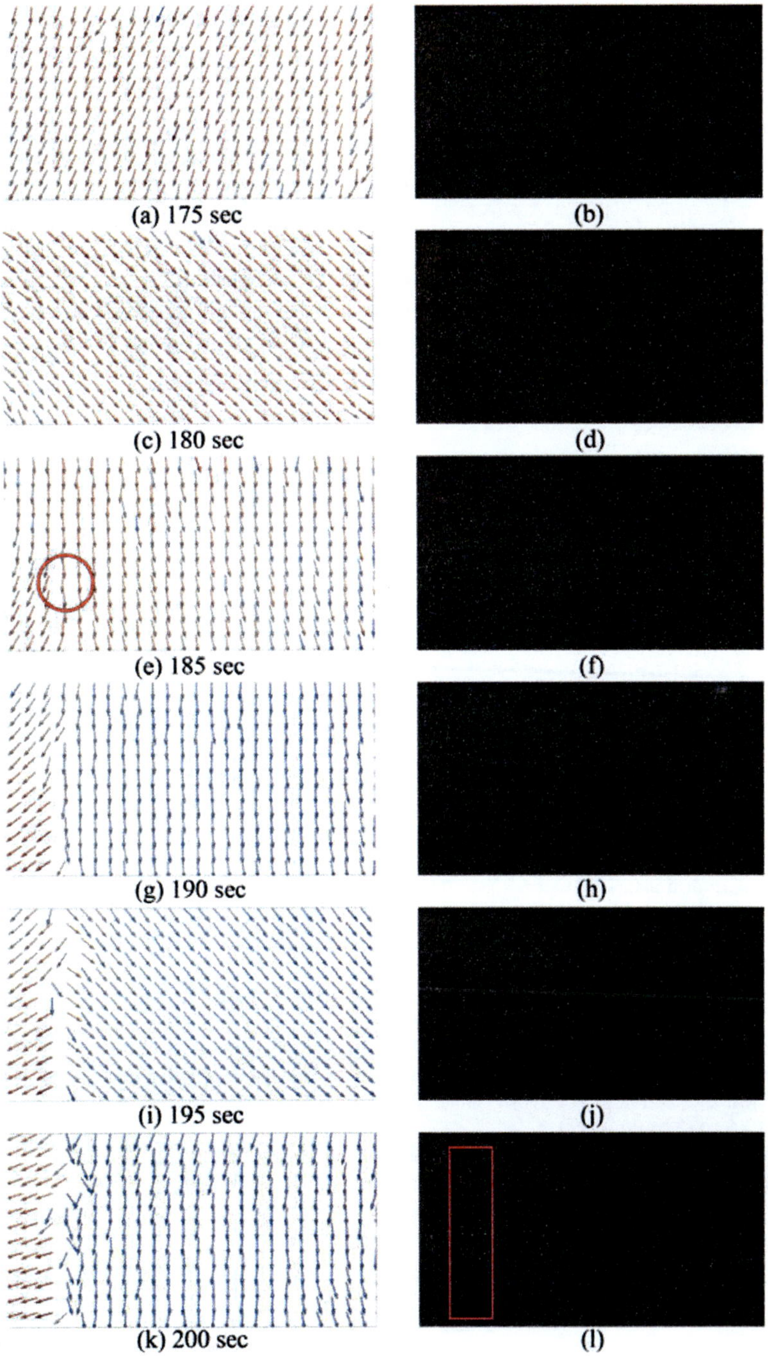

Fig. 5 Displacement plots and there corresponding images of the surface

Table 1 Root mean square error for corresponding loading conditions

Loading type	Root mean square error
Translational	0.010950
Rotation	0.006596
Uni-axial tensile	0.004876
Pure shear	0.006946

anatomy of the structures. Hence, DIC has immense applicability in structural health monitoring of aerospace, civil, mining structures and the infrastructure developed for thermal and nuclear energy generation. Future research focuses on taking DIC into the field for monitoring the deformation of field structures.

References

1. Peters W, Ranson W (1982) Digital imaging techniques in experimental stress analysis. Opt Eng 21(3):213427. https://doi.org/10.1117/12.7972925
2. Pan B (2018) Digital image correlation for surface deformation measurement: historical developments, recent advances and future goals. Meas Sci Technol 29(8):082001. https://doi.org/10.1088/1361-6501/aac55b
3. Reu P (2012) Introduction to digital image correlation: best practices and applications. Exp Tech 36(1):3–4. https://doi.org/10.1111/j.1747-1567.2011.00798.x
4. Sun Y, Pang JH, Wong CK, Su F (2005) Finite element formulation for a digital image correlation method. Appl Opt 44(34):7357–7363. https://doi.org/10.1364/AO.44.007357
5. Réthoré J, Hild F, Roux S (2008) Extended digital image correlation with crack shape optimization. Int J Numer Meth Eng 73(2):248–272. https://doi.org/10.1002/nme.2070
6. Deb D, Bhattacharjee S (2015) Optical strain for monitoring of concrete failure mechanism with discontinuity. Appl Opt 54(35):10409–10417. https://doi.org/10.1364/AO.54.010409
7. Deb D, Bhattacharjee S (2015) Extended digital image correlation method for analysis of discrete discontinuity. Opt Lasers Eng 74:59–66. https://doi.org/10.1016/j.optlaseng.2015.05.006
8. Hild F, Roux S (2006) Digital image correlation: from displacement measurement to identification of elastic Properties-a review. Strain 42(2):69–80. https://doi.org/10.1111/j.1475-1305.2006.00258.x
9. Besnard G, Hild F, Roux S (2006) Finite-element displacement fields analysis from digital images: application to Portevin-le chatelier bands. Exp Mech 46(6):789–803. https://doi.org/10.1007/s11340-006-9824-8
10. Hou HS, Andrews H (1978) Cubic splines for image interpolation and digital filtering. IEEE Trans Acoustics, Speech Sign Process 26(6):508–517. https://doi.org/10.1109/tassp.1978.1163154
11. Pan B, Hui-Min X, Bo-Qin X, Fu-Long D (2006) Performance of sub-pixel registration algorithms in digital image correlation. Meas Sci Technol 17(6):1615–1621. https://doi.org/10.1088/0957-0233/17/6/045
12. Deb D, Bhattacharjee S (2014) Method for standardizing and generating speckle pattern in digital image correlation for monitoring of deformation/strain during experimental investigation. File no: 944/KOL/2014 of 16/09/2014.

Detection of Cracks in Stainless Steel Structures Using Transient Eddy Current Oscillations Method

Sudhakar J. Ongole, Lokesh Andavarapu, and Chandra S. Angani

1 Introduction

The pipelines are most important parts of any large-scale industries like nuclear power, oil–natural gas, to transport the steam, crude oil, and several gases. Stainless steel piping systems have been the typical choice in industrial applications because of its strength and durability even in high temperature and pressure environments. However, metal piping has its own drawbacks [1]. Specifically, if the plant uses stainless steel or any other alloy to transport fluids, local wall thinning and corrosion can be a real concern [2]. Corrosion affects flow rates and efficiency. Continuous use of these components may lead change in local wall thickness due to aging. The thinned wall may not sustain the high pressures which may lead initiation of cracks that result in leakages, unwanted outages, or catastrophic failures [3]. Therefore, periodical in service inspection is necessary to monitor the structural materials to avoid the shutdowns and accidents in the industries. There are several conventional NDT methods which were developed, and they have been serving for the decades [3]. Such as Ultrasonic Testing (UT) [4], Radiography Testing (RT) [5], Magenetic Flux Leakage (MFL) [6], Barkhausen Noise (BN) [7], and Eddy Current Testing (ECT) [8]. However, the conventional methods have their own disadvantages like UT requires a coupling medium for the testing, since slight air gap between test object and the probe causes loss of energy [9], RT requires a skillful training, and RT is hazardous to the operator due to radiation [10]. The MFL requires surface preparation; it is applied only to the magnetic materials; moreover, it operates at nearly saturation magnetization, and hence, the equipment requires a high-power current source to generate a strong applied magnetic field [11]. The ECT is most

S. J. Ongole · L. Andavarapu · C. S. Angani (✉)
Department of Electronics and Physics, GIS, GITAM Deemed To Be University, Visakhapatnam, India
e-mail: cangani@gitam.edu

C. K. Mukhopadhyay and R. Mulaveesala (eds.), *Advances in Non-destructive Evaluation*, Lecture Notes in Mechanical Engineering,
https://doi.org/10.1007/978-981-16-0186-6_16

widely used method to detect cracks and inhomogeneities in the materials. ECT is based on the electromagnetic induction. In ECT, a probe consisting of an excitation coil which generates a magnetic field is used to induce the eddy currents in to the test material [8]. If there is any crack present in the test object beneath the probe causes perturbation of eddy currents that result in change in the detected magnetic field. By careful examination of this magnetic field, one can detect the cracks and thickness change in the test material. However, ECT has limited depth of penetration of induced currents due to use of single frequency excitation, also to get more eddy current density in the test object, the excitation coil should drive with more current. This high current may cause the heating of the excitation coil in the probe; it results unwanted change in coil impedance [12].

In order to detect the resultant magnetic field in the probe, several magnetic field sensors can be employed. T. Dogaru et al. used Giant Mangnetoresistance (GMR) sensors in ECT for the detection of cracks [13], Dario J. L. et al. used Anisotropic Mangeitoresistantce (AMR) sensors for the detection of defects in steel tubes using remote field ECT [14], KeijiTsukada et. al. developed ECT probes with Tunneling Magnetoresistance (TMR) sensor [11], and Tian et. al, angani et al. employed hall sensors [15, 16] in the Pulsed Eddy Current (PEC) method for the testing of steel material. Tiago et. al compared the performance of the GMR and hall sensors for the detection of defects in rails using velocity induced ECT [17]. The hall sensor electronics are simple and does not require any external magnetic field biasing like other sensors. It has liner output characteristics [18], also the commercially available hall sensors are smaller in size and cheap in price. Thus, the hall sensor has been employed in the present study to detect the magnetic field in probe.

There is always an intense research is going on the advancement of NDT methods; in this context, Angani et.al, [19] recently developed a new NDT method called Transient Eddy Current Oscillations [TECO]. The basic principle of TECO is same as ECT; however unlike the ECT, it does not used any external current source like ECT to drive the probe; instead, TECO uses a decay of oscillations generated by a combination of a capacitor and induction coil at their natural frequency [20]. The decay of oscillations induces the eddy currents in the test specimen which is beneath the probe. It is known that the eddy current technique is based on the interaction between a magnetic field source which is a coil and the test material. This interaction induces eddy currents in the test piece. One can detect the presence of very small cracks by monitoring changes in the resultant magnetic field due to perturbed eddy current flow [21]. The simple time domain features in the resultant oscillations such as amplitude, decay time can be used to identify the physical condition of the test specimen [19]. TECO has the advantages of simple electronics, less power consumption, due to short time decay of oscillations, and also coil heating can be reduced.

2 Experimental setup and Theory

The TECO system consists of a probe, arbitrary waveform generator, DC power supply, differential amplifier, and a digital oscilloscope. The schematic diagram and the photograph of the laboratory setup of TECO system are shown in Fig. 1a, b, respectively. The probe consists of an induction coil of 200 turns of 0.3 mm thick copper wire (1.43 mH) and a capacitor with 0.154 μF connected in parallel. This arrangement connected to a DC power supply through an electronic relay switch. An arbitrary function generator has been used to control the ON/OFF states of the relay. The DC power supply used to supply the current of 800 mA to the coil through the switch. A digital oscilloscope has been used to record and save the data from the probe. A differential amplifier using INA118 instrumentation op-amp with gain factor 100 was designed to amplify the output voltage of the sensor.

The function generator used to drive the relay switch with a square wave of amplitudes varies from 0 to 6 V with 50% duty cycle. During the positive half cycle of the square wave, the switch is ON, and then, the DC power supply charges the capacitor. During the 0 V half cycle of the square wave, the switch is OFF; now, the capacitor and induction coil exchange the stored energy. This physical phenomenon creates a decay of oscillations having the natural frequency equal to $1/2\pi\sqrt{LC}$. In the present study, oscillations with approximately 10 kHz frequency were generated. These oscillations create a proportional magnetic field around the induction coil in the probe. When the electrically conductive test specimen placed beneath the probe, a decay of eddy currents will be induced due to electromagnetic induction [20]. The resultant magnetic field which is the vector sum of the reflected field from the induced currents and the imposed primary field from the coil is detected by using a hall sensor. The sensor is placed in the bottom center of the coil as shown in Fig. 1a (included a small photograph of the coil). The hall sensor produces an output voltage which is proportional to the magnetic field which is perpendicular to the test surface.

The parallel combination of the capacitor and the inductor can be represented by a second-order differential equation. And the solution for the instantaneous current of the decay of oscillations can be represented as given in below Eq. (1).

$$i(t) = I_0 e^{-at} \cos(\omega t + \theta) \tag{1}$$

where *i (t)* is the instantaneous amplitude of current, I_o is the current at time t = 0, θ is the phase angle, *a* is the damping coefficient, and ω is the angular frequency of free running oscillations [19, 20]. In order to test the TECO system, a 5 mm thick steel plate test specimen having four superficial cracks with 0.5 mm width and 25 mm length with different depth of 2.5, 3, 3.5, and 4 mm was prepared using Electro Discharge Machining (EDM) process. The drawing of the test specimen is shown in Fig. 1c.

3 Experimental Results

To record the decay of oscillations, the probe is placed on the surface of the test specimen. During the measurement, the probe is exactly centered on the crack. The decay of oscillations signal is recorded in the digital oscilloscope for every negative edge of the square wave from the function generator.

Because the decay oscillations were created when the DC power supply got disconnected from the capacitor in the probe, this happens when the relay is switched OFF by the function generator. The results were obtained for all the cracks of different depths, as shown in Fig. 2a. As shown in above Fig. 2a, the decay oscillations have shown different amplitudes with different decay coefficients for different depths of the cracks. It can be observed from the figure that when there is no crack, the signal has less amplitude and decays more quickly as compared with the signals which are obtained in the presence of cracks. As shown in the figure, the amplitude is increased and the signals decay more slowly with increasing the crack depth. This can be understood as the volume of the crack increases with increasing crack depth. So the volume of the available electrically conductive test material beneath the probe decreases with increasing the crack depth. When there is no crack, then higher density of induced currents causes more reflected magnetic field results in the reduction of resultant magnetic field which is vector sum of both primary and reflected fields. Hence, the no crack signal has the less amplitude. If there is increase in the crack depth which causes reduction in induced current density, it leads to increase in the amplitude.

To get more clear view of the decay of oscillations, in Fig. 2b, the upper envelopes are plotted for different crack depths. The signal decays more quickly with decreasing the crack depth, because more volume of conductive material results in more induced currents and more resistivity; it leads to higher ohmic loss in the material. Therefore, when the probe is placed on the no crack area, the induced field decays more quickly as compared to the crack area on the test specimen. In order to get more insight and

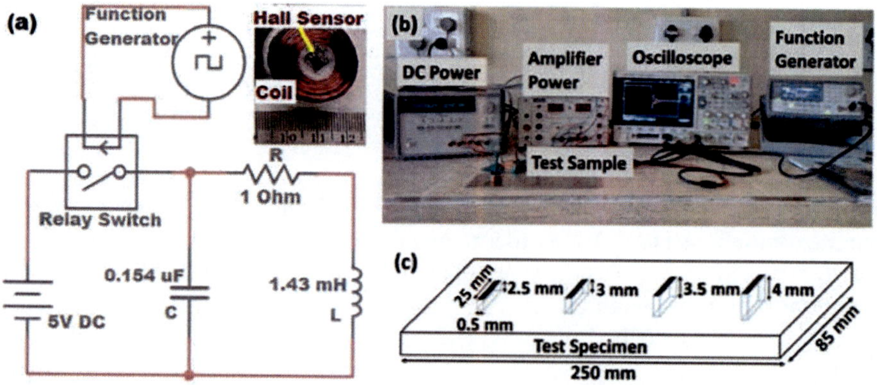

Fig. 1 na TECO system schematic **b** photograph of the system setup, and **c** drawing of the test specimen

insight in to the results, the decay coefficients of all the envelopes were calculated, by using the curve fitting. Also the energies of the oscillations are calculated using Eq. (2) given below:

$$E_s = \sum_{n=-\infty}^{\infty} |x(n)|^2 \quad (2)$$

The energy of the signal is increased with the increasing the crack depth; this is the direct result of the increase in the amplitude of the decay of oscillations with crack depth. As the area under the curve is increased with the crack depth, the energy of the signal is increased. Figure 3b shows the plot of energy and the decay coefficient of the oscillations as a function of crack depth.

4 Conclusions

Transient Eddy Current Oscillation (TECO) method is used to detect cracks in a 5 mm thick stainless steel SS 304. The probe consists of a capacitor and an induction coil connected in parallel to generate a decay of free running oscillations at resonance frequency. Those are used to induce eddy currents into the test object. Four superficial EDM notches with 0.5 mm were fabricated having different depths from the surface, in order to simulate the actual cracks. The results were obtained for different depths of cracks. The amplitude of oscillations is decreased with decreasing the crack depths, and the oscillations decay slowly with increasing the crack depth. The energy of the signal also shows the similar trend as the amplitude change with the crack depth. The results show that the selected features are quite representative. Results suggest that this method can be applied for the detection of cracks in industrial structural materials.

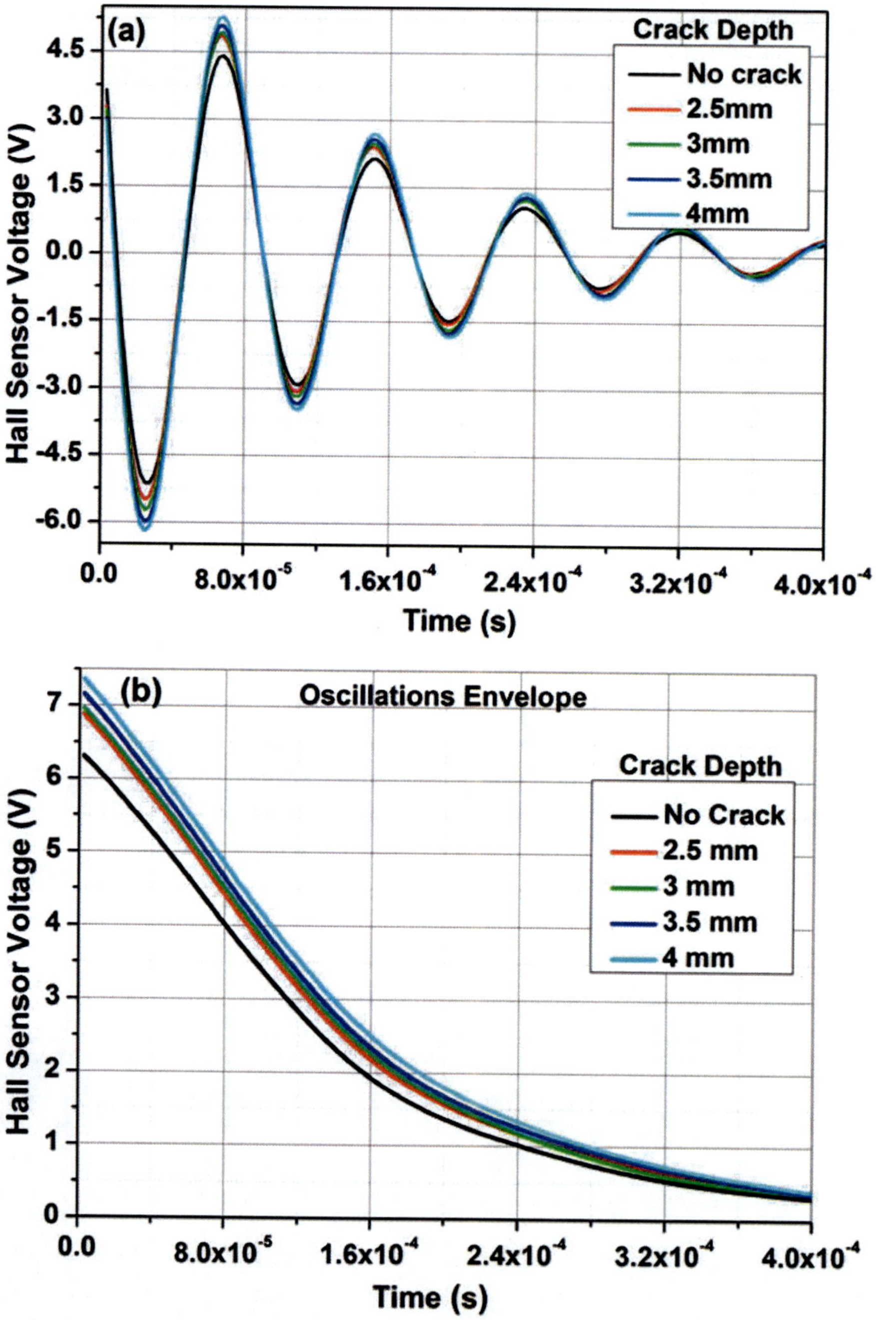

Fig. 2 **a** TECO signal for the different cracks **b** upper envelope of the oscillations

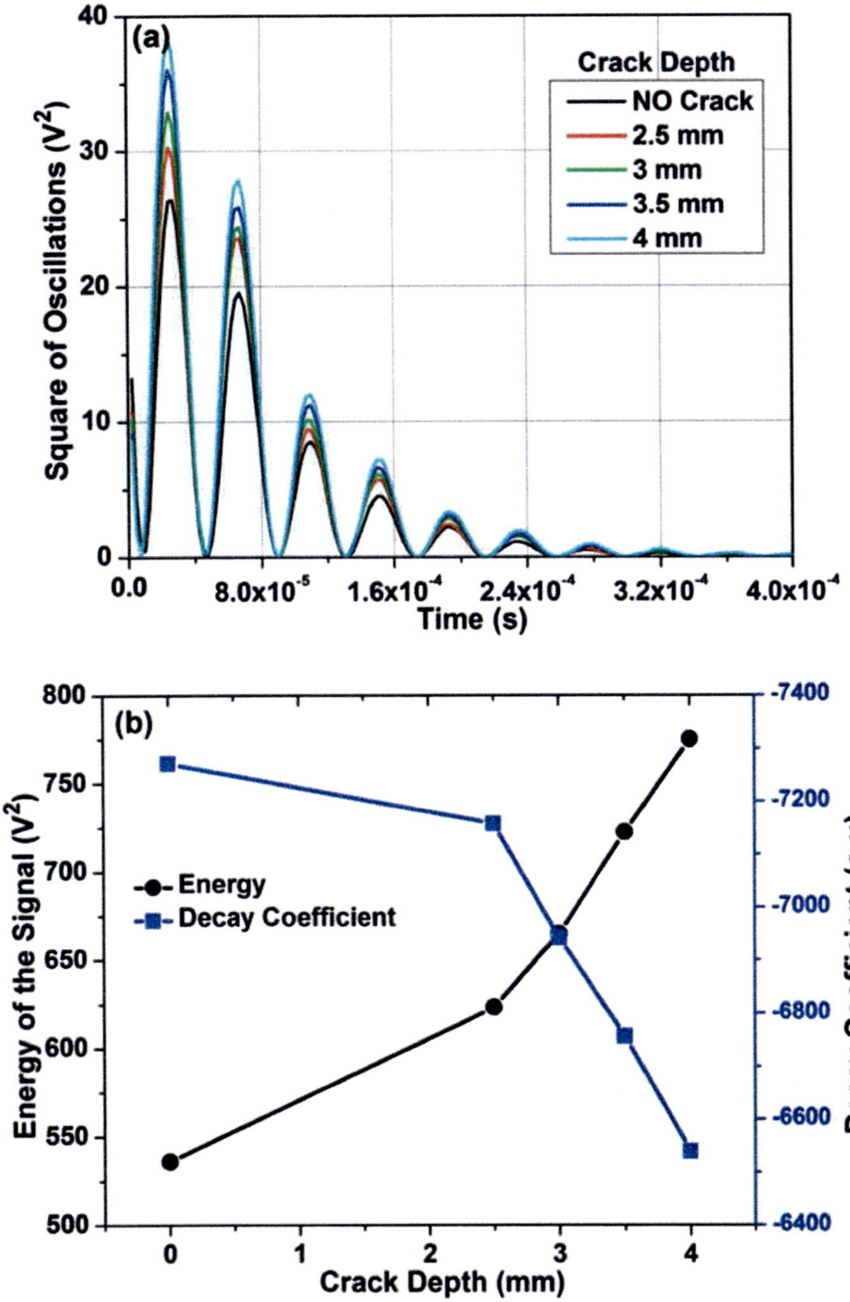

Fig. 3 a Square of the TECO signal at the different cracks **b** energy and the decay coefficient of oscillations as a function of crack depth

Acknowledgments The work has been performed in NDT laboratory at the department of electronics and physics, GITAM Institute of Science, GITAM deemed to be university, Visakhapatnam, India. This research work is supported by The Department of Science and technology (DST), Science and Engineering Research Board (SERB)—India, under Early Career Research (ECR) Award project: ECR/2016/001790. The support is greatly acknowledged.

References

1. Valdez, B., Schorr, M., Zlatev, R., Carrillo, M., Stoytcheva, M., Alvarez, L., Eliezer,A., and Rosas, N.: Corrosion Control in Industry, Environmental and Industrial Corrosion, Practical and Theoretical Aspects, Benjamin Valdez., Michael Schorr., (eds.) Intech Open, (2012).
2. Popoola LT, Grema AS, Latinwo GK (2013) Corrosion problems during oil and gas production and its mitigation. Int. J. Ind. Chem. 1–15:1–6
3. Masayuki, K., Tomohisa, Suzuki, T., Meshii.: Failure pressure of straight pipe with wall thinning under internal pressure, Int. J. of Pressure Vessels and Piping, 85, 628–634 (2008).
4. Raj B, Jayakumar T, Rao BPC (1995) Non-destructive testing and evaluation for structural integrity. Sadhana 20:5–38
5. Le M, Kim J, Kim S (2016) Nondestructive testing of pitting corrosion cracks in rivet of multilayer structures. Int. J. Precis. Eng. Manuf. 17:1433–1442
6. Moreira E, Barbosa Rabello J, Pereira M (2010) Digital Radiography Using Digital Detector Arrays Fulfills Critical Applications for Offshore Pipelines. EURASIP J. Adv. Signal Process. 2010:894643
7. Chandra Sekhar, A., Jungmin K., Minhhuy L., and Jinyi L.: Magnetic Leakage Testing Using Linearly Integrated Hall and GMR Sensor Arrays to Inspe.ct Inclusions in Cold Rolled Strips, IEEE Transactions On Magnetics. 50, 1–4 (2014).
8. Steel MBNIOCI (1997) Thomas W. Krause., Atherton, D.L., Sullivan, S.P. Nondestructive Testing and Eval. 13:309–323
9. García Martín J, Gómez-Gil J, Vázquez Sánchez E (2011) Non-Destructive Techniques Based on Eddy Current Testing. Sensors. 11:2525–2565
10. Bindal, V, N.: Water Based Couplants for General Purpose Use for Ultrasonic NDT Applications, J. of Scientific and Industrial Research. 59, 935–939 (2000). https://nopr.niscair.res.in/handle/123456789/26632.
11. Mahmoudi, A., Regragui, F.: Welding Defect Detection by Segmentation of Radiographic Images. In: 2009 WRI World Congress on Computer Science and Information Engineering, pp. 111–115. Los Angeles, CA, 2009.
12. Keiji T., Minoru, H., Yoshihiro N., Kenji Sakai, Toshihiko K.: Small Eddy Current Testing Sensor Probe Using a Tunneling Magnetoresistance Sensor to Detect Cracks in Steel Structures. IEEE Trans. on Mag., 54, 1–5 (2018), Art no. 6202205.
13. Gerhard M, Olaf H, Valentin U (2007) Deep Penetrating Eddy Currents and Probes. Materials Testing. 49:258–264
14. Giant magnetoresistance-based eddy-current sensor (2001) Dogaru, T., Smith, S, T. IEEE Trans Magn 37:3831–3838
15. Dario JL, Pasadas A, Lopes R, Helena GR, Tiago JR (2015) Defect detection in stainless steel tubes with AMR and GMR sensors using remote field eddy current inspection. Acta IMEKO 4:62–67
16. Tian GY, Sophian A (2005) Reduction of lift off effects for pulsed eddy current NDT. NDT&E Int. 38:319–324
17. Angani CS, Park DG, Kim CG, Leela P, Kollu P, Cheong YM (2010) The Pulsed Eddy Current Differential Probe to Detect a Thickness Variation in an Insulated Stainless Steel. J. Nondest. Eval. 29:248–252

18. Tiago JR, Helena GR, Lopes RA, Pasadas DJ (2015) Magnetic sensors assessment in velocity induced eddy current testing. Sens Actuators, a 228:55–61
19. Advances in Magnetic Field Sensors (2010) Pavel, Ripka., Michal, Janoˇsek. IEEE Sens J 10:1108–1116
20. Chandra SA, Helena GR, Artur LR, Tiago JR, Prashanth B (2015) Transient eddy current oscillations method for the inspection of thickness change in stainless steel. Sens. And Act. A. 233:217–223
21. Angani, C. S., Ramos, H. G., Ribeiro, A. L., Rocha T. J., P. Baskaran.: Lift-Off Point of Intersection Feature in Transient Eddy Current Oscillations Method to Detect Thickness Variation in Stainless Steel, IEEE Trans. on Mag. 52, 1–8 (2016). Article no. 6201408.
22. Nicola, Bowler.: Eddy-Current Nondestructive Evaluation, Springer-Verlag, New York (2019).

Damage Assessment of Concrete Bridge Structures Using Ultrasonic Tomography Technique

Thirumalaiselvi A. and Saptarshi Sasmal

1 Introduction

The construction of civil infrastructure such as bridges requires longer period of time, exhaustive labour and money. In such instances, failure of such structures will be catastrophic in terms economic loss along with human life. According to BS-106 report [1], in India, 42% of bridges crossed 100 years, 62% bridges crossed 80 years and 75% bridges crossed 60 years of service. Maintenance of such bridge structures is highly demanding and a proper condition assessment protocol is also of high importance considering the safety aspects [2]. This urges researchers to exploit a varied range of inspection methods for early detection and warning on critical defects [3]. Application of stress wave propagation based non-destructive techniques to assess the structural condition is increasingly emerging. Acoustic emission (AE) and ultrasonic pulse velocity (UPV) are some of the stress wave-based non-destructive techniques which are most widely used towards damage assessment of structures. Unlike metal structures, concrete structures pose the challenge in developing efficient damage assessment technique due to the inherent heterogeneity, microcracks/pores and nonlinearities from concrete. It has been reported in earlier works that predictions from AE technique are rarely quantifiable and also not straightforward that makes it to be an insufficient tool. Detailed investigation at critical locations could be made through employing ultrasonic testing [4].

The travel time of ultrasonic pulse is measured using ultrasonic testing, and the velocity is evaluated from the ultrasonic wave measurement made, which normally in the range of 3000–4000 m/s for normal concrete [5]. Since the ultrasonic velocity in air is around 350 m/s, which is lesser than that in normal concrete, damaged

Thirumalaiselvi A. (✉) · S. Sasmal
Special and Multifunctional Structures Laboratory, CSIR-Structural Engineering Research Centre, CSIR Road, Taramani, Chennai 60113, India
e-mail: selvi@serc.res.in

C. K. Mukhopadhyay and R. Mulaveesala (eds.), *Advances in Non-destructive Evaluation*, Lecture Notes in Mechanical Engineering,
https://doi.org/10.1007/978-981-16-0186-6_17

zone in concrete generally exhibits lesser velocity value. Though ultrasonic testing is effective and also simple in detecting cracks and defects present in concrete, it poses few disadvantages concerning its application in the actual field scenario. Ultrasonic testing is found to be inapplicable in inaccessible but critical locations. Further, in most of the cases, ultrasonic testing requires accessibility of two sides on the structure. and a perfect acoustic coupling between ultrasonic transducers and the concrete surface is essential. Achieving perfect coupling is sometimes not possible and labour intensive; thus, the application of the ultrasonic testing on large-scale concrete structure (for global assessment) is limited [6, 7].

Owing recent advances in computing and sensor technologies, ultrasonic tomography that utilizes elastic wave propagation is merging as one of the advanced non-destructive testing techniques for assessing the condition of structures in full scale. The advantage of ultrasonic tomography lies in its ability to visualize the interior of the structure that enables better identification of critical zones and damage progression. Although few studies such as [8–11] have been conducted to apply tomography method in damage assessment, there remain several aspects, especially for full-scale structures which need to be studied to provide a measurement protocol for damage assessment. The main objective of this study is to examine the structural condition of a laboratory scale concrete bridge model through ultrasonic tomography method. Three damage levels are defined: (1) undamaged; (2) partially damaged; (3) severely damaged, based on the maximum strain reached in longitudinal reinforcement in the bridge girder. At this pre-defined damage states, ultrasonic measurements are carried out on specimens tested in flexure, and velocity distribution is determined following the algebraic iterative reconstruction technique based on Kaczmarz method. Reconstructed velocity tomography is then verified with the experimentally observed damage pattern. The central objective of the present study is to examine the feasibility of ultrasonic tomography method for locating cracks and evaluating damage in the large concrete structures such as bridges and to provide a damage assessment methodology with limited number of ultrasonic sensors.

2 Ultrasonic Tomography

Ultrasonic tomography is the process of image reconstruction of the internal structure of a medium by examining the ultrasonic wave propagated through that [12]. This is performed by piezoelectric sensors that utilize the ultrasonic wave properties to plot the tomogram. Defects/damage, if any present in the medium, invariably causes the mechanical properties and acoustic characteristics which in turn gets reflected in the ultrasonic wave propagation. Generally, two ultrasonic wave parameters observed during damage progression, namely the amplitude attenuation of the propagated wave and the time of flight (TOF). Since the ultrasonic tomography is more concerned with the velocity distribution within the medium, determination of TOF is highly crucial. The very first wave arrivals (P-waves) should be identified to determine the TOF while computing wave velocity. However, it is often difficult to define that first arrival because sometimes amplitude of the first peak might be less, almost equal to noise levels. Hence, auto-picking algorithms, such as the Hinkley criterion (HIC), Akaike information criterion (AIC) and noise threshold criterion have been evolved in the recent years to define the TOF [13–15]. Among these, in the present study, AIC method is adopted to define TOF. The AIC method works basically on the iterative statistical analysis of ultrasonic waveform and is defined as follows:

$$\mathrm{AIC}(t_\mathrm{w}) = t_\mathrm{w}\log\left(\mathrm{var}\left(\mathrm{Sig_w}[1, t_\mathrm{w}]\right)\right) + (N - t_\mathrm{w} - 1) \\ \log\left(\mathrm{var}\left(\mathrm{Sig_w}[1 + t_\mathrm{w}, N]\right)\right) \quad (1)$$

where t_w is the fixed time range through all portion of the signal $\mathrm{Sig_w}$, and N is the final index of the windowed time series. The minimum value of computed AIC values defines the first arrival.

Several image reconstruction techniques such as synthetic aperture focusing technique (SAFT), reverse time migration (RTM), filtered back-projection (FBP) and the algebraic reconstruction technique (ART) have been applied for non-destructive testing of concrete structures based on ultrasonic wave data. ART, which was introduced in 1937, also called as Kacrmarz method is one of the most popular algorithms for tomography reconstruction [16]. ART uses linear algebraic equation.

Linear algebraic equation for velocity tomography is framed from the TOF of waves propagating along a particular ray that cross-defined pixel within the medium. The TOF set t_i is given by

$$t_i = \sum_{j=1}^{M} d_{ij} p_j (i = 1 : N) \quad (2)$$

where d_{ij} is the distance travelled by particular ray '*i*' in pixel '*j*', p is slowness or inverse velocity, t_i is TOF of wave by ray '*i*', and N and M are the total number of wave measurements and pixels, respectively. In matrix form, above equation is represented as:

$$T = DP \quad (3)$$

where T is the TOF matrix, D is the path length matrix, P is the slowness matrix. Having D and T values at hand, P is computed. Path length matrix, D, found to be a sparse one since each element in the matrix represents the path length in a pixel corresponding to the defined ray, as illustrated in Fig. 1. For example, consider 1st ray that starts from 12th pixel and ends at 121th pixel (see Fig. 1), the path length matrix, D, have only a few non-zero values as indicated by shaded region among the entire image domain. Thus, D is highly sparse, especially in the situation of analysing actual full-scale structure with larger dimensions and more number of pixels. Hence, it may not be possible to solve Eq. 3 applying conventional matrix inversion method.

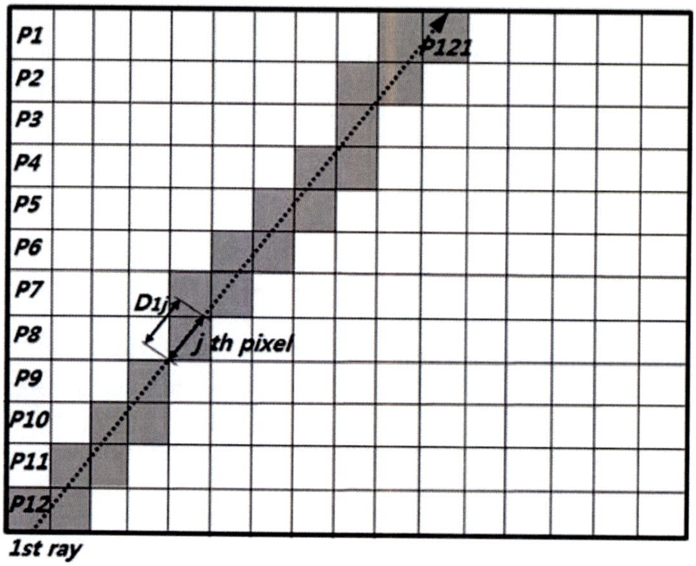

Fig. 1 Ray contribution to pixels [18]

In such instances, iterative-based matrix inversion method is needed. Algebraic reconstruction technique [16] is one such iterative solver that is followed in the present study. This iterative solver updates the residual time, and the iteration process continues till the evaluated slowness vector, which represents the inverse values of all pixels and converges to a solution with the prescribed minimum error. The iterative solver initiates by assuming the first estimate of slowness, P′. The average of slowness across all pixels from the ultrasonic measurements is generally used for the first assumption of P' in velocity tomography. The temporary TOF, T', is calculated as:

$$T' = DP' \tag{4}$$

The residual time from the measurements, dT', is defined by

$$dT' = T - T' \tag{5}$$

The calculated residual time dT' is substituted back into the back-projection equation as

$$dP' = dT' \frac{D}{\text{row}(D)^2} \tag{6}$$

where dP' is the slowness residual. The term back-projection arises from the representation of linear equations in the hyperplane, having dP′ as a normal vector to the measurement line [17] as illustrated in Fig. 2. As the solution approaches to the exact

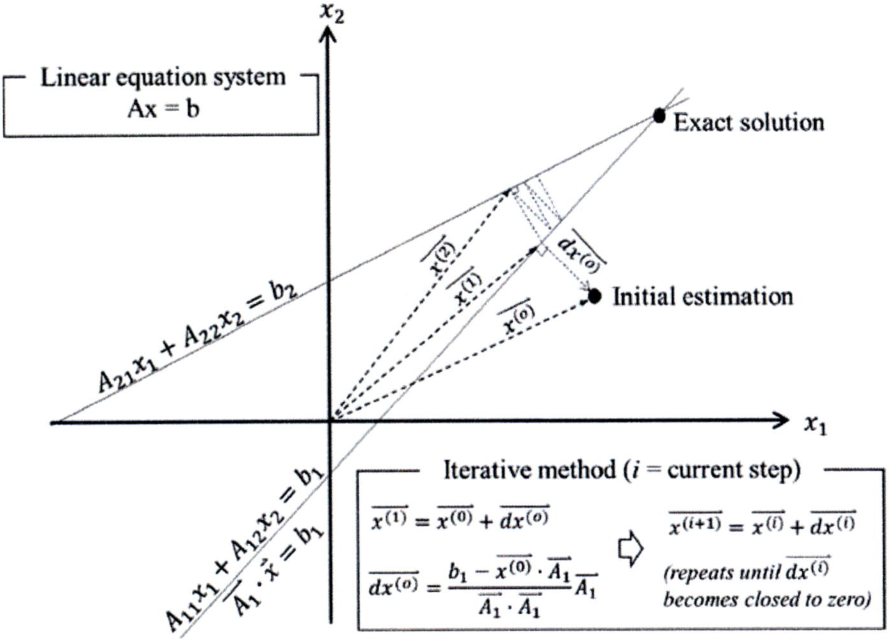

Fig. 2 Schematic illustration of ART [17]

one, among several measurement lines in the hyperplane, dP' is projected onto the lines and becomes smaller residual. The iterative solver revises the back-projection for the consecutive step as:

$$P'' = P' + dP' \tag{7}$$

where P'' is the iterative estimation of the slowness field in the next step. This process repeated till dP' converges to the minimal threshold error.

3 Experimental Investigations

A laboratory scale concrete bridge model of length 4500 mm was cast for experimental investigations, and static loading (through actuator) was applied to induce pre-defined level of damage. Geometric details about the tested beam are provided in Table 1. 3D view of the tested beam with geometry details is shown in Fig. 3.

Four-point bending test was performed with an effective span of 4200 mm to introduce different levels of damage to the beam. The loading is applied in three cycles for each level of loading, so that the damage induced in the beam stabilizes, and there will be clear distinction of different damage cases. The loading is applied in three different load steps, and crack patterns are noted after each load increment.

Table 1 Concrete mixture proportion and geometrical details

Tee-Beam geometry	
Length	4500 mm
Depth	275 mm
Flange width	600 mm
Flange thickness	75 mm
Web thickness	170 mm
Beam longitudinal bar (bottom)	2# 12ϕ
Hanger bar	2# 8ϕ
Shear stirrups	2 legged 8ϕ at 175 mm c/c
Flange longitudinal bar	6# 8ϕ
Flange transverse bar	8ϕ at 300 mm c/c
Maximum coarse aggregate size	20 mm
Concrete cover	30 mm
Concrete mix proportion	
Design compressive strength	43 N/mm^2
Cement: fine aggregate: coarse aggregate: water	1:2.25:2.35:0.5

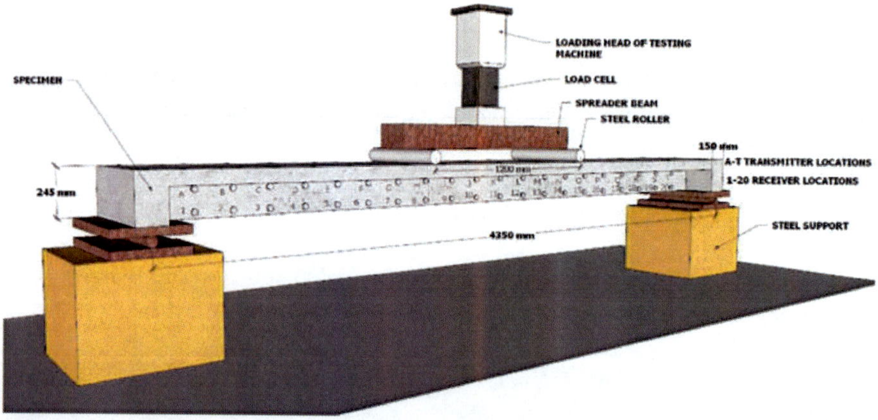

Fig. 3 3D view of the specimen kept in test pad

Load was equally distributed from the hydraulic jack to two planes on the specimen top surface each being separated at a distance of 600 mm from the specimen centre to both sides using two steel rollers and a steel beam placed at the top so that the load is symmetrically distributed to the T-beam. Load was applied in displacement controlled mode at a rate of 0.5 mm/min using a servo-hydraulic actuator with maximum capacity of 25 tonnes. Three damage levels are defined as (1) undamaged before loading, (2) partially damaged at load level of 5 kN (maximum strain in longitudinal reinforcement = 450 micro-strain) and (3) severely damaged at load level

Fig. 4 Laboratory test set-up

of 19 kN (maximum strain in longitudinal reinforcement = 2400 micro-strain). A typical laboratory set-up for the tests is shown in Fig. 4. The strain in longitudinal reinforcement is acquired using the strain gauges pasted onto the reinforcements before casting.

The wave measurements were taken at three steps: (1) before start of loading, (2) at load level of 5 kN and (3) after failure, i.e. load level of 19 kN. Figure 5 shows the instrumentation of measurement. In the measurements, stress waves were generated using a waveform generator that emits ultrasonic pulse. The wave excitation was carried out by placing ultrasonic pulser and exciting at the chosen transmitter location. The excitation was subsequently done at another transmitter location to repeat the wave generation and acquisition exercise. 20 transmitter location arrangements resulted in a total of 400 waveforms to be processed further to obtain TOF. The processed data was later used to compute velocity distributions of the measured volume using ART method.

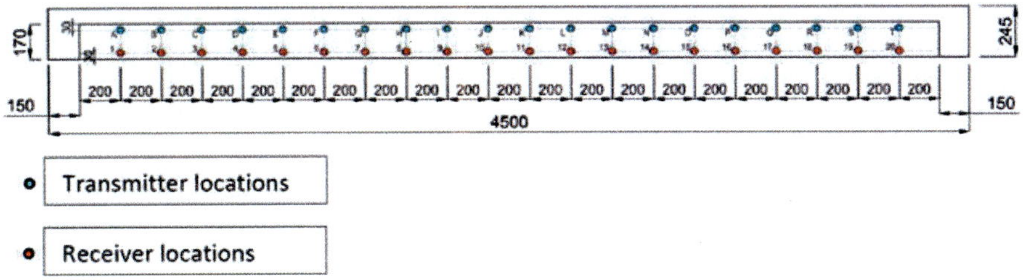

Fig. 5 Sensor locations for ultrasonic measurement

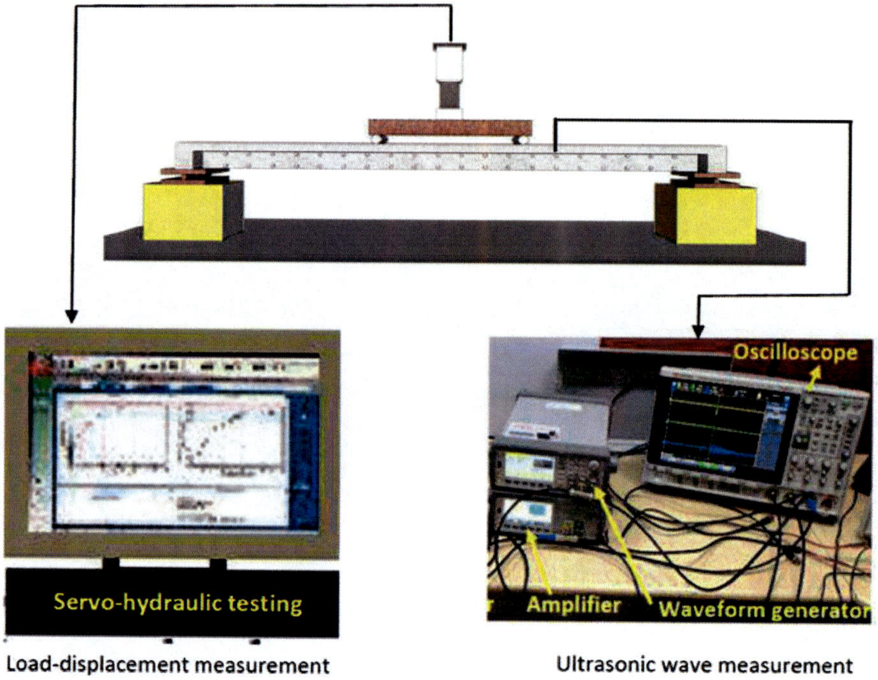

Fig. 6 Measurement protocol followed during experiments

A measurement protocol of the ultrasonic data is shown schematically in Fig. 6. The ultrasonic measurement includes the following: Waveform generator (Keysight 33500B) with maximum operating voltage range of 10 V (peak to peak, denoted as pp), two channel isolated amplifier (Keysight 33502A), two 23 mm diameter ultrasonic transducers with central frequency of 1 MHz (PA956 manufactured by Precision Acoustics Ltd.), a digital oscilloscope (Keysight InfiniiVision DSOX4024A) and personal computer with Keysight software installed. Five-cycle hanned type tone burst signal of 10 V (peak to peak) amplitude and 75 kHz was first generated, and then, it was amplified to 50 V pp and fed into the specimen through transmitting transducer. The ultrasonic signal received at the receiving locations was captured, amplified by five times and visualized using the digital oscilloscope. The input signals before and after amplification were also visualized through oscilloscope. Finally, oscilloscope was connected to a computer from where waveform generator was controlled and digitized signals were captured and stored.

4 Results and Discussion

Velocity tomograms from the measured ultrasonic data at different damage states are made based on the TOF evaluated using AIC method. Middle 2/3 specimen length, where cracks are observed in partially damages state, is considered in plotting the ultrasonic tomography. The number of pixels in the tomograms is fixed to be 22,500, and the path length matrix, D, is constructed according to the specimen dimension and pixel value. Having the measured TOF at hand, slowness field across each pixel is computed iteratively, and the velocity tomography is plotted. A two-dimensional velocity distribution at different pre-defined damage states are presented in Figs. 7, 8 and 9 for showing the transition of damage mapping. Before loading, when the specimen was intact, velocities as measured were in the range of 3000 m/s. The velocity distribution showed decline in values as damage progresses, with the severely damaged state obtained after final failure giving velocity values below 2000 m/s, in particular at sections where critical cracks were observed. Critical cracks observed at different damage states are noted with crack number (say C1, C2, etc.). Experimentally observed damage pattern at different damage state are also shown in Figs. 7, 8 and 9 along with the velocity distribution map.

To make it very clear, damage on the girder at different damage states is also schematically sketched in Figs. 7, 8 and 9. It is seen that in both damage states, velocity values are less at the locations, where cracks are noticed. In the locations

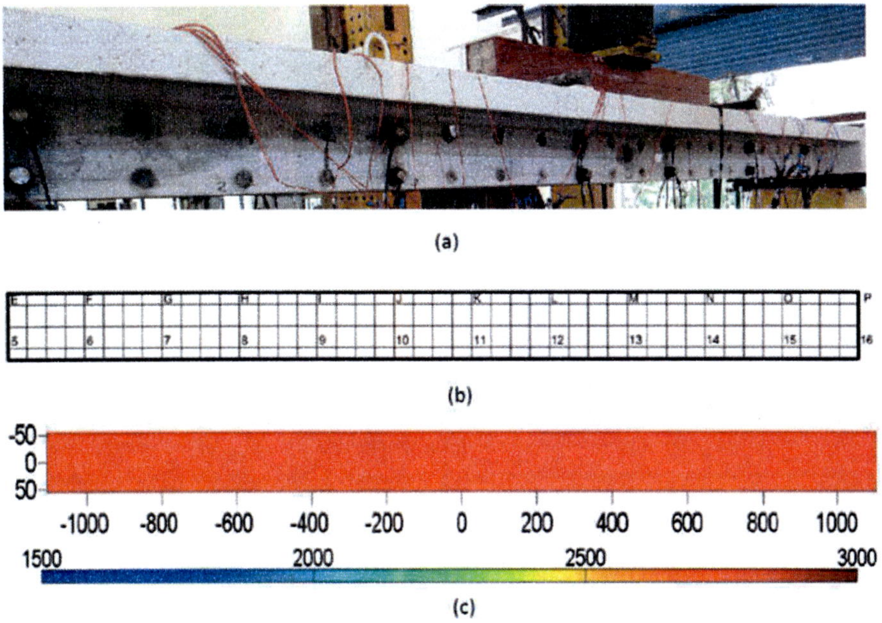

Fig. 7 Damage pattern at undamaged state (intact). **a** test photo; **b** schematically sketch; **c** tomographic image

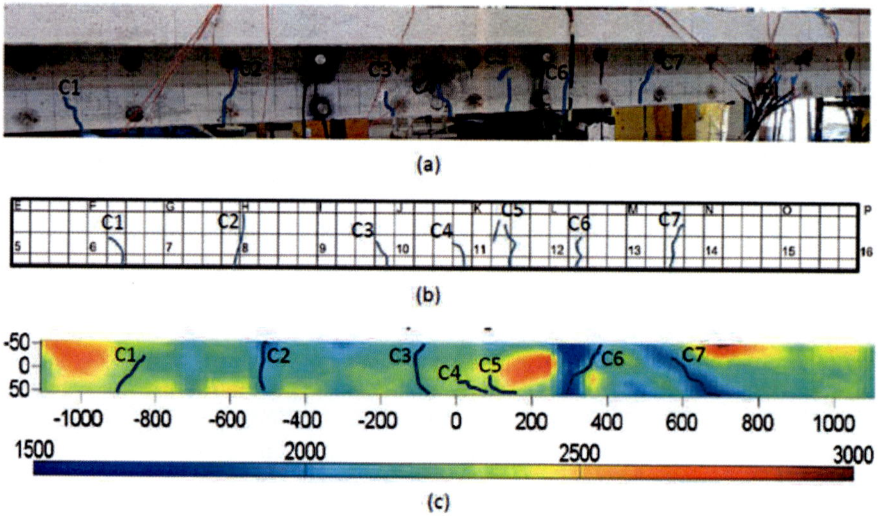

Fig. 8 Damage pattern at partially damaged state. **a** test photo; **b** schematically sketch; **c** tomographic image

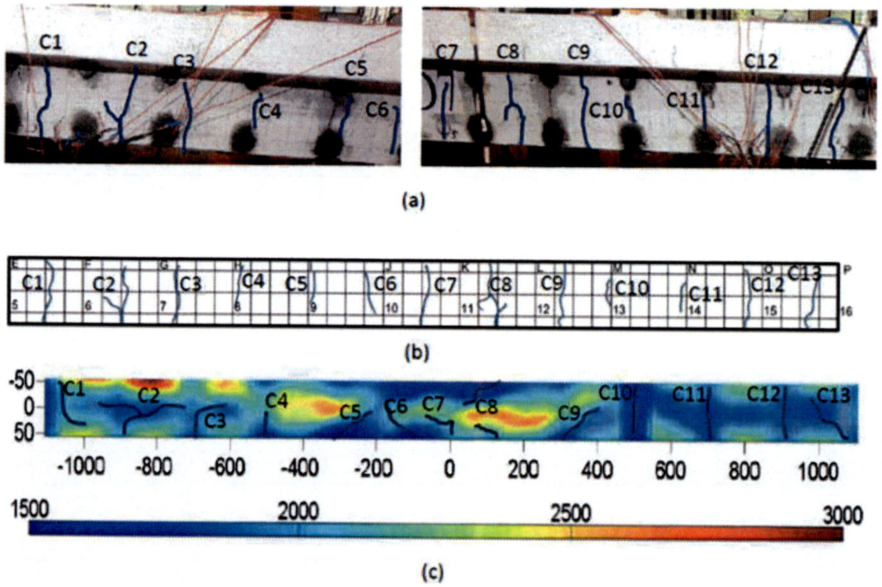

Fig. 9 Damage pattern at severely damaged state. **a** test photo; **b** schematically sketch; **c** tomographic image

where cracks formed due to flexure loading, waves generated using ultrasonic pulse generator could provide velocity distribution with noticeable difference. This demonstrates the acuity of the ultrasonic waves towards inhomogeneity brought out by cracking. In addition to the critical crack locations, few more low velocity zones are seen in tomographic image, which inherently indicates damaged zones inside the specimen and local crushing seen under loading points.

5 Conclusion

Ultrasonic testing system is applied to full-scale bridge model subjected to four-point static loading. The damage progression is monitored by ultrasonic velocity tomography. The reconstructed images provide helpful information about the damage severity and critical crack location. This study demonstrates the applicability of ultrasonic tomography for monitoring realistic damage progress within full-scale structures under static loading with limited number (two) of ultrasonic sensors available at hand and establishes tomography as an effective method to determine the condition of in situ structures. In addition, this study proves that ultrasonic tomography using ART with Kaczmarz method is a reliable method for non-destructive evaluation of concrete structure. The research group is further developing techniques and algorithms for 2D and 3D ultrasonic tomography for damage identification/condition assessment of large concrete structures with different type of damage or deterioration.

References

1. Broad guidelines for instrumentation of bridges for running higher axle loads (2016) Bridges and structures directorate. RDSO, Lucknow
2. Aggelis DG, Kordatos EZ, Strantza M, Soulioti DV, Matikas TE (2011) NDT approach for characterization of subsurface cracks in concrete. Constr Build Mater 25(7):3089–3097
3. Linderman LE, Mechitov KA, Spencer BF Jr (2013) TinyOS-based real-time wireless data acquisition framework for structural health monitoring and control. Struct Control Health Monit 20(6):1007–1020
4. Behnia A, Chai HK, Yorikawa M, Momoki S, Terazawa M, Shiotani T (2014) Integrated non-destructive assessment of concrete structures under flexure by acoustic emission and travel time tomography. Constr Build Mater 67:202–215
5. Naik TR, Malhotra VM, Popovics JS (2003) The ultrasonic pulse velocity method. In: Handbook on nondestructive testing of concrete, 2nd edn. CRC Press, pp 8–1
6. Long, R. (2000). Improvement of ultrasonic apparatus for the routine inspection of concrete (Doctoral dissertation, University of London).
7. Hall KS (2011). Air-coupled ultrasonic tomographic imaging of concrete elements. Doctoral dissertation. University of Illinois at Urbana-Champaign
8. Kobayashi Y (2007) Three-dimensional seismic tomography for existing concrete structure. In: Proceedings of second international operational analysis conference, 2007, vol. 2, pp 595–600
9. Chai HK, Momoki S, Kobayashi Y, Aggelis DG, Shiotani T (2011) Tomographic reconstruction for concrete using attenuation of ultrasound. NDT & E Int 44(2):206–215

10. Haach VG, Juliani LM (2014) Application of ultrasonic tomography to detection of damages in concrete. In: Proceedings EURODYN
11. Shiotani T, Ohtsu H, Momoki S, Chai HK, Onishi H, Kamada T (2012) Damage evaluation for concrete bridge deck by means of stress wave techniques. J Bridge Eng 17(6):847–856
12. Kak AC, Slaney M, Wang G (2002) Principles of computerized tomographic imaging. Med Phys 29(1):107–107
13. Carpinteri A, Xu J, Lacidogna G, Manuello A (2012) Reliable onset time determination and source location of acoustic emissions in concrete structures. Cement Concr Compos 34(4):529–537
14. Hu B, Kundu T, Grill W, Liu B, Toufigh V (2013) Embedded piezoelectric sensors for health monitoring of concrete structures. ACI Mater J 110(2):149
15. King D (2013) Collapse dynamics of ultrasound contrast agent microbubbles. Doctoral dissertation, University of Illinois at Urbana-Champaign.
16. Karczmarz S (1937) Approximate solution for systems of linear equations. Bull Int Acad Pol. Sic. Let. Cl. Sci. Math. Nat. 355–357
17. Kak A, Slaney M (1988) Principles of computerized tomographic imaging. IEEE Press
18. Choi H (2016) Evaluation of internal damage in reinforced concrete elements using ultrasonic tomography. Doctoral dissertation. University of Illinois at Urbana-Champaign

An Analytical and Experimental Study of Nonlinear Intermodulation in Delaminated Composite Plate

N. S. V. N. Hanuman, Subhankar Roy, and Tanmoy Bose

1 Introduction

In recent days, nonlinear ultrasonic techniques are being developed to describe the nonlinear response at the defect location of the material. Local defect resonance (LDR) is one of the nonlinear ultrasonic techniques introduced recently for the detection of damages in composites structures. When the flexural wave frequency interacts with the defect frequency, then LDR phenomenon arises at the defect region leading to increase in the interaction between the acoustic wave and defect by the continuous excitation of the defect using a particular range of frequency incorporated with LDR frequency. Corresponding nonlinear response (temperature) was developed in the defect region, which is captured by IR camera [1]. Further resonant ultrasound spectroscopy (RUS) technique was implemented on case study—flat bottom hole (FBH) of PMMA plate to detect the small damages using analytical calculation of LDR frequency [2]. In order to avoid the utilization of more input power, the ultrasonic thermography (UT) using air-coupled ultrasonic (ACU) excitation is implemented [3]. The LDR frequency provides the energy in the form of waves directly to the defect. These causes in generation of higher harmonics (HH) at the defect location, which is described as classical nonlinearity. Non-classical nonlinearity is characterized by the dynamic nature of the defect which is used to localize and size

N. S. V. N. Hanuman (✉) · S. Roy · T. Bose
Department of Mechanical Engineering, National Institute of Technology Meghalaya, Shillong, India
e-mail: nsvnhanu@gmail.com

S. Roy
e-mail: subhankar.roy@nitm.ac.in

T. Bose
e-mail: tanmoy.bose@nitm.ac.in

C. K. Mukhopadhyay and R. Mulaveesala (eds.), *Advances in Non-destructive Evaluation*, Lecture Notes in Mechanical Engineering,
https://doi.org/10.1007/978-981-16-0186-6_18

detection of the defect with high-contrast images [4, 5]. The LDR technique is suitable for damages with radius to depth ratio more than or equal to five which was proved numerically as well as well as experimentally [6]. Most of the researchers presented the importance of the LDR frequency to localize the defect in the materials. Meanwhile, some researchers tried to identify the LDR frequency at the defect location by implementing the continuous periodic signals. Hence, the broadband input signals were provided periodically for the exciting defect. Further nonlinear response evaluated by the Fourier transforms [7]. The importance of the sub-harmonic LDR excitation was presented to detect micro-voids in adhesive joints of modern aircraft structures [8]. Nonlinear wave modulation thermography (NWMT) technique was implemented to generate the high-amplitude fundamentals and higher harmonics at defect location for accurately characterize the barely visible impact defect (BVID) in composite materials [9]. The chirp signals are provided to the delamination region to identify the LDR frequency. Further fast Fourier transform (FFT) and bicoherence analysis were proposed for evaluation of the nonlinear response [10]. In the same year, double excitation and bicoherence analysis were implemented for increasing the effectiveness to detect the LDR frequency [11]. The significance of quadratic nonlinearity to characterize the delamination by comparison of analytical and experimental results is also carried out [12]. In this work, an analytical formulation is established to find out the generation of intermodulation peaks at defect location by providing dual excitation, considering quadratic nonlinearity. Simultaneously, experiments are carried out on the GFRP plate with single delamination. Further, the analysis is carried out by using FFT in both the analytical as well as experimental cases. Experimental setup, results, and discussions are explained in the next sections.

2 Methodology

The flexural wave frequency is affected by the local mass and stiffness factors of the material. These factors are decreased due to the presence of defects in a structure. When the local defect frequency is matched with flexural wave frequency, local vibrational amplitudes are produced. In this study, a GFRP plate with single delamination is considered. The delamination resonance frequency is calculated theoretically as per the analytical relation shown in Eq. (1) based on a FBH model. The residual thickness of the FBH is considered from the top surface of the plate. At a particular LDR frequency, vibrating phenomena happen at residual thickness region like vibration motion of the circular plate with pinned boundaries (width to thickness ratio is $a/d > 10$). The theoretical equation of LDR frequency is given as [2].

$$f_d = \frac{1.6d}{a^2}\sqrt{\frac{E}{12\rho(1-\nu^2)}} \tag{1}$$

Fig. 1 Generation of the nonlinear response at the flat bottom hole (FBH) when the flexural wave propagated in a plate

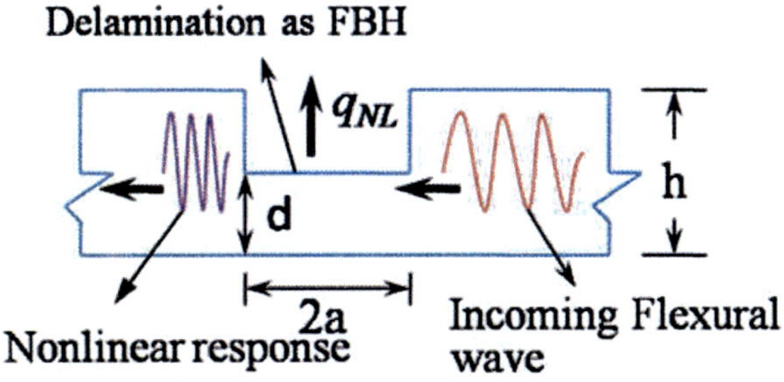

where f_d = LDR frequency, d = residual thickness, a = radius of delamination, E = Young's modulus of elasticity, ρ = density, v = Poisson's ratio (Fig. 1).

The analytical model is developed based on the classical laminated plate theory. The quadratic nonlinear approach is considered for the propagation of the wave in a plate by providing dual excitation. The nonlinear governing equation for fluxural wave propagation in a plate followed as [13]:

$$D_{11}\frac{\partial^4 w_0}{\partial x^4} + 2(D_{12} + 2D_{66})\frac{\partial^4 w_0}{\partial x^2 \partial y^2} + D_{22}\frac{\partial^4 w_0}{\partial y^4} + I_0\frac{\partial^2 w_0}{\partial t^2} - I_2\left(\frac{\partial^2 \ddot{w}_0}{\partial x^2} + \frac{\partial^2 \ddot{w}_0}{\partial y^2}\right) - q_1 - q_2 + q_{\mathrm{NL}} = 0 \tag{2}$$

where q_{NL} is the nonlinear force at the defect location (x_d, y_d), which is the opposite sign. q_1 and q_2 are point forces. The bending stiffness matrix D_{ij} is calculated by following equation [13]:

$$D_{ij} = \frac{1}{3}\sum_{k=1}^{L} Q_{ij}^{(k)}\left(z_{k+1}^3 - z_k^3\right) \tag{3}$$

Here, the total number of layers is represented by L. The stiffness matrix is reduced under the plane stress $\left(Q_{ij}^{(k)}\right)$ condition for the kth orthotropic lamina. From Eq. (2), mass moment (I_0) and second mass moment (I_2) are expressed as follows [13]:

$$I_0 = \sum_{k=1}^{L} \rho_0^{(k)}(z_{k+1} - z_k) \tag{4}$$

$$I_2 = \frac{1}{3}\sum_{k=1}^{L} \rho_0^{(k)}\left(z_{k+1}^3\right) - z_k^3 \tag{5}$$

Here, simply supported boundary condition is considered. The symbols l and b represent the length and width of the plate. The bending moment in the x and y directions is denoted by M_{xx} and M_{yy}, respectively. The total displacement (first-order perturbation theory) equation can be followed as:

$$w_0(x, y, t) = w_0^{(1)}(x, y, t) + \xi w_0^{(2)}(x, y, t) + O(\xi^2) \tag{6}$$

Here, quadratic nonlinearity is considered. $w_0^{(1)}(x, y, t)$ and $w_0^{(2)}(x, y, t)$ represent displacement related to the linear and nonlinear force (q_{NL}), respectively. The higher-order terms $O(\xi^2)$ and the nonlinear constants are ξ. The nonlinear terms represent the generation of the intermodulation terms at the defect location. Linear terms can be neglected in this part. Then, the previous equation can be rewritten as

$$w_0(x, y, t) = \xi w_0^{(2)}(x, y, t) \tag{7}$$

Substituting Eq. (7) into Eq. (2) leads to

$$\begin{aligned} &\xi D_{11}\frac{\partial^4 w_0^{(2)}(x, y, t)}{\partial x^4} + 2\xi(D_{12} + 2D_{66})\frac{\partial^4 w_0^{(2)}(x, y, t)}{\partial x^2 \partial y^2} \\ &+ \xi D_{22}\frac{\partial^4 w_0^{(2)}(x, y, t)}{\partial y^4} + \xi I_0 \frac{\partial^2 w_0^{(2)}(x, y, t)}{\partial t^2} \\ &- \xi I_2\left(\frac{\partial^2}{\partial x^2} + \frac{\partial^2}{\partial y^2}\right)\left(\frac{\partial^2 w_0^{(2)}(x, y, t)}{\partial t^2}\right) \\ &- q_1(x, y, t) - q_2(x, y, t) + q_{NL}(\xi, x, y, t) = 0 \end{aligned} \tag{8}$$

where the second-order nonlinear transverse load ($q_{\mathrm{NL}}(\xi, x, y, t)$) and displacement $\left(w_0^{(2)}\right)$ can be expressed as

$$q_{\mathrm{NL}}(\xi, x, y, t) = \sum\sum F_{mn}^{II}(\xi, t)\ \sin\alpha x\ \sin\beta y \tag{9}$$

$$w_0^{(2)}(x, y, t) = \sum\sum Y_{mn}(t) \sin\alpha x\ \sin\beta y \tag{10}$$

The elastic effect of quadratic nonlinearity can be predicted at the LDR frequency ($f_{mn} = f_d$) as follows [14]:

$$F_{mn}^{II}(\xi, t) = \overline{Q}\xi\left[\frac{\cos 2\pi f_1 t - \cos 2\pi f_d t}{f_d^2 - f_1^2} + \frac{\cos 2\pi f_2 t - \cos 2\pi f_d t}{f_d^2 - f_2^2}\right]^2 \tag{11}$$

where $\overline{Q} = \left(\frac{Q}{\pi^2 ab}\sin\alpha x_d\ \sin\beta y_d\right)^2$ Now, on substitution of Eq. (9) and (10) to Eq. (8) results in the following equation

$$\begin{aligned} &\xi Y_{mn}\left[D_{11}\alpha^4 + 2(D_{12} + 2D_{66})\alpha^2\beta^2 + D_{22}\beta^4\right] \\ &+ \xi I_0 \ddot{Y}_{mn} - \xi I_2(\alpha^2 + \beta^2)\ddot{Y}_{mn} + F_{mn}^{II} = 0 \end{aligned} \tag{12}$$

The above equation can be written in a simplified form as

$$\ddot{Y}_{mn} + 4\pi^2 f_d^2 Y_{mn} = -\overline{F_{mn}^{II}} \tag{13}$$

where

$$\overline{F_{mn}^{II}} = A^{II}\left[\cos 2\pi f_1 t\left(f_d^2 - f_2^2\right) + \cos 2\pi f_2 t\left(f_d^2 - f_1^2\right) + \cos 2\pi f_d t\left(f_1^2 + f_2^2 - 2f_d^2\right)\right]^2$$

and

$$A^{II} = \frac{\overline{Q}}{M_{mn}\left(f_d^2 - f_1^2\right)\left(f_d^2 - f_2^2\right)}.$$

The general solution of above in homogeneous ordinary differential equation can be expressed in below written form

$$w_0^{(2)}(x, y, t) = \sum\sum(V_1 + V_2 + V_3 + V_4 + V_5 + V_6)\sin\alpha x\ \sin\beta y \tag{14}$$

where the combining parts are of following form

$$V_1 = \frac{A^{II}\left(f_2^2 - f_d^2\right)(1 + \cos 4\pi f_1 t)}{8\pi^2\left(4f_1^2 - f_{mn}^2\right)}$$

$$V_2 = \frac{A^{II}\left(f_1^2 - f_d^2\right)(1 + \cos 4\pi f_2 t)}{\left(4f_2^2 - f_{mn}^2\right)}$$

$$V_3 = \frac{A^{II}\left(f_1^2 + f_2^2 - 2f_d^2\right)(1 + \cos 4\pi f_d t)}{8\pi^2\left(4f_d^2 - f_{mn}^2\right)}$$

$$V_{4,5} = \frac{A^{II}\left(f_d^2 - f_2^2\right)\left(f_d^2 - f_1^2\right)}{4\pi^2 f_{mn}} \left[\frac{(f_1 \pm f_2)\left\{\mathrm{Cos}2\pi\left(f_1 \pm f_2 - f_{mn}\right)t - \mathrm{Cos}2\pi\left(f_1 \pm f_2 + f_{mn}\right)t\right\}}{\left[(f_1 \pm f_2)^2 - f_{mn}^2\right]}\right]$$

$$V_{6,7} = \frac{A^{II}\left(f_d^2 - f_1^2\right)\left(f_1^2 + f_2^2 - 2f_d^2\right)}{4\pi^2 f_{mn}} \left[\frac{(f_2 \pm f_d)\left\{\mathrm{Cos}2\pi\left(f_2 \pm f_d - f_{mn}\right)t - \mathrm{Cos}2\pi\left(f_2 \pm f_d + f_{mn}\right)t\right\}}{\left[(f_2 \pm f_d)^2 - f_{mn}^2\right]}\right]$$

$$V_{8,9} = \frac{A^{II}\left(f_d^2 - f_2^2\right)\left(f_1^2 + f_2^2 - 2f_d^2\right)}{4\pi^2 f_{mn}} \left[\frac{(f_1 \pm f_d)\left\{\mathrm{Cos}2\pi\left(f_1 \pm f_d - f_{mn}\right)t - \mathrm{Cos}2\pi\left(f_1 \pm f_d + f_{mn}\right)t\right\}}{\left[(f_1 \pm f_d)^2 - f_{mn}^2\right]}\right]$$

Classical and non-classical nonlinear terms are obtained from Eq. 14. The classical terms are V_1, V_2, V_3 representing first and second forcing frequency and second harmonic LDR frequency, respectively. The non-classical terms are $V_{4,5}$, $V_{6,7}$ and $V_{8,9}$, $V_{4,5}$ terms representing intermodulation between the single periodic frequency (SPF) and forcing frequency (f_1), respectively. $V_{6,7}$ terms represent intermodulation between single periodic frequency (SPF) and LDR frequency (f_d). $V_{8,9}$ terms represent intermodulation between LDR frequency (f_d) and first forcing frequency (f_1).

3 Experimental Setup

Some experiments are performed on the GFRP plate for justification of analytical results. The glass fiber weave mat was used for fabricating the GFRP plate by hand layup process. The thickness of the weaving mat is 0.4 mm. The mat was cut into four layers with 180 × 180 mm^2 dimension. These four layers were binded with each other using epoxy resin (LY556) and hardener (HY951) mixture for making 2 mm thickness of plate. The mixture is taken as 10:1 by weight ratio. The mixture gives strong bonding between the layers. Later, the plate was kept at room temperature for curing process. One artificial circular delamination (Teflon tape) is inserted between the second and third layers of the plate during the fabrication process. The position of the delamination is arranged at 60 mm from x and y coordinates, which is considered from the left bottom corner of the plate as shown in Fig. 2. The material properties of GFRP plate are $E_c = 10.32$ GPa, $\nu = 0.3$, and $\rho = 2145$ kg/m^3.

Fig. 2 Schematic diagram of GFRP plate with a single delamination

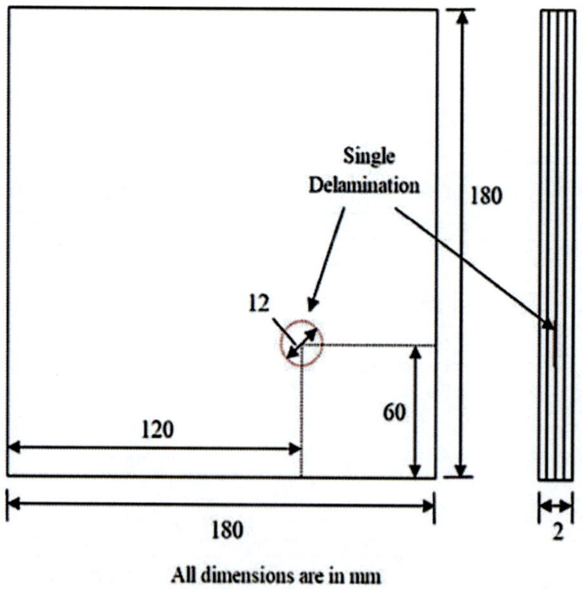

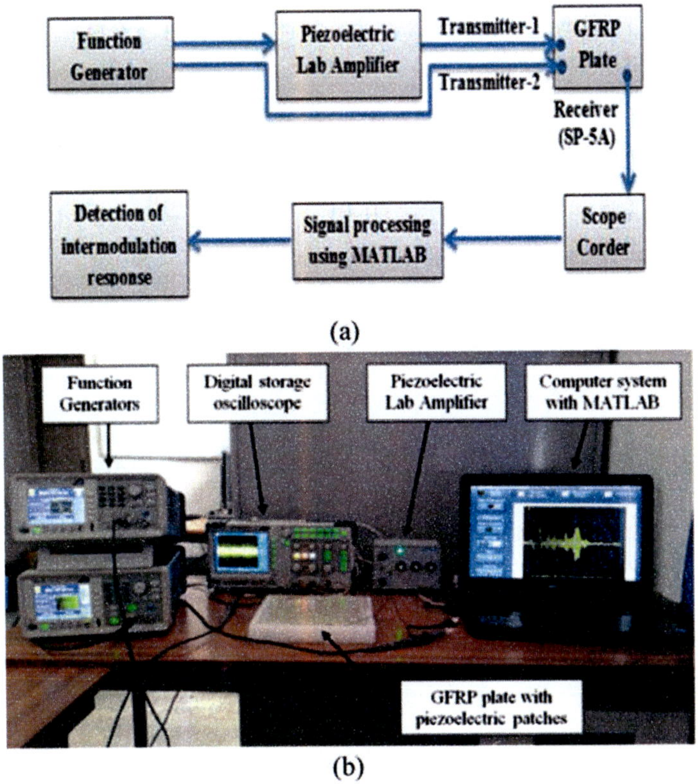

Fig. 3 Experimental setup: **a** schematic diagram and **b** pictorial view

The experimental setup consists of an arbitrary function generator, digital storage oscilloscope (DSO), piezoelectric laboratory amplifier, piezoelectric patches (SP-5A), and a computer system with MATLAB as shown in Fig. 3a, b.

Two input excitation signals such as LDR frequency (f_d) and SPF (F) are provided to the plate, which is generated by the function generator at 10 Vpp, and it is amplified to the 17 Vpp using laboratory amplifier. These input signals are passed through the piezoelectric patches (SP-5A) which is glued on the surface of the plate. One of the input signals is varied as fundamental LDR frequency, sub-harmonic, and superharmonic frequencies along with another input signal, which is fixed (SPF). The output signals are captured through the piezo electric patch to the digital storage oscilloscope (DSO). Further, these output signals are analyzed by using fast Fourier transform (FFT).

4 Results and Discussion

In this section, the analytical and experimental results of the GFRP plate with single delamination are presented. The LDR frequency (f_d) of the GFRP plate is considered as 12.9 kHz, which is calculated by the analytical formula (Eq. 1). Another excitation frequency in form of SPF is fixed at 109 kHz. The sampling frequency is taken as 500 kHz for analytical and experimental cases.

4.1 *Analytical Investigation*

Figure 4a–c represents the analytical responses of the GFRP plate when excited with sub-harmonic LDR frequency ($2f_d$), second- and third-order superharmonic frequencies ($f_d/2$, $f_d/3$), respectively. These responses are calculated using Eq. 7. The amplitude force is considered as 20 N, and nonlinear constant (ξ) is taken as 10. The plate is excited with the sub-harmonic LDR frequency ($2f_d$). Single periodic frequency intermodulation (SPFI) peaks such as $F \pm f_d, F \pm 2f_d$ *and* $F \pm 3f_d$ are obtained along with fourth harmonic LDR frequency ($4f_d$) and SPF (F) as shown in Fig. 4a. The SPFI peaks $F \pm f_d/2, F \pm 3f_d/2, F \pm 2f_d$ are recognized along with forcing frequencies, i.e., local defect resonance intermodulation (LDRI) ($3f_d/2$, $5f_d/2$) peaks, second harmonic LDR frequency ($2f_d$), and SPF (F) when the plate is excited with the second-order superharmonic frequency ($f_d/2$) as shown in Fig. 4b. Finally, the SPFI peaks $F \pm 2f_d/3$, $F \pm 4f_d/3$, and $F \pm 2f_d$ are observed along with LDRI peaks ($2f_d/3, 5f_d/3, 7f_d/3$), second harmonic LDR frequency ($2f_d$), and SPF (F) when the plate is excited with third-order superharmonic frequency ($f_d/3$) as shown in Fig. 4c. Moreover, $2F$ term can be observed in all three cases.

4.2 *Experimental Investigation*

Figure 5a–d represents experimental results when the GFRP plate is excited with sub harmonic LDR frequency ($2f_d$), fundamental LDR frequency, second- and third-order superharmonic frequencies ($f_d/2, f_d/3$), respectively. $F - f_d, F \pm 2f_d, F \pm 3f_d$ SPFI peaks are observed along with fourth and eighth harmonic LDR frequencies ($4f_d, 8f_d$) and SPF (F) when the plate is excited with the sub harmonic LDR frequency ($2f_d$) as shown in Fig. 5a. Similar peaks can be observed analytically also (Fig. 4a). Further, higher harmonic LDR frequencies such as $2f_d, 3f_d, 4f_d, 5f_d, 6f_d, 7f_d, 8f_d, 10f_d, 13f_d, 14f_d, 15f_d$peaks are obtained along with SPF (F) when the plate is excited with fundamental LDR frequency (f_d) as shown in Fig. 5b. Later the plate is excited with second-order superharmonic frequency ($f_d/2$), where corresponding SPFI peaks such as $F \pm f_d/2, F \pm 3f_d/2, F \pm 2f_d$ are recognized along with LDRI peaks ($3f_d/2$, $5f_d/2$), SPF(F), second harmonic LDR frequency ($2f_d$) and LDR frequency (f_d).

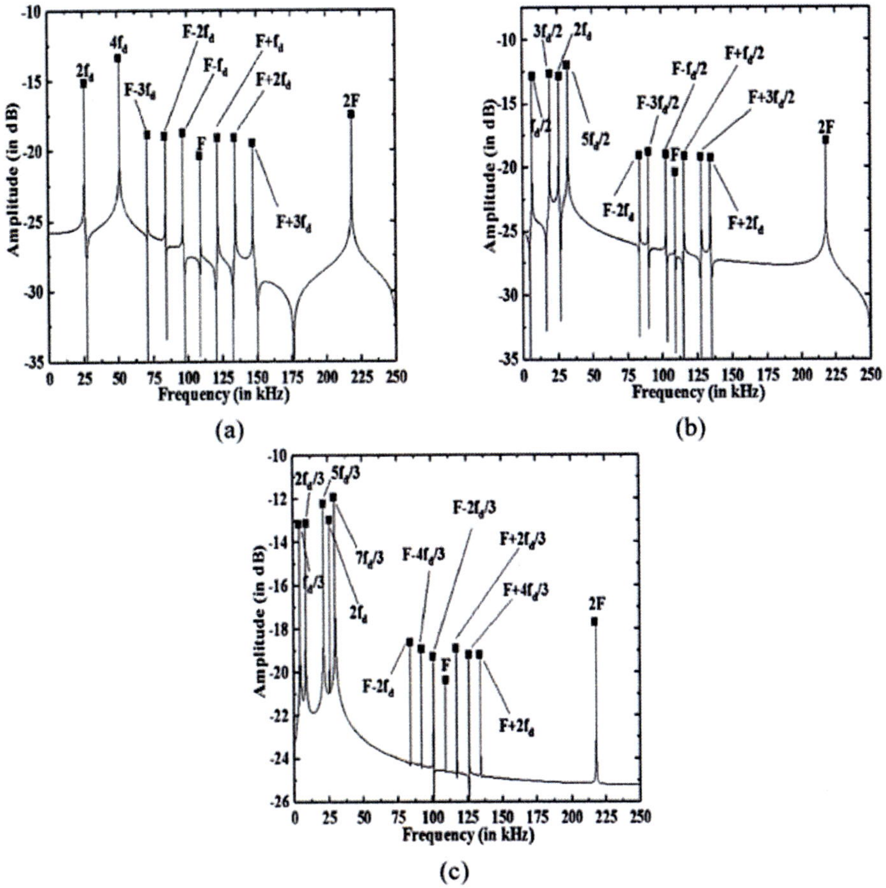

Fig. 4 Frequency spectra showing analytical nonlinear intermodulation responseof GFRP plate with excited at **a** subharmonic ($2f_d$), **b** second-order superharmonic ($f_d/2$) and **c** third-order superharmonic ($f_d/3$) excitationsin the presence of SPF(F)

Similar peaks can be recognized analytically in Fig. 4b. Further, the plate is excited with the ($f_d/3$), where SPFI peaks such as $F \pm 2f_d/3$, $F \pm 4f_d/3$, $F \pm 2f_d$ along with LDRI peaks ($2f_d/3$,$5f_d/3$, $7f_d/3$), second harmonic LDR frequency ($2f_d$), SPF (F) and LDR frequency (f_d) is obtained. Moreover, $2F$ term can also be observed in all the cases. Similar peaks can be identified analytically in Fig. 4c.

It is observed from the analytical and experimental results that both the techniques show similar kind of peaks such as LDR, LDRI, and SPFI in each case of excitation. Hence, the experimental results are validated from the analytical results. Moreover, it is observed that the experimental plots show some noise which is not found in case of analytical investigation. The noise during experimentation is due to various parameters such as equipment noise, signal noise, external noise, etc. It can be reduced by improving the sample rate of experiment and by performing averaging of signals

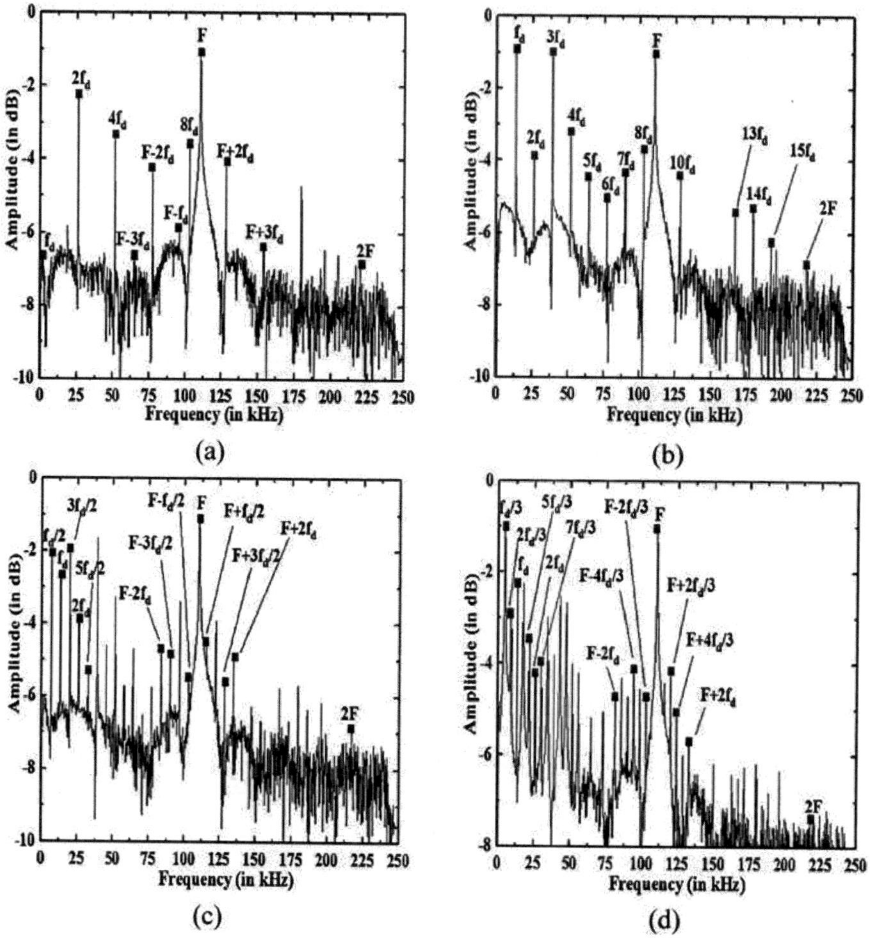

Fig. 5 Frequency spectra showing experimental nonlinear intermodulation responseof GFRP plate excited at **a** subharmonic ($2f_d$), **b** LDR frequency (f_d), **c** second-order superharmonic ($f_d/2$), and **d** third-order superharmonic ($f_d/3$) excitations in the presence of SPF(F)

during post-processing. The higher the number of signals averaged, the lesser will be the noise in experimental plots. Once the defect frequency is identified, the defect can be located by using ultrasound assisted vibro-thermography using IR camera. Therefore, this technique can be applied to detect defects in any advanced composite structure having sub-surface delamination during fabrication or during service due to low velocity impacts and thus find a vast area of repeatability.

5 Conclusions

In this paper, the analytical model validated with experiments was presented for wave propagation in a symmetric composite laminate plate. The dual-excitation frequencies, i.e., SPF (F) and variable excitation frequency of sub-harmonic frequency ($2f_d$) to second- and third-order superharmonic frequency ($f_d/2$, $f_d/3$)) were provided to develop the analytical model using quadratic nonlinear approach. Subsequently, experiments were performed on the GFRP plate with single delamination by providing dual excitation frequency using two-dimensional oscillator. Further, the output data was processed by FFT plots analytically as well as experimentally. From the FFT plots, SPFI and LDRI peaks were found in analytical and experimental cases when excitation with sub-harmonic frequency ($2f_d$), second- and third-order superharmonic ($f_d/2, f_d/3$) frequency. Excitation with LDR frequency (f_d) leads to generation of higher harmonic LDR frequencies only. Subsequently, the analytical FFT plot peaks are justified with the experimental FFT plot peaks. Finally, a good agreement between the analytical and experimental results was accomplished.

References

1. Solodov I, Bai J, Bekgulyan S, Busse G (2011) A local defect resonance to enhance acoustic wave-defect interaction in ultrasonic nondestructive evaluation. Appl Phys Lett 99:211911. https://doi.org/10.1063/1.3663872
2. Solodov I, Bai J, Busse G (2013) Resonant ultrasound spectroscopy of defects: case study of flat-bottomed holes. J Appl Phys 113:223512. https://doi.org/10.1063/1.4810926
3. Solodov I, Busse G (2013) Resonance ultrasonic thermography: highly efficient contact and air-coupled remote modes. Appl Phys Lett 102:061905
4. Solodov I (2014) Resonant acoustic nonlinearity of defects for highly-efficient nonlinear NDE. J Nondestruct Eval 33:252–262
5. SolodovI DD, Rahammer M (2015) Thermosonic Chladni figures for defect-selective imaging. Ultrasonics 60:1–5
6. Hettler J, Tabatabaeipour M, Delrue S, Abeele KVD (2017) Detection and characterization of local defect resonances arising from delaminations and flat bottom holes. J Nondestruct Eval 36(1):2. https://doi.org/10.1007/s10921-016-0380-6
7. Rahammer M, Kreutzbruck M (2017) Fourier-transform vibrothermography with frequency sweep excitation utilizing local defect resonances. NDT&E Int 86:83–88. https://doi.org/10.1016/j.ndteint.2016.11.012
8. Ginzburg D, Ciampa F, Scarselll G, Meo M (2017) SHM of single lap adhesive joints using-subharmonic frequencies. Smart Mater Struct 26:105018. https://doi.org/10.1088/1361-665X/aa815c
9. Dionysopoulos D, Fierro GPM, Meo M, Ciampa F (2018) Imaging of barely visible impact damage on a composite panel using nonlinear wave modulation thermography. NDT E Int 95:9–16
10. Roy S, BoseT DK (2018) Detection of local defect resonance frequencies using bicoherence analysis. J Sound Vib 443:703–716. https://doi.org/10.1016/j.jsv.2018.12.006
11. Roy S, Bose T (2019) Efficient determination of local defect resonance frequencies from bicoherence plots using double excitations. Mech Sys and Sig Proc 127:595–609. https://doi.org/10.1016/j.ymssp.2019.03.017

12. Hanuman NSVN, Roy S, Bose T (2019) Detection of local defect resonance intermodulation peaks using bicoherence analysis. J Mech Sci 163:105092
13. Reddy JN (1997) Mechanics of laminated composite plates: theory and analysis. CRC Press
14. Landau LD, Lifshitz EM (1986) Theory of elasticity. Elsevier Ltd.

Ultrasonic Inspection of Weld Defects Using Total Focusing Method

Lalith Sai Srinivas Pillarisetti and Subramanian Anandamurugan

1 Introduction

The advent of ultrasonic phased array transducers with beam steering and focusing capabilities have made the ultrasonic inspections faster than those performed using a single-element probe [1]. With suitable time-delay laws given to each element of the array, the ultrasonic beam can be steered and focused at various depths [2]. Array-based inspection can perform multiple scans such as a linear scan (B-scan) and a sectorial scan (S-scan) at a single placement of the probe. However, these conventional phased array imaging techniques generate images with poor resolution, as focusing during reception cannot be achieved throughout the image due to the unavailability of individual A-scan data of all the receiving elements during data acquisition [3].

The total focusing method (TFM) is an advanced ultrasonic imaging technique that relies on the data acquired using independent parallel reception channels. The TFM generates high-resolution images by focusing on both transmission and reception on every image pixel by utilizing the complete response of the array, called full matrix capture (FMC) [3]. FMC data generation is possible if the data acquisition system has an independent transmission and reception channels. The TFM algorithm can be extended for imaging in multiple modes because of the mode conversions that are possible inside the specimen when the probe is excited [4]. In the past two decades, TFM has gained prominent importance in the ultrasonic inspection for showing

L. S. S. Pillarisetti (✉)
NDT & E Lab, Department of Mechanical and Aerospace Engineering, IIT Hyderabad, Hyderabad, India
e-mail: me15b19m000002@iith.ac.in

S. Anandamurugan
BHGE Inspection Technologies, JFWTC, 122, EPIP, Ph 2, Whitefield, Bangalore, India
e-mail: anandamurugan.s@bhge.com

C. K. Mukhopadhyay and R. Mulaveesala (eds.), *Advances in Non-destructive Evaluation*, Lecture Notes in Mechanical Engineering,
https://doi.org/10.1007/978-981-16-0186-6_19

improved imaging performance with better defect detectability [5–7]. Though the TFM generates a highly focused image, it also faces some challenges for angle-based weld inspections. In this paper, the time of flight framework for multimode TFM imaging is presented for an angular wedge configuration to perform weld inspections. The challenges in implementing TFM for weld inspection, such as the area of coverage and mode conversions, are discussed in detail. Finally, to show the efficacy of TFM, the imaging performance of TFM in terms of defect detectability and image resolution is compared with the phased array sector scan images by performing inspection over weld specimens with defects.

The paper is arranged as follows. The first section presents the theory of FMC and multimode TFM for an angular wedge configuration. The second section illustrates the challenges of wedge and probe selection and the mode conversions in the specimen for TFM image reconstruction. Finally, experimental results of TFM and phased array sector scan on different weld defects using a shear wedge are compared to show the efficacy of TFM.

2 Theoretical Background

This section describes the theoretical background of the full matrix capture (FMC) data acquisition and the total focusing method (TFM). The TFM algorithm is also extended for imaging using multiple wave modes by accounting for the wave mode conversions in the specimen. Further, the time of flight calculations required for post-processing the image output for an angular wedge configuration is presented for multimode TFM.

2.1 Full Matrix Capture

FMC data generation involves the excitation of each array element with simultaneous independent reception from all the array elements. Unlike the conventional phased array imaging, where only the summed up response of all the array elements is stored, FMC data involves capturing the response of individual array elements separately during the reception. This receiving capability is possible if the array controller has an independent set of parallel reception channels. As shown in Fig. 1, FMC data can be visualized as an A-scan data matrix with transmitting and receiving elements on rows and columns. The ith row and j^{th} column box in Fig. 1 refer to the A-scan data A_{ij} received by the jth array element when the ith element is excited. If there are N elements in an array, the FMC data comprises N^2 A-scan data, as shown in Fig. 1.

Fig. 1 FMC A-scan data matrix

Receiving Elements (j) →

Transmitting Elements (i) ↓

A_{11}	A_{12}	A_{13}	A_{14}	…	A_{1J}	…	A_{1N}
A_{21}	A_{22}	A_{23}	A_{24}	…	A_{2J}	…	A_{2N}
A_{31}	A_{32}	A_{33}	A_{34}	…	A_{3J}	…	A_{3N}
A_{41}	A_{42}	A_{43}	A_{44}	…	A_{4J}	…	A_{4N}
⋮	⋮	⋮	⋮	⋮	⋮	…	⋮
A_{I1}	A_{I2}	A_{I3}	A_{I4}	…	A_{IJ}	…	⋮
⋮	⋮	⋮	⋮	⋮	⋮	…	⋮
A_{N1}	A_{N2}	A_{N3}	A_{N4}	…	…	…	A_{NN}

2.2 *Total Focusing Method*

TFM is an image post-processing algorithm that reconstructs the image using the entire FMC data of the array probe by focusing on both the transmission and reception on each image pixel. The region of interest (ROI) and the grid resolution of the image are defined before performing TFM. The algorithm then sums up all the transmit–receive element pair A-scan data in the FMC data to synthesize a focus at each pixel point by calculating the time of flight values corresponding to the pixel point from all the array elements.

Figure 2 schematically illustrates the concept of TFM for an angular wedge contact configuration. If P is a pixel point in the defined ROI, the image value, $I(P)$ corresponding to the pixel is given by:

Fig. 2 Schematic of the TFM concept

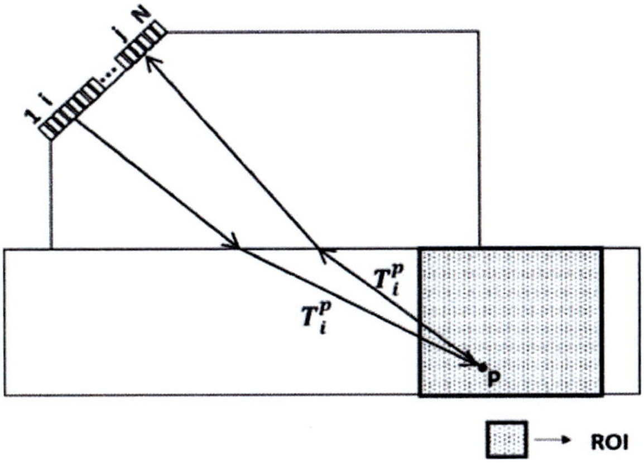

$$I(P) = \sum_{i}^{N} \sum_{j}^{N} A_{ij} \left(T_i^P + T_j^P \right) \tag{1}$$

T_i^P and T_j^P represent the time of flight values between the pixel point P and transmitting element i, and between the pixel point P and the receiving element j, respectively.

2.3 Multimode Total Focusing Method

Wave mode conversions take place when a propagating ultrasonic wave hits the free surface of the specimen. For example, when a transverse or longitudinal wave is incident on a free surface, it splits, producing both longitudinal and transverse waves [8]. These wave mode conversions possible when the specimen is excited with an ultrasonic probe lead to many options of post-processing TFM using multiple wave modes. This multimode imaging capability of TFM is clearly illustrated in this section.

For all the subsequent subsections, consider an angular wedge configuration, where the ROI is in the y–z plane with the specimen top surface at $z = 0$, as shown in Fig. 3. The origin O is taken as the intersection point on the specimen top surface where the normal drawn from the center of the array probe intersects the surface. Let the coordinates of a transmitting–receiving element pair are (y_i, z_i) and (y_j, z_j), respectively, and $P(y_\text{p}, z_\text{p})$ be a pixel point in the ROI. The velocity of the ultrasonic wave in the wedge and specimen be v_w and v_s, respectively.

Direct Mode TFM Imaging In the direct mode TFM imaging, the transmission and reception paths for each pixel point are evaluated without considering any skip from the specimen back surface, as shown in Fig. 3a.

Let $(y_1, 0)$ be the intersection point on the specimen top surface where the transmitted ultrasonic wave hits the surface before focusing on the pixel point P. The time of flight in transmission T_i^P can be written as:

$$T_i^P(y_1) = \frac{\sqrt{(y_i - y_1)^2 + (z_i)^2}}{v_\text{w}} + \frac{\sqrt{(y_\text{P} - y_1)^2 + (z_\text{P})^2}}{v_\text{s}} \tag{2}$$

According to Fermat's principle, the wave path between any two points, either in the same medium or different medium, is the path that corresponds to the minimum time of flight between those two points. Therefore, the time of flight can be estimated by calculating the zeros of the first derivative of T_i^P with respect to y_1. There are analytical solutions such as the Ferrari method [9] and iterative techniques such as Newton–Raphson to calculate the zeros of a function.

The time it takes for the wave to travel from the pixel point to the receiving element is the same as the time it takes to travel to the pixel point if the wave is

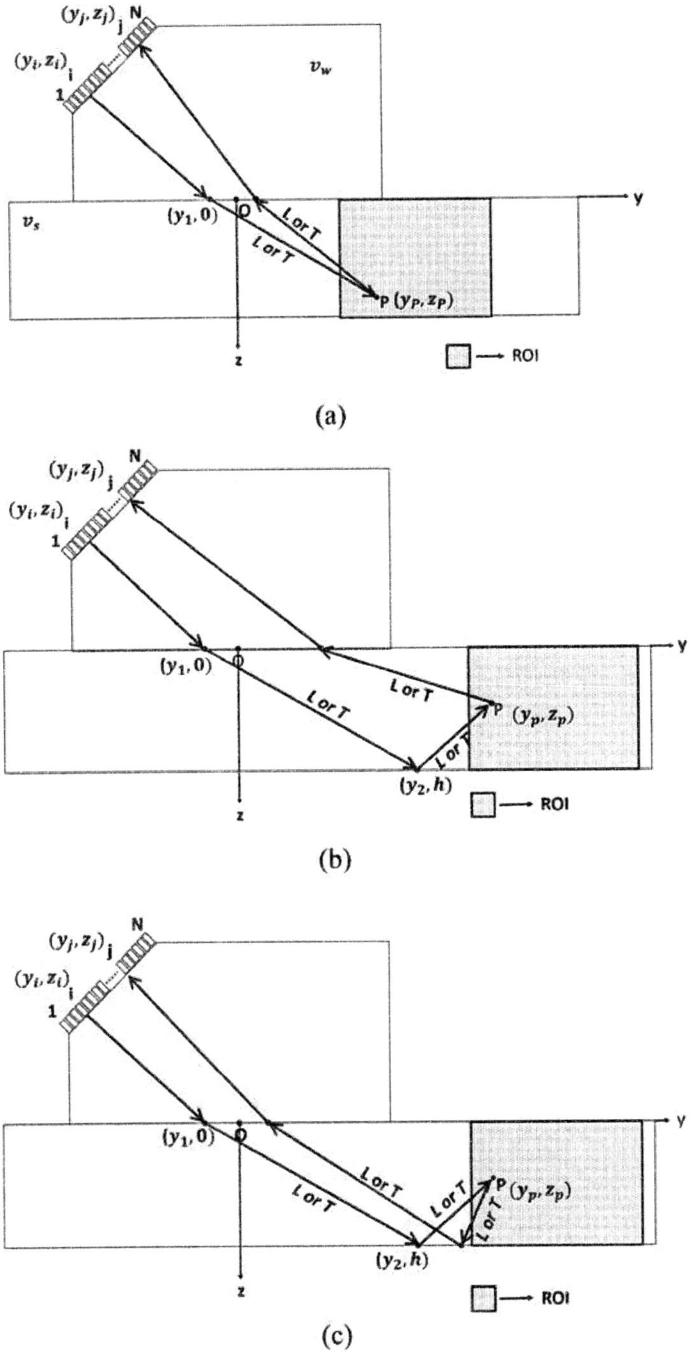

Fig. 3 Schematic of multimode TFM imaging: **a** direct mode, **b** corner mode, **c** indirect mode

excited from the same element under reception. This analogy indirectly states that Eq. 2 on replacing y_i with y_j and z_i with z_j can be used for calculating the time of flight in the reception phase. The TFM image value at each pixel point can then be conveniently calculated using Eq. 1.

The wave nature of the ultrasonic wave entering and leaving the specimen interface can be longitudinal (*L*) or transverse (*T*) because of the mode conversions inside the specimen. These mode conversions lead to the possibility of TFM imaging using 2^2 direct modes (L-L, L–T, T-L, and T-T). These multiple TFM direct modes are schematically represented in Fig. 3a. T-L symbolically means transverse wave path in transmission to pixel point and longitudinal wave path in reception from the pixel point to the receiving element. Corresponding to the TFM mode of interest, the wave speed in the specimen (v_s) in Eq. 2 is adjusted as transverse wave speed (v_T) or longitudinal wave speed (v_L).

Corner Mode TFM Imaging In corner mode imaging, either the transmission wave path or the reception wave path for each pixel point is evaluated considering a skip from the specimen back surface. For the case of weld inspection, as the wedge is positioned away from the weld region, the skip during transmission path would be more relevant as the specimen is inspected in half-skip configuration, as shown in Fig. 3b.

The time of flight computation for the transmission phase is different from what is stated in the direct mode TFM imaging, as it involves refraction and reflection at the specimen top and back surfaces, respectively. Let *h* be the thickness of the specimen and $(y_1, 0)$ and (y_2, h) be the intersection points during the transmission phase on the specimen top and back surfaces, respectively. The time of flight in the transmission phase can be written as:

$$T_i^{\mathrm{P}}(y_1, y_2) = \frac{\sqrt{(y_i - y_1)^2 + (z_i)^2}}{v_{\mathrm{w}}} + \frac{\sqrt{(y_2 - y_1)^2 + (h)^2}}{v_{s1}} + \frac{\sqrt{(y_{\mathrm{P}} - y_2)^2 + (z_{\mathrm{P}} - h)^2}}{v_{s2}} \tag{3}$$

Depending on the TFM mode of interest, v_{s1} and v_{s2} can be either longitudinal or transverse wave speed. The unknown variables (y_1, y_2) can be calculated using the Newton–Raphson method for finding the zeros of multivariable functions. As the reception wave paths of both the corner and direct TFM modes are the same, the reception time of flight equation of the direct mode can be used for calculating the time of flight values during the reception for the corner mode.

The mode conversions in the specimen result in the possibility to image 2^3 TFM corner mode images (LL-L, LL-T, LT-L, LT-T, TT-T, TL-T, TT-L, and TL-L), as represented in Fig. 3b. TL-T symbolically means the transverse wave path in transmission to the specimen back surface, the longitudinal reflected wave path from specimen back surface to the pixel point, and the transverse wave path in reception from the pixel point.

Indirect Mode TFM Imaging Indirect mode imaging involves a skip from the specimen back surface in both the transmission and reception phases between the pixel point and each transmit–receive element pair, as represented in Fig. 3c. The time of flight calculation during the transmission and reception is therefore calculated using the transmission time of flight calculations presented for the corner mode imaging.

In this case, the mode conversions result in the possibility to capture 2^4 different TFM indirect mode images. The indirect TFM modes can be represented as (WW-WW), where W can be either longitudinal wave (L) or transverse wave (T), as illustrated in Fig. 3c. For example, considering the TL-LT mode, the transmission phase TL corresponds to the transverse wave path in transmission to the specimen's back surface and longitudinal reflected wave path from the specimen's back surface to the pixel point. The reception phase LT corresponds to the longitudinal wave in reception from the pixel point to the specimen's back surface and transverse wave path from the specimen's back surface to the reception array element.

3 TFM for Weld Inspection

The focusing capability of TFM at each pixel in the ROI results in better resolution images compared to the conventional phased array scans. TFM capabilities are therefore utilized for inspecting defects in the welds. This section presents the challenges in selecting the wedge and probe for TFM inspection of welds and the effects of mode conversions on the TFM image.

3.1 Wedge and Probe Selection

Weld defects can occur in any portion of the weld, such as the root crack, which occurs at the root of the weld, and the toe crack near the top weld bead. In phased array sector scanning, the full aperture of the phased array is excited using the appropriate time delays to steer and focus the ultrasonic beam at various angles covering the entire weld region, whereas the beam spread in TFM is limited to the directivity of the array probe because of single element excitation. Though the ROI defined in TFM is not limited to the number of probe elements and array pitch like linear scan (B-scan), only the image pixels of ROI that fall within directivity of the array probe are effectively imaged. Therefore, in TFM, the array probe and wedge must be selected to cover the entire weld region within the beam directivity of probe in either the direct or half-skip configuration for better detectability of weld flaws. Therefore, for all the simulations and experiments presented in this paper, ray-tracing software is used to select the array probe and wedge to make sure that the entire weld region is covered under the beam spread of the array probe.

Unlike the longitudinal wedge, which allows both longitudinal and transverse waves into the specimen, shear wedge restricts the entry of the longitudinal waves because of its wedge angle lying between the longitudinal and transverse wave critical angles. Shear wedge, therefore, implicitly limits the number of mode conversions possible inside the specimen resulting in reducing the image artifacts in the TFM image. Transverse wave also has better detectability towards small defects because of shorter wavelength compared to longitudinal waves. Therefore, an array probe mounted on a shear wedge, with a probe aperture wider enough to cover the weld region within its beam directivity, is considered optimum for weld inspection.

3.2 *Effect of Mode Conversions on TFM Images*

The multiple TFM mode images that can be generated at a single probe position calls into question the modes that can be considered for inspection purposes. Simulations are performed in CIVA [10] on a steel specimen of 20 mm thickness, for inspecting a 3 mm vertical notch to study the effects of mode conversions on the TFM mode images using a shear wedge, as stated in Fig. 4. The specifications of the specimen, array probe, and the shear wedge are shown in Table 1. The wedge Z-offset and wedge front dimensions mentioned in Table 1 are illustrated in Fig. 4, along with the position of the ROI and the dimensions of the specimen. The array probe and shear wedge specifications presented in Table 1 are used for both the numerical and experimental studies presented in this paper as the beam spread of the probe can cover all the specimen defects inspected in this study.

Though only the transverse wave is transmitted into the specimen from the shear wedge, multiple TFM mode images can be generated because of the mode conversions at the specimen back surface and the defects. Therefore, the TFM mode images for which the wave nature of the refracted wave into the specimen during transmission is transverse are considered and imaged, as shown in Figs. 5, 6, and 7. The simulation results suggest that most of the TFM mode images have misleading flaw indications also called imaging artifacts. These imaging artifacts occur because of the nearness of the time of flight values between the actual TFM mode of interest and the other TFM modes [11, 12]. The TFM mode images, therefore, comprise all

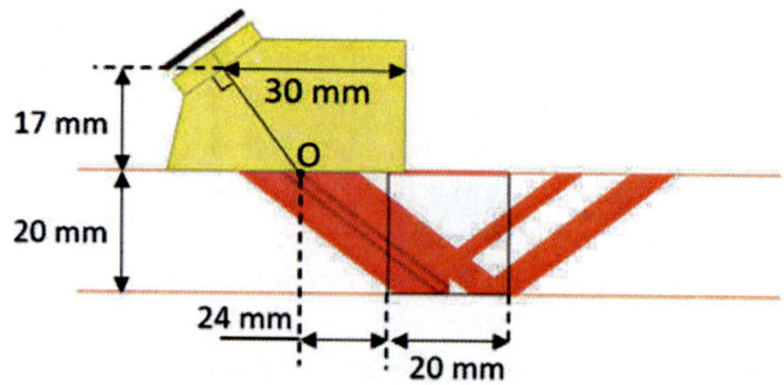

Fig. 4 TFM setup for notch inspection

Table 1 Specimen, array probe, and wedge specifications

Specimen (steel)	Longitudinal velocity	5900 m/s
	Transverse velocity	3230 m/s
Array probe	Number of elements	32
	Array pitch	0.5 mm
	Element length	0.42 mm
	Aperture length	15.5 mm
	Frequency	5 MHz
Wedge	Longitudinal velocity	2380 m/s
	Transverse velocity	1150 m/s
	Wedge Z-offset	17 mm
	Wedge front	30 mm
	Wedge angle	36°
	Nominal refraction angle	53°

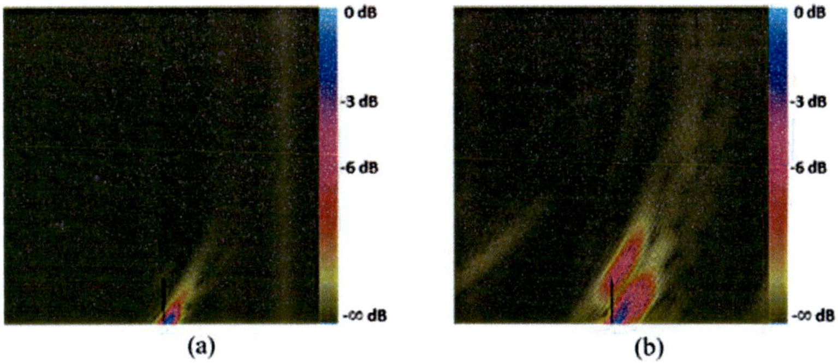

Fig. 5 Direct mode TFM images: **a** T-T, **b** T-L

the image artifacts that are having the time of flight values in the vicinity of those corresponding to the TFM mode used for image reconstruction. These artifacts may mislead the user and lead to erroneous interpretation and identification of the actual defect.

The direct mode TFM T-T mode image is shown to capture the echo only from the corner of the notch, whereas the T-L mode image shows echoes from both the tip and corner of the notch with a significant increase in background noise due to image artifacts. Among the corner mode TFM images, only the TT-T mode is shown to image the complete notch with no image artifacts. Similar to the direct mode T-T image, the TT-TT mode images only the corner of the notch with other imaging modes exhibiting significant noise due to the image artifacts.

The simulation results of T-T and TT-T TFM mode notch inspection can be validated by the experimental results presented in [13]. In all the simulated results, the completely transverse modes (T-T, TT-T, and TT-TT) are shown to process the

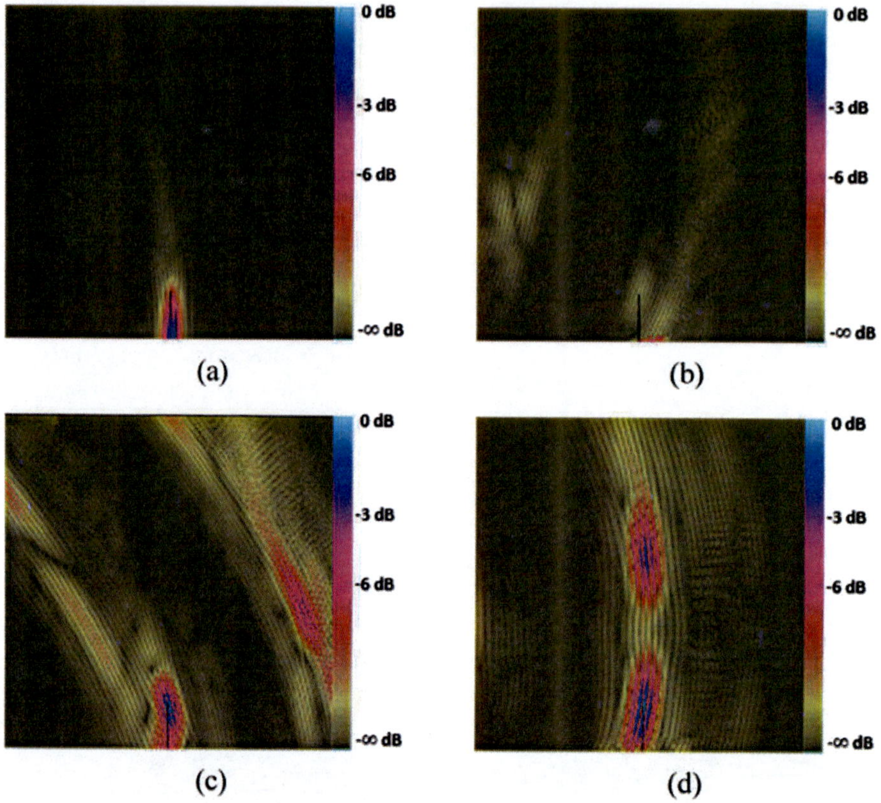

Fig. 6 Corner mode TFM images: **a** TT-T, **b** TL-T, **c** TT-L, **d** TL-L

images with no image artifacts throughout the ROI. They, therefore, are considered for experimental investigation of weld specimens using the same probe and wedge.

4 Weld Specimens and Scan Setup

The multimode TFM technique and phased array sector scan are used to investigate two weld samples of thickness 12 and 18 mm, having a total of five different types of defects combined, as shown in Table 2 and Fig. 8. Two different specimen thicknesses are studied to illustrate the challenges of the TFM technique over the sector scan for higher specimen thickness. The specifications of the probe, wedge, and specimen used for experimental investigation are the same as stated in Table 1. With the help of ray-tracing software, the distance required from the wedge front of the shear wedge to the weld centerline is estimated for both the inspection techniques for the known defect positions.

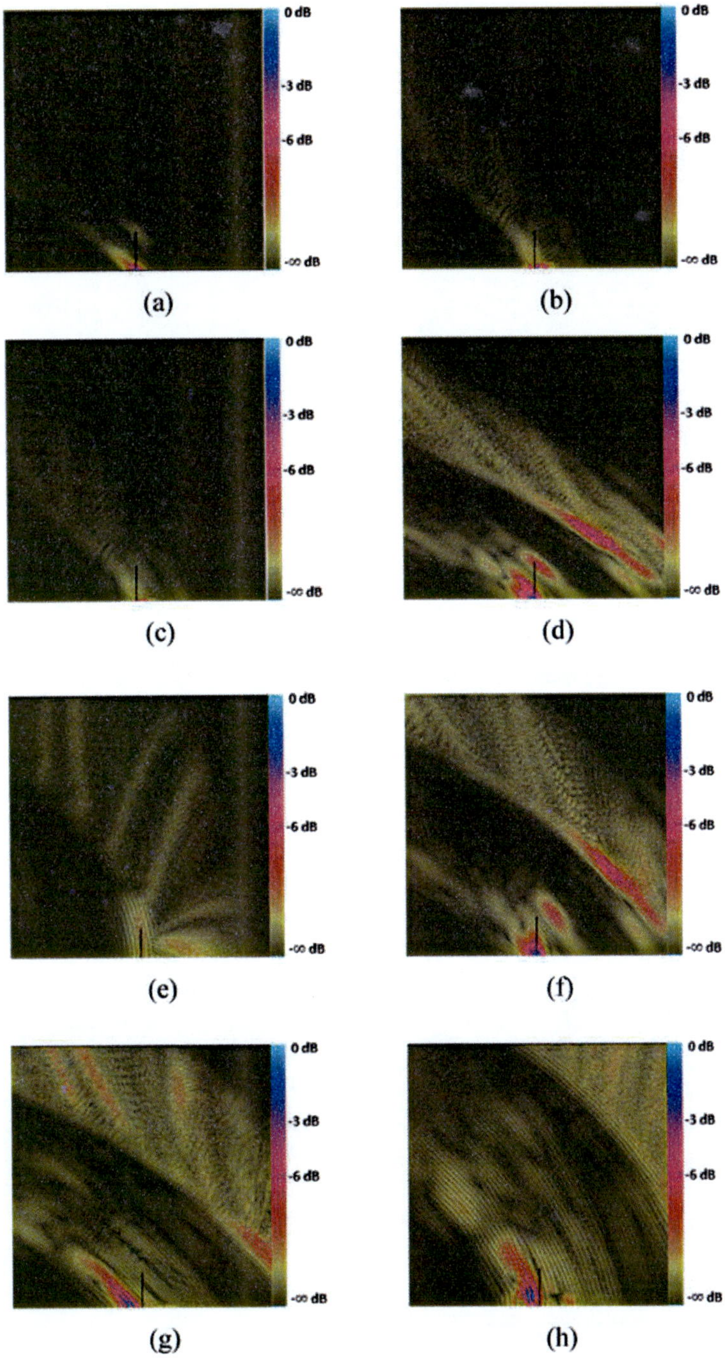

Fig. 7 Indirect mode TFM images: **a** TT-TT, **b** TL-TT, **c** TT-LT, **d** TT-TL, **e** TL-LT, **f** TT-LL, **g** TL-TL, **h** TL-LL

Table 2 Specifications of weld specimens having different types of weld defects

Specimen no.	Weld material	Thickness (mm)	Length (mm)	Defects
1	Carbon steel	12	300	Toe crack, lack of sidewall fusion, slag
2	Carbon steel	18	300	Root crack and porosity

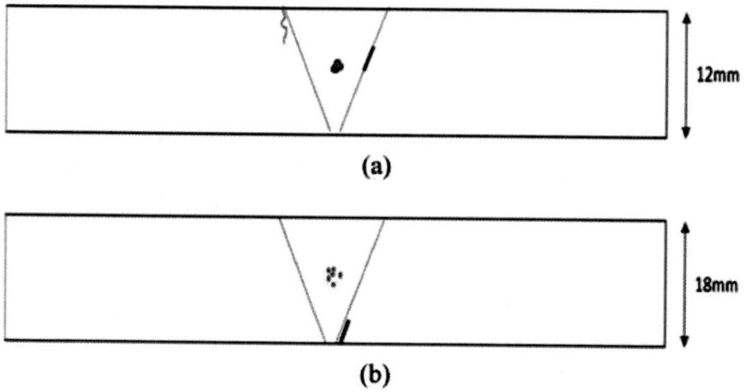

Fig. 8 Schematic of the weld specimens with defects: **a** specimen 1, **b** specimen 2

The FMC data is captured using the FMC ultrasound data acquisition unit on weld samples, and the TFM is processed using the MATLAB for image reconstruction. The block diagram of the TFM inspection involving data acquisition and post-processing is shown in Fig. 9.

For sector scans, the refracted angular range of 35°–70° with the specimen top surface and angular resolution of 1° is selected to cover the root and cap portions of the weld region in direct and half-skip configurations, respectively, whereas in TFM, beam steering is not possible due to single element excitation, and therefore, the beam spread is limited to the probe directivity. As the location of defects in the weld specimen is known, the probe placement is decided using the ray-tracing software. In

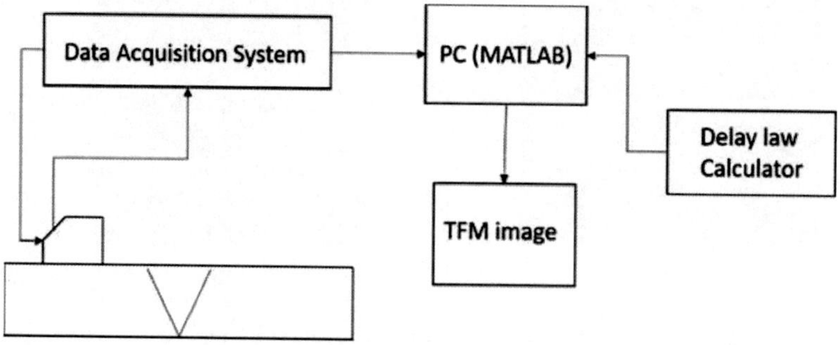

Fig. 9 FMC-TFM setup block diagram

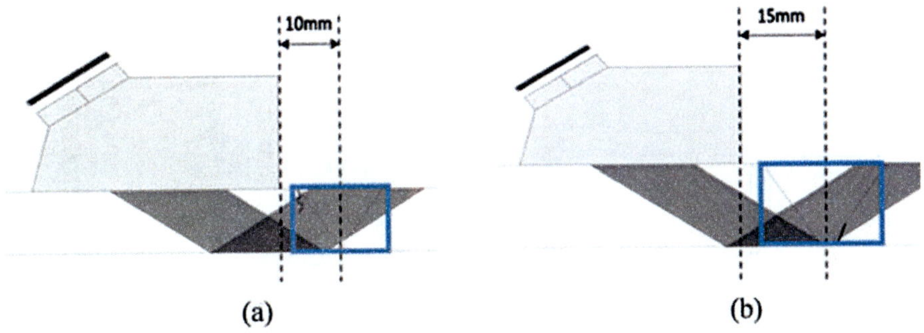

Fig. 10 Scan setup for FMC data capture: **a** specimen 1, **b** specimen 2

contrast to the specimen 1, it can be observed from Fig. 10 that the 32 element probe aperture is not sufficient to cover the entire weld region of specimen 2. Therefore, the TFM inspection over larger weld thickness specimens requires multiple probe position inspections to cover the whole weld region. However, for specimen 2, as the defects positioning is known, the probe is placed by covering the weld defects within the beam spread of the probe, as shown in Fig. 10b. The grid resolution of 512×512 is selected for TFM with the ROI covering the weld region. The TT-T and TT-TT TFM modes are used for post-processing the TFM images as the specimens can only be inspected in the half-skip configuration because of the presence of top weld bead.

5 Experimental Results and Discussion

In this section, the TT-TT TFM mode imaging results of toe crack, sidewall fusion, slag, root crack, and porosity defects are compared with phased array sector scan images. The weld overlay is drawn on the TFM and sector images to visualize the defect's location. In sector imaging, the weld overlay is repeated to facilitate visualization of defects for half-skip modes, whereas, in TFM, as the image processing is performed only in the ROI, a single weld overlay is sufficient for visualization. The results of both the imaging techniques are shown on different scales as the image area of the sector scan is much higher than the TFM results.

The presented imaging results prove the efficacy of the TFM technique in capturing all the defects with accurate positioning and high resolution. The toe crack, sidewall fusion, and root crack sector images, as shown in Figs. 11b, 12b and 14b show comparable imaging performance to their corresponding TFM results. In contrast, the sector images suffer from degradation in image quality for capturing smaller defects such as the slag and porosity, as can be observed in Figs. 13b and 15b, whereas the TFM images show high defect detectability even for smaller defects (Figs. 13a and 15a). This better imaging performance of TFM can be attributed to its ability to focus on each pixel in both transmission and reception. TFM for weld inspection also

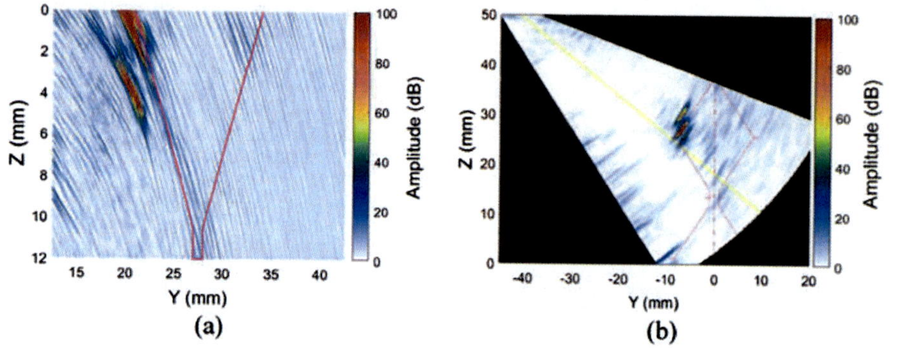

Fig. 11 Imaging results of specimen 1 with toe crack defect: **a** TFM image, **b** sector image using cap inspection

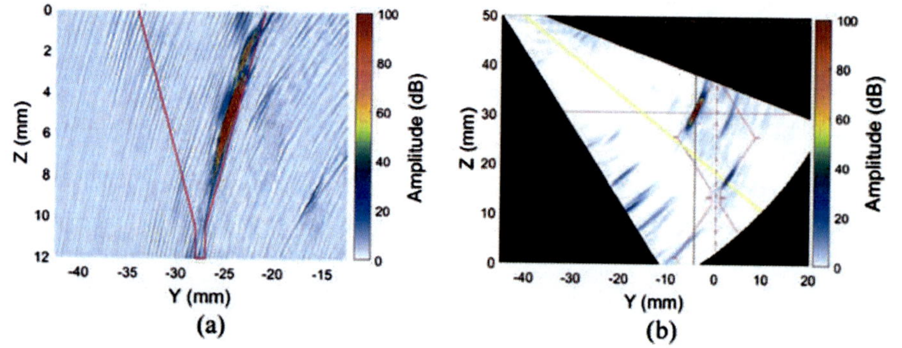

Fig. 12 Imaging results of specimen 1 with lack of sidewall fusion defect: **a** TFM image, **b** sector image using cap inspection

provides better visualization than the sector scan as the image is post-processed in the user-defined ROI, thereby generating the half-skip mode images within a single weld overlay.

6 Conclusion

The multimode TFM results of different weld defects prove the efficacy of the imaging technique in providing high-resolution images with improved defect detectability. TFM results also suggest an increased efficiency in detecting smaller defects than the phased array sector scan. However, TFM has challenges such as the weld coverage area compared to the phased array sector scan because of single element excitation. Therefore, TFM inspections require using more array elements or probing the weld specimen in multiple positions, resulting in increased processing

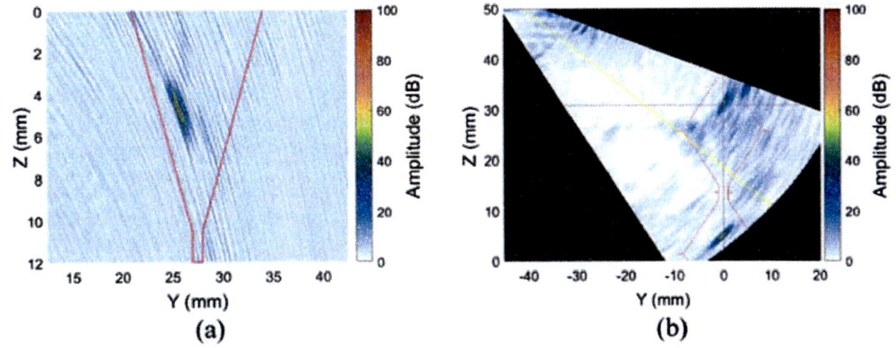

Fig. 13 Imaging results of specimen 1 with slag defect: **a** TFM image, **b** sector image using cap inspection

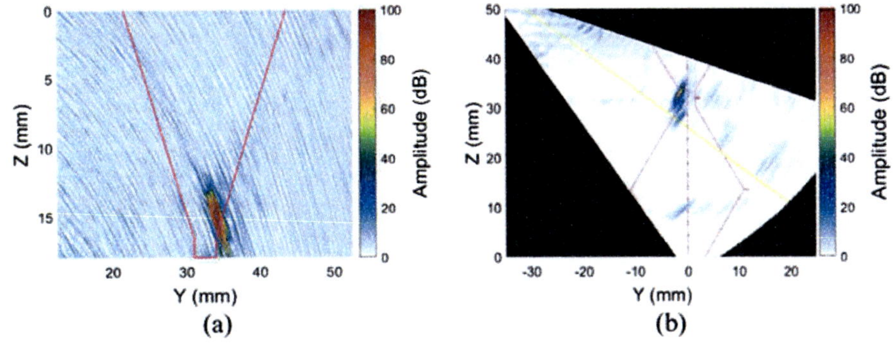

Fig. 14 Imaging results of specimen 2 with root crack defect: **a** TFM image, **b** sector image using root inspection

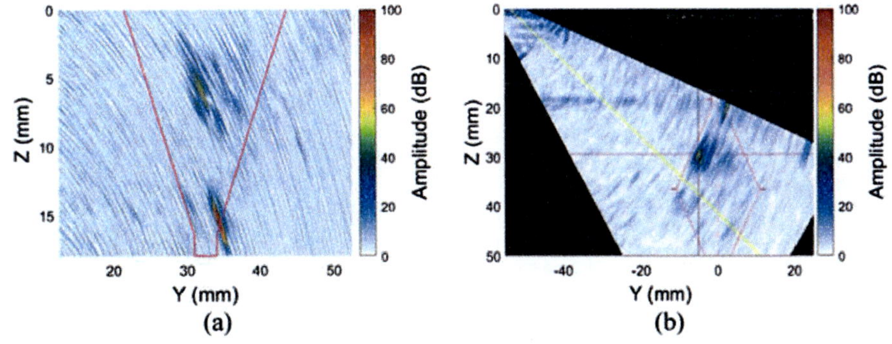

Fig. 15 Imaging results of specimen 2 with porosity defect: **a** TFM image, **b** sector image using cap inspection

time. Hence, the user must decide on the tradeoff between real-time processing speed versus the area of coverage and resolution. Besides, single element excitation limits the acoustic power transmitted into the medium, decreasing the signal-to-noise ratio [3]. The wave mode conversions in the specimen also lead to the formation of imaging artifacts in the TFM images, requiring more training and expert advice to avoid the false positives of weld defects. Nonetheless, the FMC-TFM method could be a promising technique for weld inspection with an increased probability of defect identification owing to its focusing capability in both the transmission and reception.

References

1. Drinkwater BW, Wilcox PD (2006) Ultrasonic arrays for non-destructive evaluation: a review. NDT & E Int 39(7):525–541
2. Schmerr LW Jr (2015) Fundamentals of ultrasonic phased arrays, 1st edn. Springer, Switzerland
3. Holmes C, Drinkwater BW, Wilcox PD (2005) Post-processing of the full matrix of ultrasonic transmit–receive array data for non-destructive evaluation. NDT & E Int 38(8):701–711
4. Zhang J, Drinkwater BW, Wilcox PD, Hunter AJ (2010) Defect detection using ultrasonic arrays: the multi-mode total focusing method. NDT & E Int 43(2):123–133
5. Wilcox PD, Holmes C, Drinkwater BW (2006) Enhanced defect detection and characterisation by signal processing of ultrasonic array data. In: 9th European conference on NDT. ECNDT, Berlin, pp 1–9
6. Wilcox PD, Holmes C, Drinkwater BW (2007) Advanced reflector characterization with ultrasonic phased arrays in NDE applications. IEEE Trans Ultrason Ferroelectr Freq Control 54(8):1541–1550
7. Zhang J, Drinkwater BW, Wilcox PD (2008) Defect characterization using an ultrasonic array to measure the scattering coefficient matrix. IEEE Trans Ultrason Ferroelectr Freq Control 55(10):2254–2265
8. Rose JL (2014) Ultrasonic guided waves in solid media, 1st edn. Cambridge University Press, Cambridge, New York
9. Uspensky JV (1948) Theory of equations, 1st edn. McGraw-Hill, New York
10. Calmon P, Mahaut S, Chatillon S, Raillon R (2006) CIVA: An expertise platform for simulation and processing NDT data. Ultrasonics 44:e975–e979
11. Portzgen N, Gisolf D, Verschuur DJ (2008) Wave equation-based imaging of mode converted waves in ultrasonic NDI, with suppressed leakage from nonmode converted waves. IEEE Trans Ultrason Ferroelectr Freq Control 55(8):1768–1780
12. Iakovleva E, Chatillon S, Bredif P, Mahaut S (2014) Multi-mode TFM imaging with artifacts filtering using CIVA UT forwards models. In: AIP conference proceedings, vol 1581, no. American Institute of Physics, pp 72–79
13. Sy K, Bredif P, Iakovleva E, Roy O, Lesselier D (2018) Development of the specular echoes estimator to predict relevant modes for Total Focusing Method imaging. NDT & E Int 99:134–140

Modelling the Propagation of Partial Discharge Signals Inside Gas Insulated Transmission Line Sections

Yugandhara Rao Yadam, Sarathi Ramanujam, and Kavitha Arunachalam

1 Introduction

Gas insulated transmission lines (GITL) are safe alternative to overhead power lines. A GITL consists of two concentric conductors insulated by a gas mixture mainly consisting of nitrogen and sulphur hexafluoride (SF_6) [1]. The inner conductor is supported by a dielectric spacer material. The main advantages of GITL are safety, less space occupancy, reliability and ability to transmit high power [2]. But its reliability reduces due to internal defects such as partial discharges (PDs) through sharp protrusions, contaminated metal particles, floating conductors, voids in spacers, etc. A PD is a localized dielectric breakdown of the electrical insulation which does not bridge the gap between the line and ground electrodes. Real-time monitoring and detection of PDs inside GITL are necessary to avoid complete breakdown of the insulation. PD signals are commonly detected using ultra high frequency (UHF) sensors operating over 0.5–3 GHz frequency band [3]. A typical GITL runs for few km and has straight sections, bends, joints and turns, which alters the characteristics of the PD signals received by the UHF sensor. Thus, it is important to understand the influence of the GITL sections on PD signal characteristics for sensor positioning and signal interpretation. Numerical simulations provide flexibility to study PD signal characteristics before construction. But modelling a large and complex system is numerically expensive and time consuming [4]. EM wave propagation inside a 3D model of a GITL was reported using full wave analysis over 1 MHz to 3 GHz with step size of 10 MHz. The computation time was reported as 3 days and 10 h on a computer with 128 GB RAM and 8 cores. Thus, there is a need to simplify the

Y. R. Yadam · K. Arunachalam (✉)
Department of Engineering Design, Indian Institute of Technology Madras, Chennai, India
e-mail: akavitha@iitm.ac.in

S. Ramanujam
Department of Electrical Engineering, Indian Institute of Technology Madras, Chennai, India

C. K. Mukhopadhyay and R. Mulaveesala (eds.), *Advances in Non-destructive Evaluation*, Lecture Notes in Mechanical Engineering,
https://doi.org/10.1007/978-981-16-0186-6_20

computational burden. In this work, a hybrid technique is proposed to model PD signal propagation inside a GITL using full wave EM simulations and transmission line network analysis to minimise the computational burden. Full wave analysis is carried out for the 90° L-bend section, and the results are coupled to transmission line network using system transfer functions. PDs due to Corona were propagated inside straight GITL, L-GITL using the hybrid model, and simulation results were compared with measurements.

2 Hybrid Modelling of GITL

2.1 Numerical Modelling

Hybrid modelling proposed in our study is based on transfer function analysis, which is a combination of transmission line analysis and full wave numerical simulation using Ansys HFSS® software. Figure 1 illustrates the method of hybrid modelling for L section of a GITL. L-GITL can be visualized as three separate sections: Section I and III are straight GITLs, whereas section II is 90° bend GITL. The total length of L-GITL is 3 m (each section is 1 m long) with inner diameter of 0.4 m. To reduce the computational burden without losing the accuracy, the L-GITL was divided into different sections. Straight GITL is relatively simple, and hence, sections I and III were analysed using transmission line theory [5]. Section II was analysed numerically using Ansys HFSS®. The transfer function of the straight section I, $H_1(f)$ was obtained for a unit impulse excitation over 300 MHz–3 GHz. The transfer function,

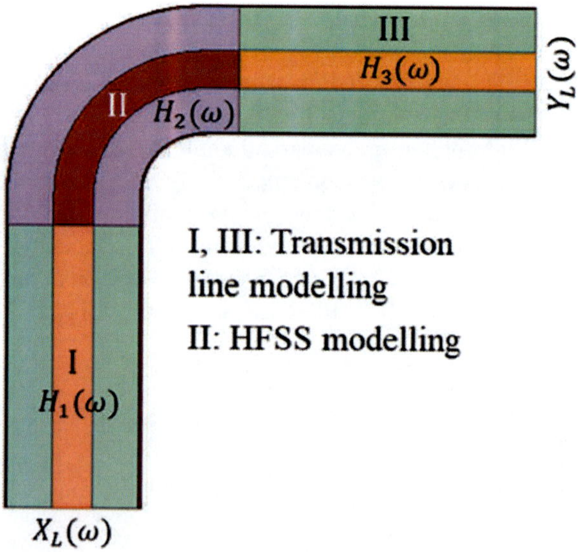

Fig. 1 Illustration of hybrid modelling technique on L-GITL

$H_3(f)$ for section III is same as $H_1(f)$. The transfer function of section II obtained from HFSS® simulations is $H_2(f)$. The complete transfer function of the L-GITL, $H_L(f)$ is given by,

$$H_L(f) = \frac{Y_L(f)}{X_L(f)} = H_1(f)H_2(f)H_3(f) \tag{1}$$

In Eq. (1), $X_L(f)$ and $Y_L(f)$ are the frequency responses of the input and output PD signals, respectively.

Corona PDs that occurred in a compressed SF_6 gas insulation was measured and fed as input to the L-GITL model. Measurements were gathered for high voltage AC (HVAC) of power frequency (50 Hz) at 4 bar pressure using a commercial UHF sensor connected to a digital storage oscilloscope (DSO). The input signal was propagated through straight section, and the delay introduced by the GITL was used to calculate $H_1(f)$. The signal received at the other end of the section I is given by,

$$Y_1(f) = X_1(f)H_1(f) \tag{2}$$

The frequency domain signal $Y_1(f)$ was used as input for full wave analysis of L-section carried out in Ansys HFSS® simulation software. The signal at the other end of L-section $Y_2(f)$ was used to calculate $H_2(f)$. The signal $Y_2(f)$ was multiplied by the transmission line transfer function $H_2(f)$ to obtain $Y_3(f)$. IFFT of $Y_3(f)$ yields the PD signal that propagated through the L-bend.

The numerical model for section II of L-GITL (90° L-bend) is shown in Fig. 2. The length of the section is 1 m; diameter of the inner conductor is 0.12 m, and the diameter of the outer conductor is 0.4 m. The inner and outer conductors were assigned as aluminium. The area between the outer and inner conductors was filled with SF_6 gas, which has dielectric constant of 1.002, conductivity ($\sigma = 0$, loss less medium) and density 23.73 kg/m^3 [6]. A wave port, which assumes a semi-infinite long waveguide was assigned to Ports 1 and 2 shown in Fig. 2. Port 1 was defined as input port with 1 W power over 300 MHz–3 GHz, and Port 2 was defined as output port. An iterative solver with modal driven solution was used in the simulation. The ratio of the output power to input power was used to calculate the transfer function of the 90° L-bend.

The transfer function for sections I/III and II are shown in Fig. 3. It is observed that the straight section I is exhibiting lossless nature (0 dB), whereas the section II which is a 90° bend, introduces loss in the transmitted signal.

3 Generation and Measurement of Corona Signals

Figure 4 shows the test setup that was used to generate and acquire Corona induced PD signals. It consists of HVAC source, test cell for simulating Corona discharges,

Fig. 2 Numerical modelling of the section II of L-GITL

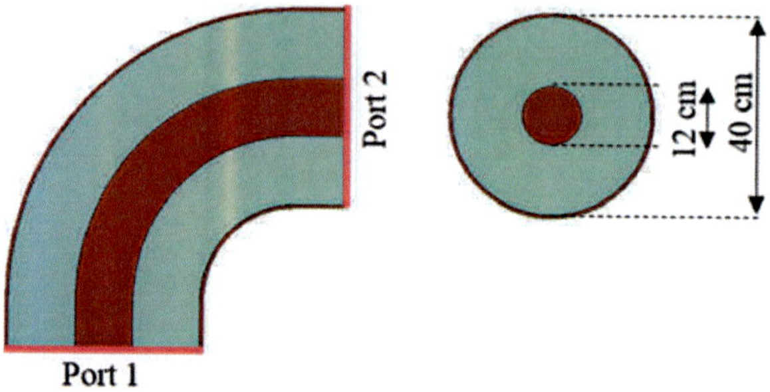

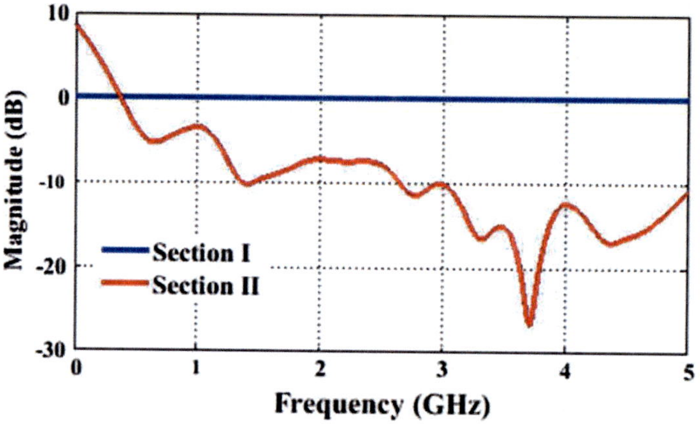

Fig. 3 Comparison of transfer functions for transmission line sections I and II of L-GITL

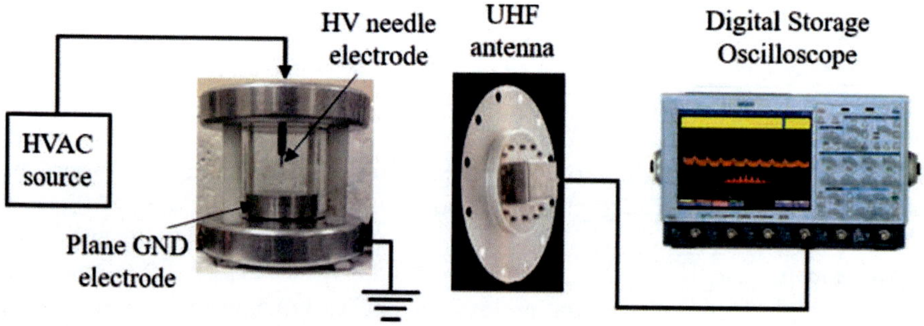

Fig. 4 Test setup for measurement of Corona type of partial discharges

UHF sensor (antenna), and oscilloscope for acquiring the PD signals. The HVAC is of power frequency (50 Hz) maintained by a transformer (5 kVA, 50 Hz, 100 kV). The applied AC voltage was measured using capacitance voltage divider circuit [7]. The test cell consists of top needle electrode, which is supplied with HVAC and bottom plane ground (GND) electrode are separated by a Perspex™ tube. SF_6 gas

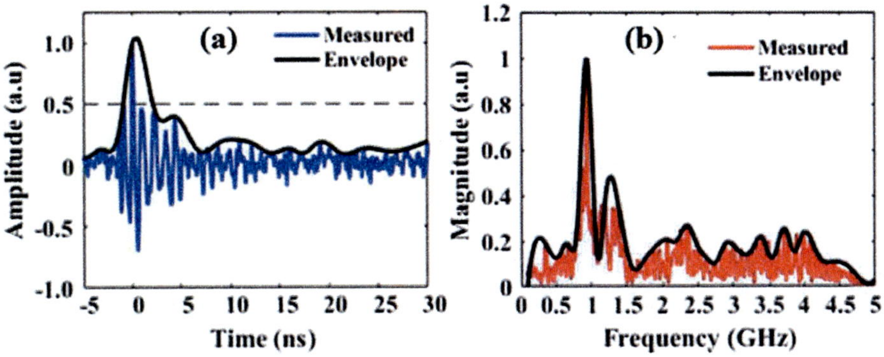

Fig. 5 (**a**) Typical Corona PD signal acquired in air and (**b**) its spectrum

was filled inside the Perspex™ chamber through a valve, and 4 bar pressure was maintained using pressure gauge that was connected to the test cell. The distance between the needle tip to GND plane was around 1 cm. The HVAC was increased at a rate of 300 V/s to the test cell until Corona PDs were generated. Once Corona was initiated, it occurs continuously and PD pulses radiated in SF_6 medium pass through Perspex™ chamber and radiate in the air. A UHF sensor placed in the vicinity of the test cell received the PD signals. The PD signals were digitized by a real-time digital storage oscilloscope (LeCroy, four-channel digital real-time oscilloscope, 3-GHz bandwidth, operated at 20 GSa/s) with an input impedance of 50 ohms and displayed on the monitor. Since the PD signals are random in nature, a minimum of 200 signals were acquired and then averaged to get a typical Corona PD signal in air. Figure 5 shows the typical Corona PD signal in air and its spectrum. The spectrum of the PD signal was used as $X_1(f)$ in the hybrid model of the GITL. To validate the technique, a Corona PD signal was measured at the output of the 1 m long straight GITL with the same cross section. The measured time domain signal at the output of the straight GITL is represented as $y_{SM}(t)$, and corresponding spectrum is represented as $Y_{SM}(f)$. This signal was compared with the signal obtained from the transmission line model

$y_L(t)$, which is the time domain signal corresponding to the frequency domain signal $Y_L(f)$.

4 Results and Discussion

Figure 5 shows the PD signal measured in air and its envelope. The pulse width determined from full width half maximum (FWHM) of the envelope was measured as 3.1 ns. The dominant spectral peaks are listed in Table 1. For subsequent PD signals, signal envelope alone is presented.

Table 1 Dominant spectral peaks of Corona PD signals in a GITL

PD signal	Spectral peaks (GHz)
Straight GITL_measured	0.66, 0.93, 1.21, 1.39, 1.66, 1.88, 2.08, 2.34, 2.56, 2.78, 3.1, 3.34, 3.57, 4.06, 4.28
Straight GITL_modelled	0.76, 0.94, 1.18, 1.36, 1.92, 2.09, 2.3, 2.5, 2.65, 2.88, 3.06, 3.37, 3.43, 3.72, 3.94
L-GITL_measured	0.5, 0.87, 1.4, 2.0, 2.24, 2.6, 3.04, 3.31, 3.56, 3.82, 4.09, 4.31
L-GITL_modelled	0.54, 0.77, 0.93, 1.15, 1.33, 1.92, 2.09, 2.3, 2.5, 2.64, 2.89, 3.03, 3.45, 4.01

Figure 6 shows the time and frequency domain signals for a 1 m long straight GITL obtained using transmission line model. PD signals from measurement and modelling shown in Fig. 6 are in good agreement with each other. The pulse width of the measured PD signal is 4.4 ns, whereas pulse width of modelled PD signal is 3.4 ns. The dominant frequencies in the spectrum of the measured and modelled Corona signals lie around 1 GHz, and the shape of the spectrums is also similar. In both cases, the spectrum spans from 0.5 to 4 GHz. Spectral peaks in the measured and modelled PD signals for straight GITL are given in Table 1.

Figure 7 shows the envelope of Corona signals and their spectrum at the end of L-GITL gathered from measurement and hybrid modelling technique. It can be observed that the measurement and modelled signals are in good agreement with each other. The pulse width of the Corona signal after passing through 3 m long L-GITL is measured as 2.7 ns, and the dominant frequencies remain about 1 GHz. The pulse width of the measured and modelled PD signals is 2.7 ns and 2.8 ns, respectively. Unlike the PD signals in straight section, spectral content of the L- spanned over 0.5 to 3 GHz. Similar spectral content was observed in the simulated output of the hybrid model. Spectral peaks of measured and modelled PD signals through L-GITL occur at different frequencies and are listed in Table 1.

5 Conclusion

In this work, a hybrid technique to model PD signal propagation inside GITL is proposed. The technique is based on transfer function analysis. The hybrid transfer function is a composite transfer function that is derived from full wave EM simulations and transmission line network analysis. The hybrid technique considerably reduced the computational burden (10 h on 128 GB RAM computer and 8 cores) as only the L-section was modelled. Corona PD signal propagation through straight and L-GITLs calculated by the model were compared with the measured PD signals. The measured and modelled signals are in good agreement suggesting that the proposed technique can be used for analysing PD signal propagation in GITL network.

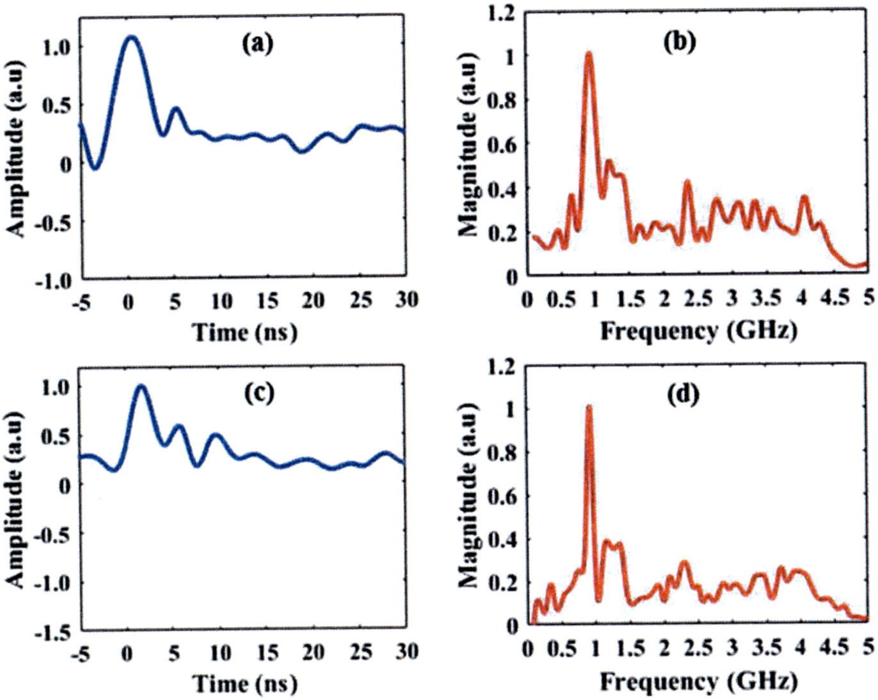

Fig. 6 Comparison of Corona signals. (**a**) Measured signal and (**b**) its spectrum, (**c**) modelled signal and (**d**) its spectrum for 1 m long straight GITL

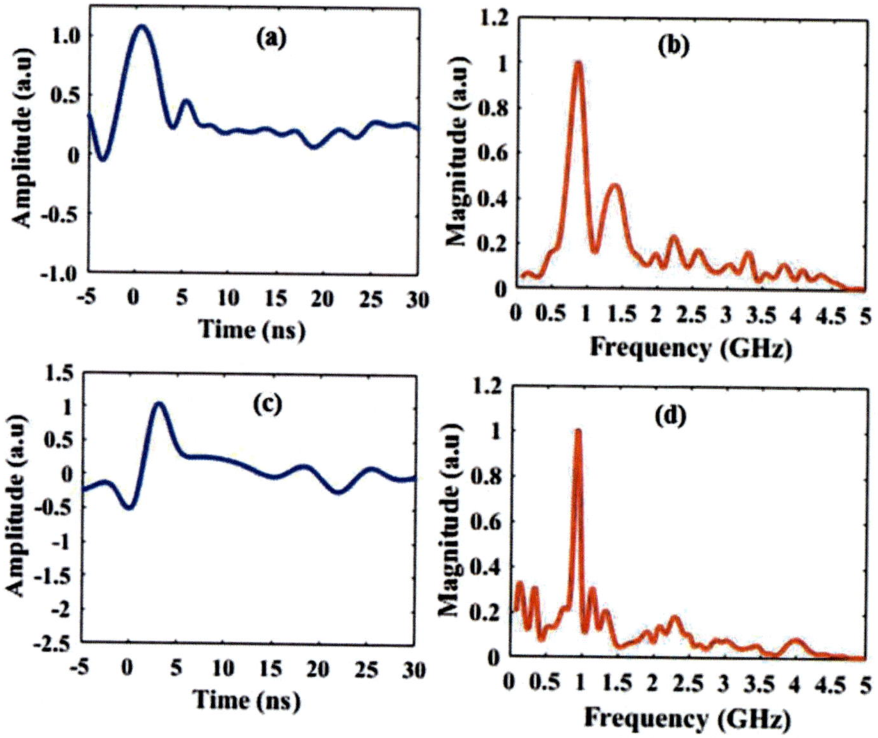

Fig. 7 Comparison of Corona signals (**a**) Measured signal and (**b**) its spectrum, (**c**) modelled signal and (**d**) its spectrum for 3 m long L-GITL

References

1. Judd MD, Farish O, Hampton BF (1996) The excitation of UHF signals by partial discharges in GIS. IEEE Trans Dielectr Electr Insul 3(2):213–228
2. Kaneko S, Okabe S, Yoshimura M, Muto H, Nishida C, Kamei M (2009) Detecting characteristics of various type antennas on partial discharge electromagnetic wave radiating through insulating spacer in gas insulated switchgear. IEEE Trans Dielectr Electr Insul 16(5):1462–1472
3. Hoshino T, Maruyama S, Sakakibara T (2008) Simulation of propagating electromagnetic wave due to partial discharge in GIS using FDTD. IEEE Trans Power Delivery 24(1):153–159
4. Behrmann G, Smajic J (2016) RF PD signal propagation in GIS: Comparing S-parameter measurements with an RF transmission model for a short section of GIS. IEEE Trans Dielectr Electr Insul 23(3):1331–1337
5. Pozar, David M (2009) Microwave engineering. John Wiley & Sons
6. Naidu MS (2008) Gas insulated substations: GIS. IK International Pvt Ltd.
7. Umamaheswari R, Sarathi R (2011) Identification of partial discharges in gas-insulated switchgear by ultra-high-frequency technique and classification by adopting multi-class support vector machines. Electric Power Components Syst 39(14):1577–1595

Assessment of Sensitivity of Radioisotope-Based Radiometry Data During Inspection of Large Sized Solid Rocket Motors

S. Khan Lubna, Amali Umesh, N. K. Ghosh, S. C. Bhattacharyya, and S. Srinivasan

1 Introduction

1.1 Brief

NDT of SRMs is carried out to find out flaws such as voids, foreign particle inclusion, cracks, interfacial defects, etc., in the motor. These flaws may get introduced in the grain during processing or during handling of the SRM. There are large numbers of NDT methods, which are followed worldwide [4], but most suitable methods for inspection of SRMs are X-ray radiography and ultrasonic testing (UT). X-ray radiography technique is generally employed in inspection of propellant grains, whereas UT is used for inspection of interfaces [5].

NDT inspection of large sized SRMs by RT method involves dividing the motor in different sections and orientations and subsequently their scanning. Full inspection of a large sized SRM is not possible due to practical difficulties of excessively large number of exposures and equally longer duration of time for inspection. Therefore, sampling inspection of the SRM is performed by formulating suitable radiography plans based on the motor geometry, criticality of flaws, and their characterization [6] -all this in conformance with the acceptance criteria as per the mission requirement[6]. Radiography plans mostly have more than 70% coverage.

A non-conventional methodology of inspection of propellant grains is developed to inspect propellant grain fully and in quick time [7]. The completely automated technique uses a radioisotope-based source and PMT-based detector. The activity of the radioisotope plays an important role in the quality of the results obtained; the effect of depleting activity of the radioisotope on sensitivity of the radiometry data is studied through various experiments and is presented here.

S. Khan Lubna (✉) · A. Umesh · N. K. Ghosh · S. C. Bhattacharyya · S. Srinivasan
Advanced Centre for Energetic Materials, Nasik, India
e-mail: lubna.khan@acem.drdo.in

C. K. Mukhopadhyay and R. Mulaveesala (eds.), *Advances in Non-destructive Evaluation*, Lecture Notes in Mechanical Engineering,
https://doi.org/10.1007/978-981-16-0186-6_21

2 Experimental Set-up

The experimental set-up consists of a radioisotope source (60Co), an array of PMT detectors (ref Fig. 1), the platform scanner on which the object under inspection is placed and a cantilever boom for the movement of radioisotope from the camera to destination position. Gamma rays (of energies 1.17 and 1.33 meV) are emitted from the 60Co radioisotope source which is positioned at the cantilever boom, in the port area of motor. All PMT detectors are aligned with tip of the cantilever boom where the radioisotope is positioned to detect the transmitted gamma photons coming from object. During scanning of radiometry, cantilever boom accesses inside the port area. The boom position is adjusted such that when boom goes inside the port of motor, it must not scratch the propellant surface. Schematic diagram of the experimental set-up is as shown below in Fig. 2 .

The rocket motor is mounted on a platform scanner which is used for longitudinal and rotational movements of the SRM. The large size rocket motor is divided into

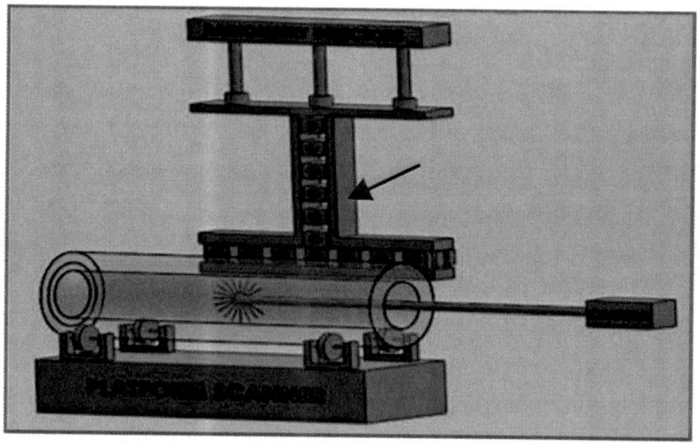

Fig. 1 Array of PMT detectors

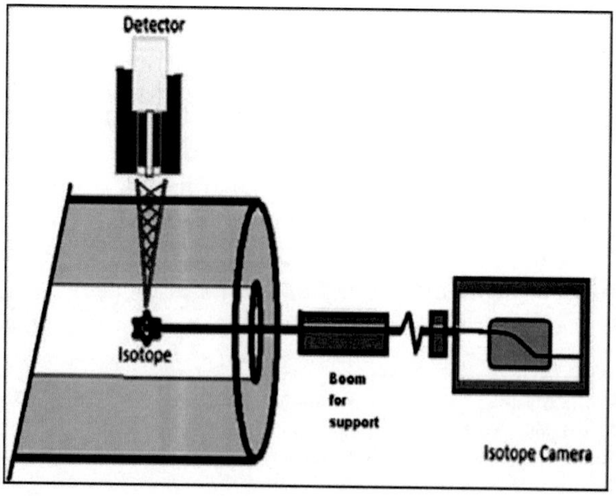

Fig. 2 Schematic set-up

Fig. 3 Image quality indicator (IQI)

zones, and position of the probe on the rocket motor is calibrated fitting to the computer screen. Artificial defects (as shown in Fig. 3) were put on the surface of the motor and were treated as image quality indicators. Various parameters used during the study are listed below in Table 1.

3 Procedure

Radiometry inspection is based on the penetration of gamma radiation through the material, and subsequently, the differential absorption as a function of thickness and density of material is recorded. The transmitted beams are detected which carries the information of volumetric flaws in the material. The transmitted beam is then analyzed to determine the presence, size, and location of flaws and discontinuities. The basic electronic instruments used in radiometry are mainly an array of PMT detectors, signal amplifier, ADC, and a display mechanism that permits interpretation of the transmitted beams Tables 2, 3 and 4.

Intensity of the beam depends upon the activity of the radioisotope. After each half-life, activity of the source is halved, which affects the quality of the radiometry data in terms of sensitivity. Experiments were performed by varying scan speed, step of scanning and collimator opening width of the PMT-based detector, and sensitivity in terms of SNR was then recorded. The results were analyzed, and the suitability of the use of technique with optimized parameters under depleted activity condition was arrived at (Figs. 4 and 5) .

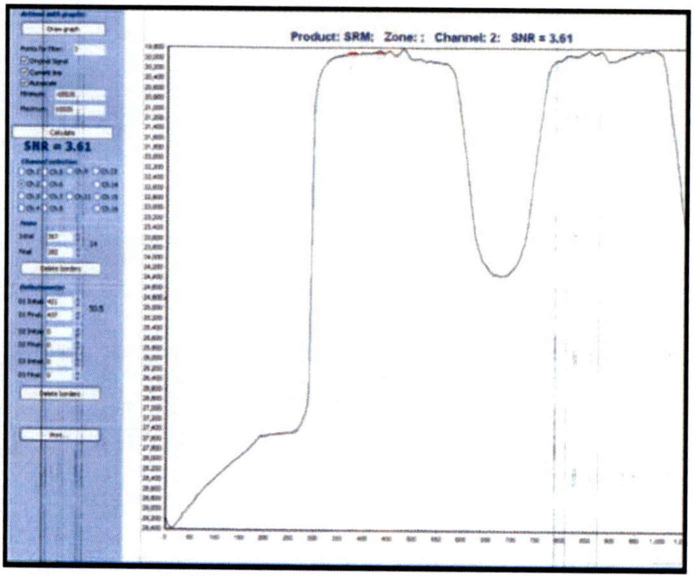

Fig. 4 A graph of photon counts versus length under scan

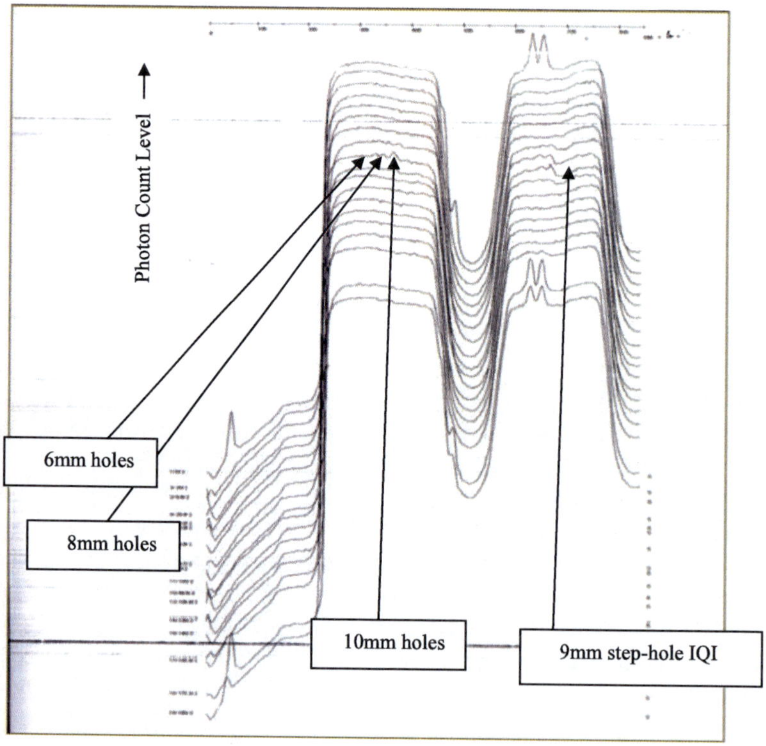

Fig. 5 A graph depicting full profile of zone under inspection

4 Observations

(1) In the first set, radiometry experiment was performed keeping fixed collimator opening width and step of scan and by varying scan speed. It was found that reducing the speed below 10 mm/s did not give any significant improvement in the SNR. The size of the smallest flaw detected was a 6 mm cube.
Step of scanning: 16mmCollimator size: 10 × 20 mm.

(2) In the second set, radiometry experiment was performed keeping fixed collimator opening width, scan speed and by varying step of scan. It was found that variation in step of scan did not give any significant improvement in the SNR. The size of the smallest flaw detected was a 6 mm cube.
Scanning Speed: 15 mm/secCollimator size: 10 × 20 mm.

(3) In the third set, two different size collimators were used by keeping scan speed and step of scan fixed.
Scanning speed: 15 mm/secStep of scanning:16 mm.

(4) Thickness sensitivity of 1.2% can be achieved with optimized parameters at an activity of 91 Ci during radiometry inspection of large sized propellant grain with web thickness (300–500 mm).

5 Conclusion

(a) Radioisotope-based radiometry technique has a unique advantage in propellant grain inspection, as full inspection of the motor is possible in a short span of time. This is very useful particularly in large sized rocket motors where conventional RT is a time-consuming process.
(b) The thickness sensitivity achieved in the inspection of large sized SRMs by radiometry technique is comparable with the results obtained using conventional RT.
(c) Inspection of large sized SRM by radiometry technique has resulted in almost 75% reduction in time of inspection as compared to conventional film-based RT.
(d) With the present parameters, fairly good sensitivity signals are obtained for the propellant grain at a depleted activity of the radioisotope for a large sized rocket motor.

6 Limitations

The technique mainly offers qualitative information about the flaw. Though the coordinate of the flaw can be obtained by radiometry, the exact nature and type of the

Table 1 Parameters and details

Radioisotope activity	91 Ci	Radioisotope activity
Inspection technique	Single wall grain inspection	Inspection technique
Scan method	Continuous line/ring/spiral	Scan method
Detector	Photo multiplier tube	Detector
Sensitivity (% thickness)	1.2%	Sensitivity (% thickness)
Radioisotope activity	91 Ci	Radioisotope activity

Table 2 Output result with varying scan speed

Parameter	Defect size (in mm)					
Scanning speed↓	2	3	4	6	8	10
5 mm/sec	x					
10 mm/sec	x	X	x	√	√	√
15 mm/sec	x	X	x	√	√	√

Table 3 Output result with varying step of scan

Parameter	Defect size (in mm)					
Step of scanning ↓	2	3	4	6	8	10
10 mm	X	X	‘	√	√	√
16 mm	X	x	X	√	√	√

Table 4 Output result with varying collimator size

Parameter	Defect size (in mm)					
Collimator size ↓	2	3	4	6	8	10
20 × 10 mm	X	X	X	√	√	√
14 × 11 mm	X	X	X	√	√	√

flaw can be ascertained by using an imaging technique only. Very tight cracks which require a high sensitivity technique may go undetected but such probabilities exist in the conventional RT too.

References

1. George PS, Oscar B (2011) Rocket propulsion elements, Eighth edition, John Wiley & Sons
2. Ghose B (2016) Evaluation of probability of detection and minimum number of exposures required for detection of planar flaws in cylindrical object by radiographic NDE method. https://www.ndt.net

3. Peter JS (2002) Non-destructive evaluation theory techniques and applications 1 Marcel Dekker
4. Halmshaw R Chapman, Hall: Industrial Radiology
5. Jack B, Geoff S (1996) Ultrasonic methods of non-destructive testing1 Ed. Chapman and Hall
6. Ghose B, Kankane DK (2008) Estimation of location of defects in propellant grain by X-ray radiography vol 41 pp 125–128 Elsevier
7. Sai Suryanarayana PV, Viswanathan K Non destructive evaluation of solid rockets and missile systems Chapter-12, Advances in NDE of Solid Rocket Motors and Future Scenario, DESIDOC, New Delhi (in Press)

Planar Printed E-Field Sensor Array for Microwave NDE of Composites

Jayaram Kizhekke Pakkathillam, Nitheesh M. Nair, Parasuraman Swaminathan, and Kavitha Arunachalam

1 Introduction

Non-destructive evaluation (NDE) deals with testing of a material or a component without compromising its utility. Different techniques are used in NDE based on the physical and material properties of the interrogating specimen and its geometry and environment. The commonly used energy sources for non-destructive testing (NDT) include acoustic, thermography, eddy current, X rays, and radiography [1]. Acoustic emission and ultrasonic testing (UT) are widely used for detecting cracks in wide range of metallic specimens [1, 2]. Thermography and optical methods are employed for surface imaging [1, 2]. Eddy current testing is usually carried out for metallic components [3]. X rays and radiography have their own advantages for NDT, but the radiation hazard is the main problem faced during testing [4]. Unlike the conventional NDT methods which are primarily designed for inspection of metallic components, microwave NDE (MNDE) finds importance in inspecting dielectrics and non-metallic composites without affecting its utility [5, 6]. In MNDE, high-frequency (300 MHz–300 GHz) electromagnetic (EM) energy is used to interrogate dielectric materials, engineered composites, and metal surfaces. Of late, the use of composite materials has increased in aerospace, automotive, and medical applications. This gives rise to the need for NDE of large flat panels. In microwave, NDE near and farfield probes can be used. The probe classification is based on the operating distance between the probe and specimen and specimen geometry. Generally, open-ended coaxial probe

J. K. Pakkathillam · K. Arunachalam (✉)
Department of Engineering Design, IIT Madras, Chennai, India
e-mail: akavitha@iitm.ac.in

N. M. Nair · P. Swaminathan
Department of Metallurgical and Materials Engineering, IIT Madras, Chennai, India

N. M. Nair
Department of Electrical Engineering, IIT Madras, Chennai, India

C. K. Mukhopadhyay and R. Mulaveesala (eds.), *Advances in Non-destructive Evaluation*, Lecture Notes in Mechanical Engineering,
https://doi.org/10.1007/978-981-16-0186-6_22

and open-ended waveguide probe operating in monostatic condition are used as near field probes [7]. Horn antenna with and without spot focusing lens is used as far field probe [8]. An array of probes enables inspection of large area of the sample and reduces measurement time. In the literature, a two-dimensional (2D) sensor array was proposed for electric field measurement in the very low frequency (VLF) band (3–30 kHz) [9]. A D dot sensor (D dot stands for the time derivative of electric flux density) was employed to measure the current which in turn was related to the time derivative of the electric field. Another work was reported on 2D sensor array for measuring electric field radiated by an antenna [10]. Theoretical modeling of magnetic field sensor arrays was proposed in [11]. In this work, an electrically short dipole antenna array inkjet printed on a photo paper is proposed used for sensing the electric field perturbation in the near field of dielectric specimens. In the proposed linear sensor array, the array acts as a passive receiver operating over 8-12 GHz and is used to measure the electric field maintained by dielectric composites illuminated by spot focusing horn operating in the X band (8–12 GHz). The design of the sensor and sensor utility for NDE of dielectric composite are demonstrated for an engineered defect in the sample.

2 Sensor Array Design

2.1 *Sensor Working Principle*

Figure 1 shows the schematic of an electrically short electric (E) field sensor. This sensor consists of an electrically short dipole antenna, a Schottky diode, a resistive transmission line, and an amplifier circuit to amplify the voltage induced on the dipole receiver.

The resistive transmission line acts as a low-pass filter for the incoming signal and reduces the field perturbation. The dipole antenna in Fig. 1 can be replaced with its open circuit voltage (V_{oc}) and antenna impedance, $Z_a = 1/j\omega C_A$, where C_A depends on the length and width of the printed dipole. The electric field incident on the dipole antenna is proportional to the open circuit voltage of the dipole and is given by,

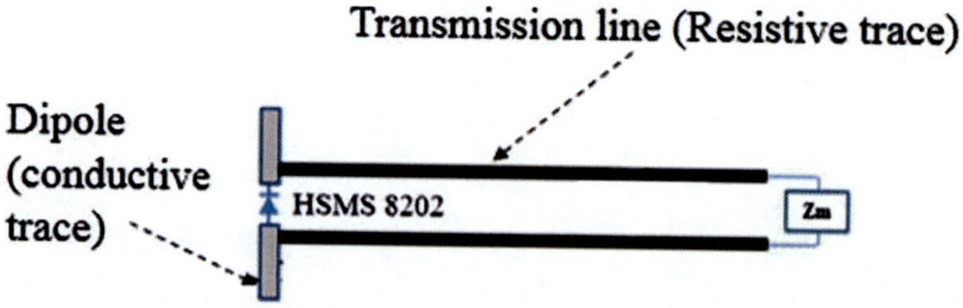

Fig. 1 Schematic of the unit sensor element

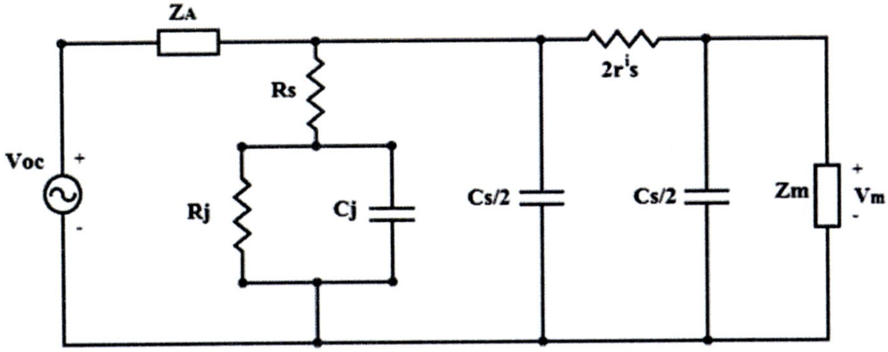

Fig. 2 Equivalent circuit of the unit sensor element

$$V_{oc}(\omega) = E_z^i(0,\omega)\int_{-h/2}^{h/2}\vec{E}.\mathrm{d}z \tag{1}$$

The diode and the transmission line can also be replaced by their equivalent circuits. The equivalent circuit of the sensor element is shown in Fig. 2. The measured output voltage V_m is given as [12],

$$V_m = \frac{1}{2R_j}\left(\frac{|V_{oc}|^2(\omega R_j C_j)^2 R_m}{(\omega R_j(C_A + C_j) + 1)^2(R_m + 2r^i s)}\right). \tag{2}$$

In Eq. (2), R_S is the diode series resistance; R_j and C_j are the junction resistance and capacitance of the diode, respectively; $r^i s$ and C_s are transmission resistance and capacitance for a given length; s and Z_m are the input resistance of the instrumentation amplifier. Thus, the voltage at the input of the instrumentation amplifier is proportional to the square of the incident electric field.

2.2 *Numerical Modeling*

Figure 3 shows a single dipole modeled as a unit cell element in HFSS [13], Ansys EM simulation software using Floquet port excitation. Floquet port imposes periodic boundary (master-slave boundary) on all sides of the unit cell. Thus, array analysis can be simplified using single element. The appropriate choice of the length (b) and width (a) of the dipole, dipole spacing in the array, and the conductivity of the dipole arms was numerically studied for the numerical model in Fig. 3. The purpose of the numerical simulations is to identify sensor configuration for better signal transmission and less reflection of the incident wave. The unit cell was modeled on a 0.2 mm thick photo paper with dielectric constant of 2.6, and the dipole arm was modeled as 2D sheet of silver.

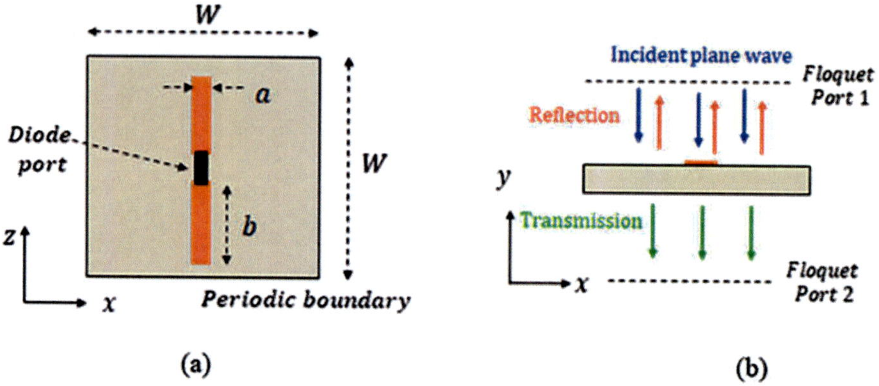

Fig. 3 Numerical modeling. **a** unit cell element schematic, **b** Floquet port arrangement for reflection and transmission coefficient analysis

The dipole simulations were carried out over 8–12 GHz. The unit cell width was fixed as 30 mm which was equal to one wavelength at the center frequency. The dipole length, a was varied over 1–3 mm, and width, b was varied over 5–7 mm. The dimensions were chosen such that the dipole is electrically small at the center frequency. Unit cell reflection ($|S_{11}|$) and transmission ($|S_{21}|$) coefficients calculated at the center frequency (10 GHz) are shown in Table 1 for varying dipole dimensions. It can be identified that longer and thicker the dipole arm, higher is the reflection. Hence, $a = 1$ mm and $b = 5$ mm were chosen for the dipole arms.

Figure 4 shows the reflection and transmission coefficients at 10 GHz for varying unit cell width (spacing between the dipoles) and fixed dipole arm length $a = 1$ mm and width $b = 5$ mm. Simulation results are shown for 1.25λ, 1λ, 0.75λ, 0.5λ and 0.25λ spacing between the adjacent dipoles in the array. It can be identified that wave reflection lowers with increase in array spacing. 0.5λ spacing was chosen for array design as reflection and transmission coefficients were in the acceptable range.

Table 1 Reflection and transmission coefficients of the 30 mm wide unit cell at 10 GHz

Dimension	$\|S_{11}\|$ at 10 GHz (dB)	$\|S_{21}\|$ at 10 GHz (dB)
$a = 1$ mm, $b =$ 5 mm	− 33	0.005
$a = 2$ mm, $b =$ 5 mm	− 28	0.009
$a = 3$ mm, $b =$ 5 mm	− 25	0.010
$a = 1$ mm,$b =$ 6 mm	− 33	0.005
$a = 1$ mm, $b =$ 7 mm	− 30	0.008

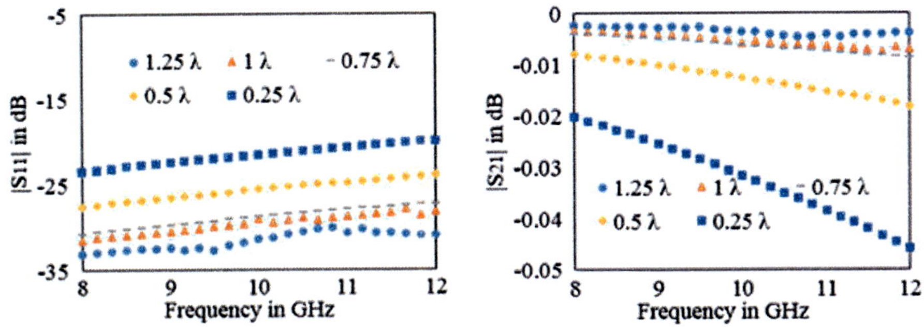

Fig. 4 Influence of spacing between array elements. (a) unit cell reflection and (b) transmission coefficients for, $a = 1$ mm and $b = 5$ mm.

Further, 0.5λ spacing provided enough space to print the transmission lines used for sensor output voltage measurements.

Unit cell simulations were carried out by varying electrical conductivity of the dipole arm and fixed spacing of 0.5λ to study field perturbation caused by the dipole antenna. The material for the dipole arm was assigned as an ohmic sheet ($\sigma = 777$ S/m) and silver ($\sigma = 6.3 \times 10^7$ S/m). Figure 5 shows the reflection and transmission coefficients for silver and ohmic dipole arms. Simulation results of photo paper without the printed dipole is also shown in Fig. 5 to understand the perturbation. It can be observed that the ohmic dipole arms provided low reflection (Fig.5a) but is lossy in nature as seen in the transmission coefficient data (Fig. 5b).

In order to quantify the reflection and transmission coefficients in the presence of the transmission lines, numerical simulations were carried out for the numerical model in Fig. 6. Here, the photo paper containing the resistive transmission lines was placed in a vacuum cylinder. A plane wave source kept at $3\lambda_c$ distance from the substrate illuminated the photopaper, where λ_c is the wavelength at the center frequency. The reflection and transmission coefficients were calculated using the incident and scattered electric field, at reference planes 1 and 2 situated at $2\lambda_c$ distance

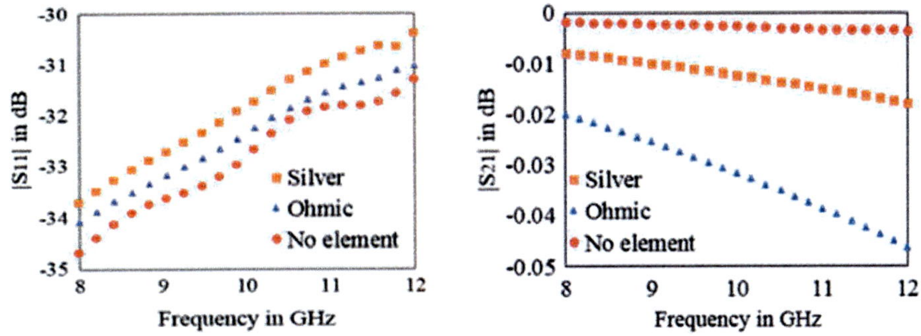

Fig. 5 Varying dipole arm conductivity. (a) Reflection and (b) transmission coefficients for $a = 1$ mm, $b = 5$ mm and 0.5λ sensor spacing

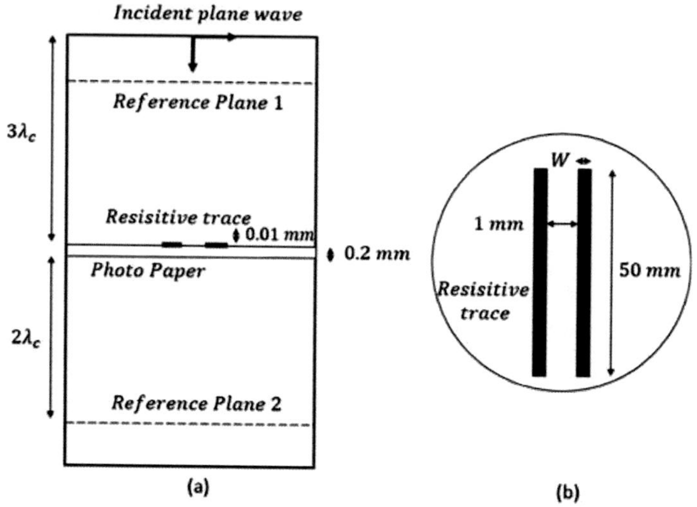

Fig. 6 Numerical model of the transmission line on the substrate. (a) Side and (b) top views

from either sides of the substrate. The length of the transmission line was fixed as 50 mm so that the contact pads for external wire connection was outside the focal spot of the spot focusing horn antennas. Numerical simulations were carried out for different trace widths, and the results are tabulated in Table 2.

Trace width $W = 0$ mm corresponds to no trace condition. Here, the reflection is very low, and transmission is very high. This is taken as reference for normalizing the S parameters. As the trace width was increased, the reflection coefficient was increased and the transmission coefficient was reduced. The 0.5 mm trace offers better reflection, but the overall line resistance is very high. When $w = 2$ mm, the line resistance is low but the reflection to the incident field caused by the line is very high. Hence, $w = 1$ mm was chosen as trace width for optimum resistance and lesser field perturbation.

Table 2 Reflection and transmission coefficients of transmission line at 10 GHz

Trace width, W (mm)	Normalized $\lvert S_{11}\rvert$ at 10 GHz (dB)	Normalized $\lvert S_{21}\rvert$ at 10 GHz (dB)	Line resistance $M\Omega$
0	0	0	–
0.5	1	−0.05	12.8
1	1.2	−0.07	6.43
2	2.7	−1.2	3.21

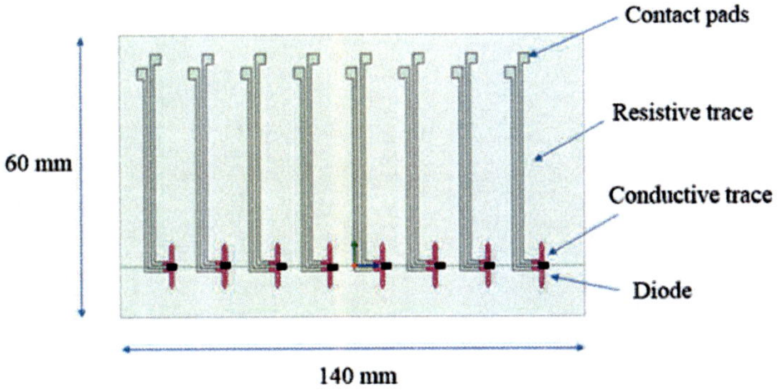

Fig. 7 Schematic of the linear sensor array

3 Measurements

3.1 Measurement Setup

Figure 7 shows the schematic of the linear sensor array that was printed using an in-house inkjet printer and commercially available resistive and conductive inks. The dipole arms and spacing between the dipole antennas were based on numerical simulations. The sensor array was fabricated using the inkjet printer on 0.2 mm thick photo paper. The dipole arm was printed using a silver ink. Transmission line part was printed using resistive ink. Zero bias Schottky diode (HSMS 8202, Avago technologies, USA) was attached to the printed transmission line and dipole arms using conductive epoxy. Figure 8 shows the measurement setup and the fabricated sensor array.

A spot focusing horn antenna was excited at 10 GHz and the incident locally plane wave illuminated a dielectric composite sample positioned on a non-metallic holder. A fiber reinforced glass composite (FR-4) of dimension 300 mm × 300 mm × 3 mm was tested for the electric field distribution adjacent to the sample. A defect of 20 mm diameter and depth of 0.5 mm was machined on one side of the sample. Sensor was placed behind the defective composite sample as illustrated in Fig. 9, and the sample was illuminated by the spot focusing horn at 10 GHz. The output voltage of each sensor element was amplified and recorded using a multiplexer and digital multi-meter controlled by a computer. Multiple trials were carried out for defect free and defective composite samples.

3.2 Sensor Measurements

The sensor voltage recorded for the defect free sample (calibration sample) was subtracted from the measurements for the defective sample. Figure 10 shows the

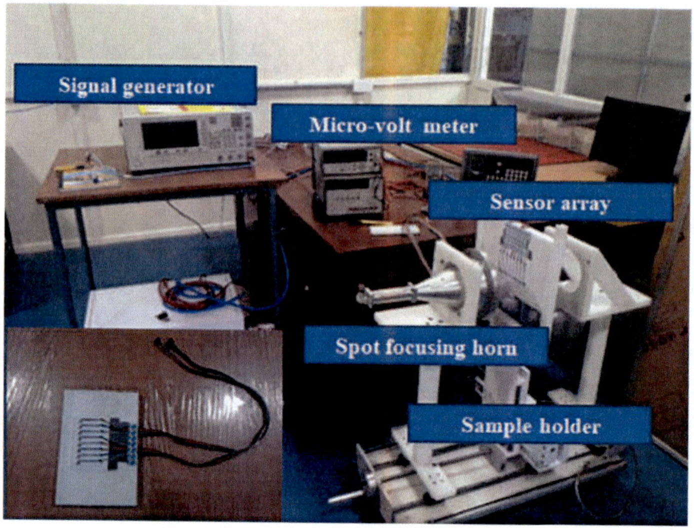

Fig. 8 Measurement setup and the fabricated E-field sensor array

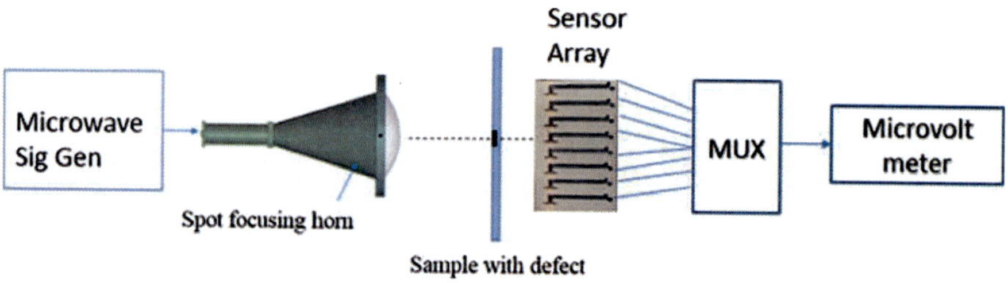

Fig. 9 Schematic illustration of the measurement setup

difference in the sensor array measurements gathered across the defect. It can be observed that the difference in the sensor voltage is relatively higher in the defective region (sensors 4 and 5) of the sample than the defect free regions (sensors 1–3, 6–8). The significantly higher voltage difference measured for sensors 4 and 5 is due to the dielectric variation introduced by the air pocket in the specimen. The nonzero difference in voltage measurements for sensors 1–3 and 6–8 is due to the localized spatial variation in the dielectric property of the defective specimen when compared to the measurements recorded for the defect free calibration sample. The noticeably small deviation in the voltage difference measured for three trials confirms that the nonzero voltage measured for sensors 1–3 and 6–8 is due to the localized spatial variation in the dielectric property of the composite. E-field sensor measurements gathered for three trials are repeatable and confirm that the proposed E-field sensor array could be used to measure perturbation in the electric field caused by defects in dielectric composites.

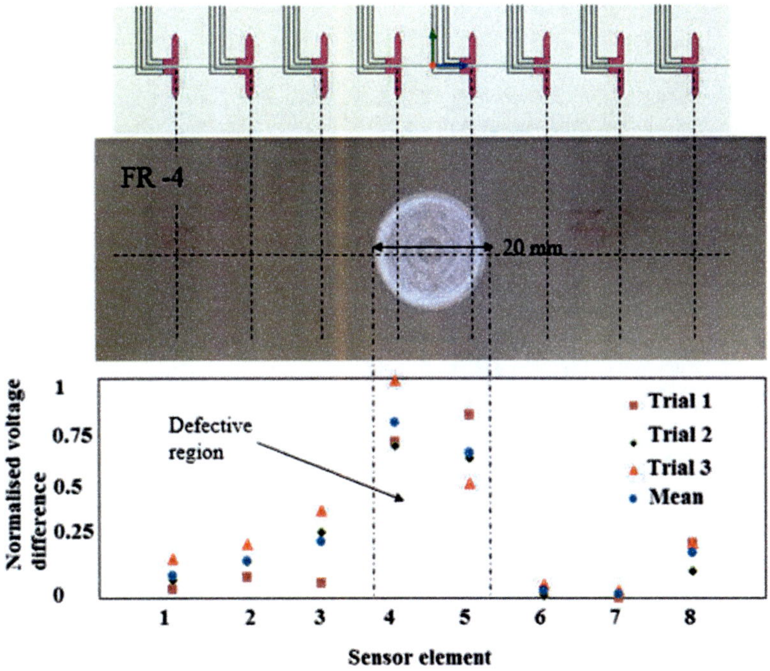

Fig. 10 Defective sample and the normalized voltage difference as a function of sensor element position

4 Conclusion

Design, fabrication, and operation of a linear array of electric field sensors were proposed for microwave NDE of dielectric composites. Numerical simulations were carried out to choose the dipole dimensions, electrical conductivity of the dipole arms, and spacing between the adjacent sensors such that the field perturbation introduced by the sensor is as low as possible. The utility of the sensor was demonstrated for a plane illumination wave maintained by a spot focusing horn antenna with 60 mm focal length and 28 mm focal spot size operating in X band (8–12 GHz) to avoid diffraction from sample edges from affecting the sensor measurements. A machined defect on FR-4 sample was probed using the sensor array. Array response for a defect free sample was used to calibrate the sensor voltage for defective sample. Preliminary results of the linear sensor array appear promising for microwave NDE of dielectric composites and require further investigation for sensor optimization.

References

1. Wang B, Zhong S, Lee T-L, Fancey KS, Mi J (2020) Non-destructive testing and evaluation of composite materials/structures: a state-of-the-art review In: Advances in mechanical engineering
2. Raj B, Jayakumar T, Thavasimuthu M (2002) Practical non-destructive testing. Woodhead Publishing, Cambridge, pp 1–27
3. Zhang N, Wang S, Ning S, Wang S (2017) Study on planar coil with multi-frequency stimulations applied to an eddy current non-destructive testing. In: 2017 20th international conference on electrical machines and systems (ICEMS), Sydney, pp 1–4
4. Halmshaw R (1996) Introduction to the non-destructive testing of welded joints. Woodhead Publishing, Cambridge, pp 40–47
5. Marc DN, Ram MN (2019) Microwave imaging of multilayered structures using ultrawideband noise signals. NDT and E Int 104:19–33
6. Li Z, Haigh A, Soutis C, Gibson A (2018) Principles and applications of microwave testing for woven and non-woven carbon fibre-reinforced polymer composites: a topical review. Appl Compos Mater 25(4):965–982
7. Zoughi R (2000) Microwave non-destructive testing and evaluation. Kluwer Academic Publishers, Dordrecht, Netherlands
8. Hajisaeid E, Dericioglu AF, Akyurtlu A (2018) All 3-D printed free-space setup for microwave dielectric characterization of materials. IEEE Trans Instrum Meas 67(8):1877–1886
9. Chung HE et al (2017) Active two-dimensional electric field imaging at very low frequencies. IEEE Sens J 17(21):7123–7130
10. Gimm Y, Ju Y, Kahng S, Lee Y, Lee S (2013) Quick SAR measurement system by 2D array E-field sensors. In: 2013 Asia-Pacific microwave conference proceedings (APMC), Seoul, pp 383–385
11. Abou-Elnour A, Abo-Elnor O, Mohamed EY, Ibrahim MM (2006) Modeling of a precise MOS sensor array system for 2D detection of magnetic fields. In: 2006 IEEE international conference on information acquisition, Weihai, pp 732–736
12. Bassen H, Smith G (1983) Electric field probes—a review. IEEE Trans Antennas Propag 31(5):710–718
13. HFSS (2019) Ansys Corporation. http://www.ansys.com/products/electronics/ansys-hfss. Last accessed 21 Nov 2019

Studies on Using FBG Sensor as a Tool for Identifying Corrosion in Pipelines

B. Arun Sundaram, S. Parivallal, and K. Kesavan

1 Introduction

Pipelines exist for the transport of crude and refined petroleum, fuels—such as oil, natural gas and biofuels. In India, the total length of gas pipeline network is 16232.6 km and currently 9509 km of pipelines are under construction. These pipelines running across several kilometers are prone to man-made and natural damages. Therefore, monitoring and maintaining its integrity and reliability become essential to the nation's energy requirements. Pipelines are susceptible to a different kinds of damage scenarios and defects due to aging. Some of the most common causes of failure in pipelines are corrosion, stress-cracks, seam weld cracks, material flaws, and externally induced damage by excavation equipment and also due to natural disaster such as earthquakes, landslides and also man-made disasters. Leakages of oil and gas from pipeline are dangerous for people and environment. Oil leakage leads to contamination of groundwater and explosions may be caused due to the leakage of gas. Leakage detection along the in-service pipeline network is an essential part of the inspection and monitoring activity. In the recent days, health assessment and real-time monitoring is gaining popularity among the researchers as it provides quality information on the behavior of real-life structures during its operation. Integrated health monitoring of structures with latest sensor and communication technology can improve reliability, system performance and safety against natural hazards, and intentional attack [1, 2]. To ensure continuous safe operation of the transmission pipelines, continuous monitoring or periodic assessment of the integrity of the pipelines becomes very important. In the field of pipeline monitoring and inspection, the main aim is to identify the exact locations that have defects, as

B. Arun Sundaram (✉) · S. Parivallal · K. Kesavan
Structural Health Monitoring Laboratory, CSIR-Structural Engineering Research Centre, Taramani, Chennai, India
e-mail: arunsundaram@serc.res.in

C. K. Mukhopadhyay and R. Mulaveesala (eds.), *Advances in Non-destructive Evaluation*, Lecture Notes in Mechanical Engineering,
https://doi.org/10.1007/978-981-16-0186-6_23

these pipeline run for several hundreds of kilometers. Once the location of defect is identified, the operators can take appropriate actions to prevent further damage and thereby avoid any accident. The development of multi-sensor tools is essential for identifying the location of defects and condition assessment of pipelines [3]. Many researchers have carried out investigations on the development of health monitoring techniques for buried pipelines using different sensing techniques [4]. Experimental investigations using electrical sensing methods, magnetic sensing, and acoustic emission were carried out to monitor structural damage in segmental concrete pipelines during earthquakes [5]. A lot of optical fiber sensor-based monitoring systems have been developed for continuous measurement and real-time assessment of various engineering structures such as bridges, buildings, tunnels, pipelines, wind turbines, railway bridges, and geotechnical structures [6, 7]. The optical fiber sensors have been widely used in life-cycle monitoring of civil infrastructure and are capable of measuring various structural parameters such as strain, temperature, acceleration, deflection/displacement, cracks, and corrosion. Fiber optic sensors offer a relatively new technology for the monitoring and evaluation of pipeline integrity and performance [8, 9]. Fiber optic sensors are used for measuring various parameters and are cost effective. Fiber optic sensor's high resolution, insensitivity to electromagnetic interference, real-time monitoring capabilities, and relatively low cost provide potential benefit to the industry.

Fiber Bragg gratings are latest class of fiber optic sensors made by exposing the core of a single-mode fiber to intense laser light. The exposed portion of the fiber core will have a permanent change in the refractive index compared to other regions and is referred as Bragg grating. These grating portions with a change in refraction will reflect some part of light passing through it. This reflection is dependent on strain and temperature variations. The pipeline under operation are subjected to hoop stresses due to the operating pressure inside the pipeline. The magnitude of the hoop stress will depend on the operating pressure inside the pipeline and wall thickness of the pipeline. Internal corrosion in oil and gas pipelines will lead to reduction in the wall thickness of the pipelines. The operating pressure inside the pipeline is almost constant along the entire stretch of the pipeline. Hence, the change in wall thickness of the pipeline due to internal corrosion will have higher hoop stresses developed in the pipeline for the same operating pressure [10, 11]. This paper presents, in detail, the studies carried out toward development of a non-destructive method to monitor corrosion by measuring the hoop strain variations using FBG-based sensors. Finite element simulation studies were also carried out for validation of hoop stress variation in pipelines due to corrosion.

2 Theory of Fiber Bragg Grating Sensors

Fiber Bragg grating (FBG) sensors are the most promising optical fiber sensors based on the state-of-the-art technologies. The advantages of FBG over conventional electrical resistance sensors are small size and lightweight, immune to electromagnetic noise, compatibility with new composite materials like glass or carbon fiber reinforced composite, etc. Bragg gratings are periodic structures created in single-mode optical fibers by subjecting them to intense ultraviolet radiations. The refractive index at these locations will vary with respect to other locations of the optical fiber core. These Bragg gratings act as wavelength reflectors and make them suitable for sensing applications. At each periodic refraction change, a small amount of light is reflected. All the reflected light signals integrate coherently to one large reflection at a specific wavelength in the grating period. This is known as the Bragg condition, and the wavelength at which this reflection happens is referred to as the Bragg wavelength. During FBG writing process, the intensity of the ultraviolet illumination is made to occur in a periodic fashion along the fiber core. At a particular high power level, local variations are created in the core having a periodic change in the local refractive index. This change in refractive index (RI) created is permanent and sensitive to a number of physical parameters, such as pressure, temperature, strain, and vibration. Thus, by monitoring the resultant changes in reflected wavelength, FBG can be used for sensing applications to measure various physical quantities.

Fiber optic Bragg grating sensor response arises from two sources (Fig. 1), namely the induced change in pitch length (Λ) of the grating and the perturbation of the effective core refractive index (n_{eff}). The wavelength of the reflected spectrum band is defined by the Bragg condition

$$\lambda_{\text{B}} = 2n_{\text{eff}}\Lambda \tag{1}$$

When an FBG is strained, the Bragg wavelength (λ_{B}) changes and the relation is given by

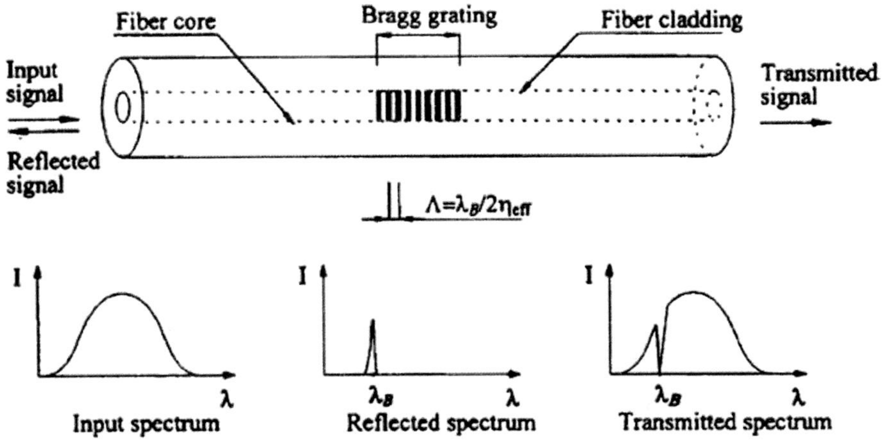

Fig. 1 Reflection inside a fiber Bragg grating sensor

$$\text{Strain,}\ \varepsilon = \frac{\Delta\lambda_B/\lambda_B}{(1 - p_e)} \tag{2}$$

where $\Delta\lambda B$ = change in Bragg wavelength ($\lambda - \lambda B$).

λB = initial Bragg wave length.

λ = Bragg length after straining/loading.

p_e = effective photo elastic constant for the fiber (~0.22).

3 Finite Element Simulation Studies

Finite element simulation studies were carried out to understand the hoop strain variation in the pipeline subjected to internal pressure. A pipeline model is created in ABAQUS, and pressure was applied on the inner surface. The pipeline is modeled using three dimensional deformable solid elements. The material properties of steel are given as input to the developed model. The developed pipeline model was meshed, and the internal pressure was applied to the pipeline. The hoop strain and pipeline thickness are inversely proportional. As the thickness reduces due to corrosion, the hoop strain increases. The hoop strain variation on the wall of the pipeline is shown in Fig. 2. Corrosion in pipeline will lead to reduction of the pipe wall thickness. Hence, the model was modified with varying thickness, and pressure was applied. Three pipeline models with varying thickness were created and are merged along the edges to form a single pipeline model. The hoop strain variation for varying thickness of pipeline is shown in Fig. 3. The model was modified with a reduction in thickness for a particular portion of the pipeline, and the hoop strain variation is shown in Fig. 4. From the finite element investigations, it can be seen that the operating pressure

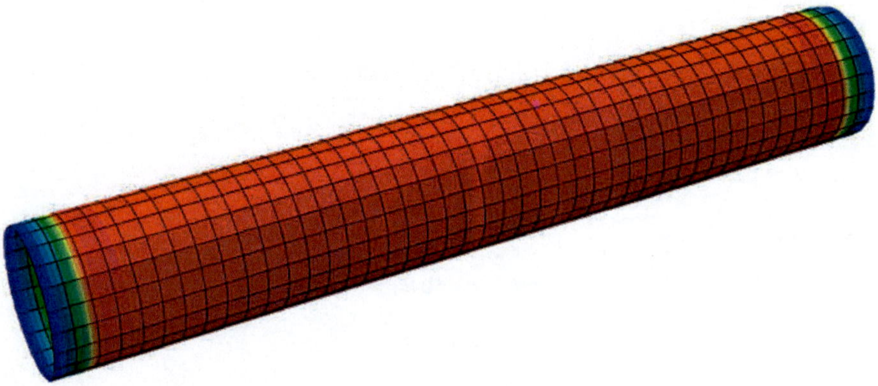

Fig. 2 Hoop strain variation in pipeline with uniform wall thickness

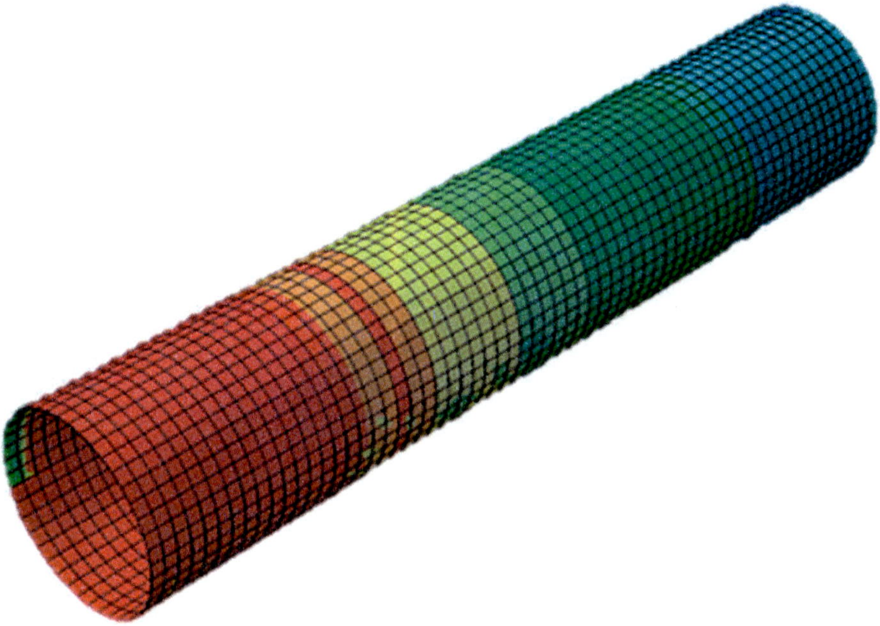

Fig. 3 Hoop strain variation in pipeline with varying wall thickness

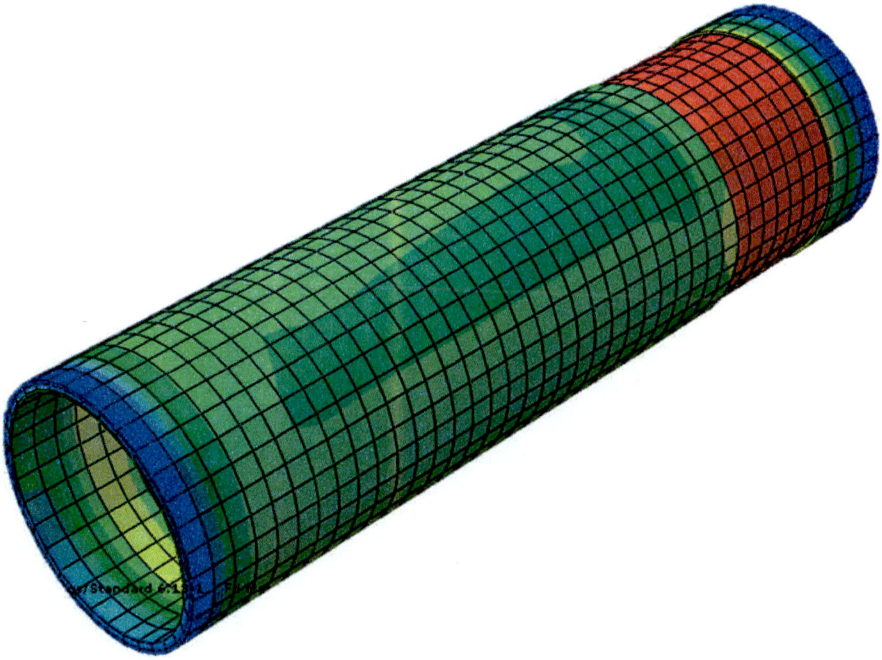

Fig. 4 Hoop strain variation in pipeline with varying thickness at a particular region

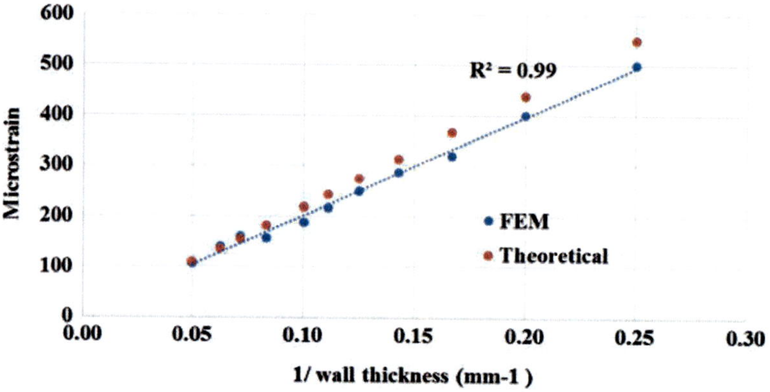

Fig. 5 Hoop strain variation for various pipe wall thickness

inside the pipeline creates a hoop strain on the pipe wall and thickness changes in the pipeline wall will have variation in the hoop strain measured. The hoop stress at any cross section of a pipeline is directly proportional to its internal pressure and diameter and inversely proportional to the thickness. The diameter of the pipeline remains constant throughout the pipeline network, and the operating pressure is also maintained at a constant level. Hence, the hoop stress varies with respect to the wall thickness of the pipeline. Thus, by measuring the hoop strain in a pipeline, we can ascertain the initiation of corrosion in a pipeline. Hence, by measuring the hoop strain variation in the pipeline, we can identify the initiation of corrosion. The hoop strain variation for different thickness of the pipeline was obtained from the finite element analysis and plotted in Fig. 5. From the plot, it can be seen that the hoop strain is inversely proportional to the pipe wall thickness.

4 Hoop Strain Measurement Using FBG Sensors

From the finite element studies, it can be seen that corrosion in pipeline leads to hoop strain variations and the hoop strain is inversely proportional to the wall thickness. By measuring the hoop strain variation, we can easily identify the corrosion activity in pipelines. Hence, for measuring the hoop strain in the pipeline, experimental investigations were carried out using FBG sensors in pipelines. A pipe specimen with a diameter of 150 mm, length as 600 mm and wall thickness as 5 mm is chosen and both ends are closed with flange plates. A perforated GI sheet of 1 mm thick and diameter 50 mm is fixed inside the pipe. The pipe specimen is fitted with an inlet for filling water. A pressure gage is fixed on the pipe specimen to monitor the pressure. It is planned to conduct accelerated corrosion test by filling 3.5% NaCl solution. In this accelerated corrosion test, steel pipe specimen act as anode and perforated GI sheet act as a cathode (Fig. 6). In order to have a controlled corrosion on the pipeline, the

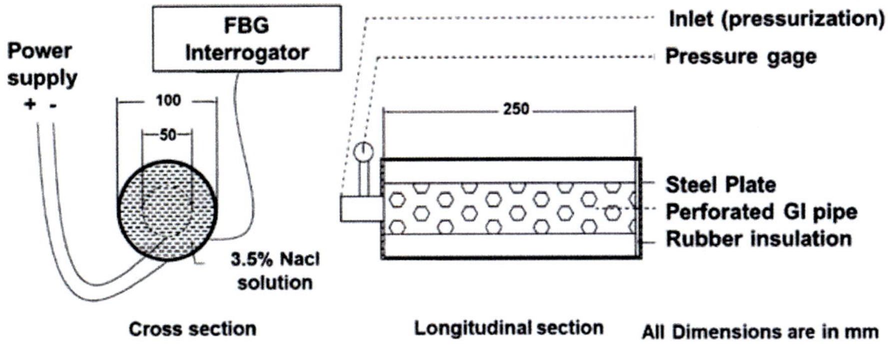

Fig. 6 Schematic representation for accelerated corrosion in pipeline

inner surface of the pipe was painted prior to testing with epoxy by leaving middle 50 mm band. The pipe specimen was instrumented with one FBG strain sensor (FBG 1) circumferentially at the 50 mm band portion, and another FBG sensor (FBG 2) was instrumented at other location (painted). The FBG sensors were instrumented after proper surface preparation on the pipeline. The sensors were bonded by following standard procedures. The sensors were connected to the data acquisition system. The pipe specimen is filled with 3.5% NaCl solution, and known pressure (P) is applied (Fig. 7). The hoop strain was measured from FBG sensors at regular intervals. By keeping this constant pressure, accelerated corrosion test is carried out.

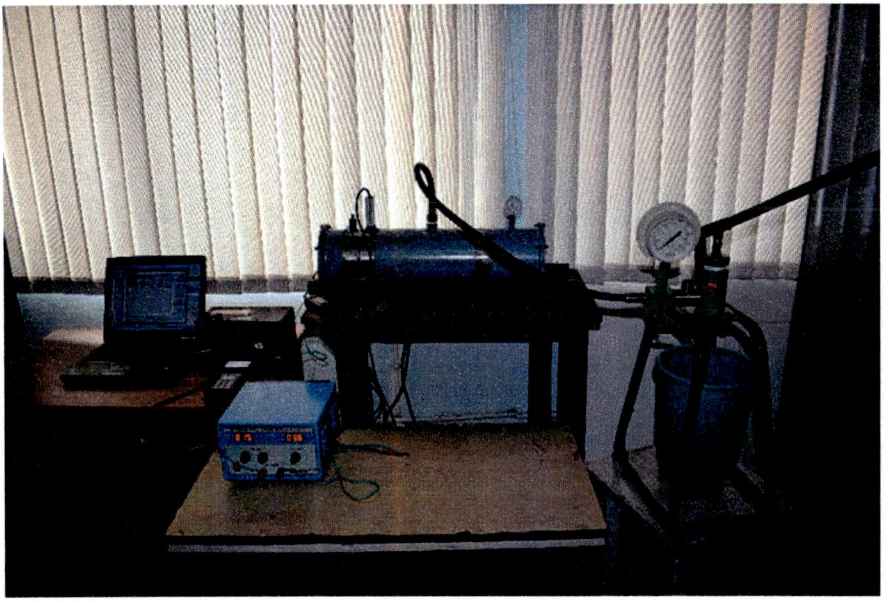

Fig. 7 Test setup for accelerated corrosion

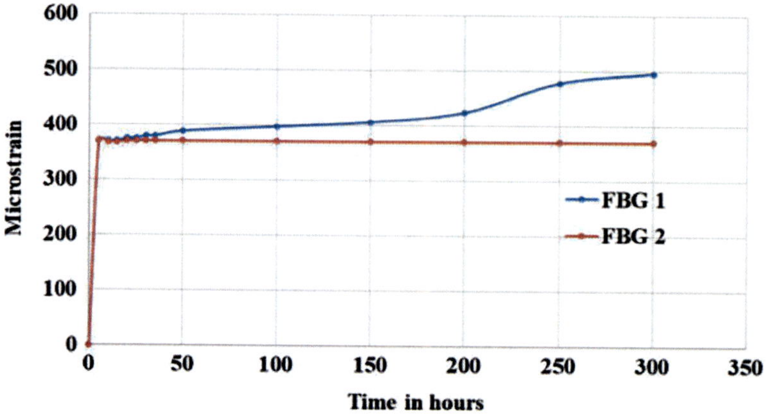

Fig. 8 Hoop strain response for accelerated corrosion

Strain response from instrumented FBG strain sensors is monitored during accelerated corrosion test. The applied pressure is maintained throughout the experiment. Additionally, electrical strain gages were also instrumented for comparing the hoop strain measured from FBG sensors. Strain response was measured from all the instrumented sensors at regular intervals (Fig. 8). From the measured responses, it can be seen that due to the corrosion effect, the strain response from FBG1 at corrosion location increases, whereas the strain response from the other FBG sensor is constant. This sudden increase in strain indicates the initiation of corrosion in the pipeline. Further studies are planned by relating the rate of corrosion, pipe wall thickness, and strain change.

5 Conclusion

Experimental studies carried out on hoop strain monitoring using FBG sensors during accelerated corrosion test is described in this paper. Finite element studies were also carried out to verify the effect of corrosion on the hoop strain variation. From the studies, it can be seen that bare FBG sensors can be used for monitoring the buried oil and gas pipelines against corrosion. Thus, FBG sensors act as a NDT tool for monitoring the initiation and propagation of corrosion. Further study needs to be carried out by relating the hoop strain variations and corrosion rate for efficient real-time monitoring of pipelines.

References

1. Balageas D, Claus-Peter F, Alfredo G (2007) Structural health monitoring. ISTE Ltd., London

2. Kesavan K, Ravisankar K, Parivallal S, Sreeshylam P, Sridhar S (2007) Fiber Bragg Grating (FBG) sensors for health monitoring of civil engineering structures. J Non-Destruct Test Evaluat 6(3):9–21
3. Costello SB, Chapman DN, Rogers CDF, Metje N (2007) Underground asset location and condition assessment technologies. Tunn Undergr Sp Technol 22:524–542
4. Arun Sundaram B, Kesavan K, Parivallal S (2018) Recent advances in health monitoring and assessment of in-service oil and gas buried pipelines. J Inst Eng India Ser A 99:729
5. Pour- M (2011) Using electrical, magnetic and acoustic sensors to detect damage in segmental concrete pipes subjected to permanent ground displacement. Cement Concr Compos 33:749–762
6. Ye XW, Su YH, Han JP (2014) Structural health monitoring of civil infrastructure using optical fiber sensing technology: a comprehensive review. Sci World J
7. Tennyson RC, Morison WD (2006) Long gage-length optic sensors for monitoring pipeline integrity. In: Proceedings of SPIE.
8. Tapanes E (2001) Fibre optic sensing solutions for real-time pipeline integrity monitoring. Australian Pipeline Indus Assoc Natl Conv 3:27–30
9. Rajeev P, Kodikara J, Chiu WK, Kuen T (2013) Distributed optical fibre sensors and their applications in pipeline monitoring. Key Eng Mater Trans Tech Publ 558:424–434
10. Wang Y, Zhang ZL, Zhang HX, Li MH (2008) Natural gas pipeline corrosion failure analysis. Int J Chem Equip Technol 29(5):53–54
11. Ren L, Jia Z-G, Li H-N, Song G (2014) Design and experimental study on FBG hoop-strain sensor in pipeline Monitoring. Opt Fib Technol 20:15–23

Numerical Modelling of Remote Field Eddy Current Testing of Steam Generator Tubes

Ranjani Jayaraman, J. Selva Solomon, N. Sridhar, Chitti Venkata Krishnamurthy, and Kavitha Arunachalam

1 Introduction

Steam generator (SG) used in prototype fast breeder reactor (PFBR) in IGCAR is one of the most critical components where sodium and water exist close by. They are vertically placed at a height of 26 m, and there are eight SGs in the PFBR where each SG has 557 tubes [1]. It operates under radioactive, high pressure, temperature, and sodium environment. Hence, proper consideration on operating conditions has to be inferred in selecting the tube material. Modified 9Cr–1Mo steel (with the addition of stabilizing elements) is chosen as the material for SG tubes because of its excellent creep and fatigue resistant properties [1]. Periodic testing and evaluation are necessary to ensure structural stability of the SG tube and to prevent failure and damage of SG [2]. Among the available electromagnetic non-destructive evaluation (ENDE) techniques, eddy current testing is widely used for detection of defects in electrically conducting specimen up to 6 mm thickness. For test specimens of thickness greater than 10 mm, optimization of exciter coil and test frequency are necessary for defect detection [3]. Evaluation of magnetic materials is different from non-magnetic materials as magnetic permeability is one of the primary factor influencing the skin depth [4]. During tube manufacturing process, pre-service inspection is done by saturation eddy current technique where magnetization and demagnetization of the tube is feasible. Once installed, due to practical limitations magnetization and demagnetization of individual tubes is not feasible. An alternative approach has to be used

R. Jayaraman · K. Arunachalam (✉)
Department of Engineering Design, Indian Institute of Technology Madras, Chennai, India
e-mail: akavitha@iitm.ac.in

J. S. Solomon · N. Sridhar
Electronics and Instrumentation Division, EIG, IGCAR, Kalpakkam, Tamil Nadu, India

C. V. Krishnamurthy
Department of Physics, Indian Institute of Technology Madras, Chennai, India

C. K. Mukhopadhyay and R. Mulaveesala (eds.), *Advances in Non-destructive Evaluation*, Lecture Notes in Mechanical Engineering,
https://doi.org/10.1007/978-981-16-0186-6_24

for the inspection of SG tubes [1]. Remote field eddy current (RFEC) technique is widely used for inspection of ferromagnetic tubes since conventional eddy current technique has limitations to magnetic permeability variations and skin depth [1, 4].

RFEC is a low-frequency technique which uses two coils for inspection: one to induce magnetic field and another to detect the defects in the tube [5–7]. The exciter coil generates a strong uniform magnetic field and the pickup coil positioned apart gathers the magnetic field in the remote zone which is typically 2–3 tube diameters away from the exciter [4]. The magnetic field behaviour inside the tube can categorized as follows: (i) direct zone: due to interaction of exciter coil with the tube, (ii) remote zone: secondary magnetic field is created due to the generation of eddy currents which extends radially and axially and coupling between the coils take place indirectly (iii) transition zone: interaction between direct and indirect field occurs where the phase angle changes abruptly, thereby no useful information can be acquired [4, 5]. The pickup coil is positioned in the remote region to avoid direct coupling field from the exciter coil [4]. RFEC is a through wall technique where the magnetic field passes through the tube twice: from inside near the exciter coil and when it flows from outside towards the pipe near the pickup coil [8, 9]. In addition to that, RFEC detects the inner wall and outer wall defects with equal sensitivity [8, 10]. Dual peaks are obtained in the RFEC signal response: primary peak when the defect is under the pickup coil and the secondary peak when the defect is under the exciter coil [11, 12].

In this paper, numerical modelling of RFEC inspection of SG tubes was carried out using finite element method (FEM)-based software ANSYS Maxwell. It is low-frequency electromagnetic field simulation software for analysing and solving 2D and 3D problems using electrostatic, magnetostatic, eddy current and transient solvers. Velocity effects can be incorporated using transient solver which aids in signal characterization for defect detection in high speed inspection systems [13]. The organization of the work is as follows: Sect. 2 presents the 3D numerical model for RFEC simulations and simulation results for estimating the electrical properties of the SG tube with machined defects [14]. Numerical modelling to study the influence of magnetic field coupling between the neighbouring SG tubes is reported in Sect. 3. Simulations are reported for a 2D groove in the centre SG tube inspected using RFEC probe in the centre tube, and a 2D groove in the adjacent tube and RFEC probe in the centre tube.

2 Numerical Modelling of RFEC Technique

2.1 Model Construction

The SG tube modelled in this study has an outer diameter of 17.2 mm and wall thickness of 2.3 mm. The exciter and pickup probes are bobbin coils wound on a

Table 1 Modelling parameters for RFEC inspection

S. No.	Parameter	Exciter coil	Pickup coil	SG tube
1	Inner diameter (mm)	5.9	5.9	12.6
2	Outer diameter (mm)	10.45	10.45	17.2
3	Height (mm)	8	5	200
4	Inter-coil spacing (centre–centre) mm	42.7		

former. The modelling parameters used for RFEC inspection of SG tube are listed in Table 1.

RFEC measurements gathered from a SG tube with ASTM standard machined defects were used to validate the model and to estimate the unknown electrical properties of the SG tube. The standard describes distinct artificial flaws of varying sizes to assess flaw detection sensitivity [15]. Among those, two flaws were chosen for this study. Table 2 lists the defects modelled in this work and their dimensions. Figure 1a shows the 3D model of the SG tube with defects A and B inspected using the RFEC probe. To avoid magnetic field perturbation at tube edges, defects were moved instead of the RFEC probe. Figure 1b shows the mesh plot used to capture the weak magnetic field perturbation caused by the pinhole defect. The SG tube was assigned skin depth mesh with at least 4 layers, whereas the coils and defects were assigned length-based meshes. A total of 741,775 tetrahedra were generated for the pinhole, defect A. The boundary of the air region surrounding the SG tube was kept at a distance 10 times the coil outer diameter. Neumann boundary condition was assigned on the faces of the air box surrounding the SG tube [13]. Simulations were carried out for 50 mA excitation current applied to the exciter coil at 800 and 1100 Hz. The solver computes induced eddy current and displacement currents in conducting medium. Eddy current effects were applied to the SG tube, and displacement currents were neglected in the model. The numerical simulations were carried out on a computer with the following configuration: Intel Xeon E5 v4 (dual), 2.4 GHz, 64 GB RAM. The coil inductance in air was measured as 156 μH for the exciter coil and 350 μH for the pickup coil. The number of turns in each coil was optimized numerically by enclosing the coils in an air box in the absence of the SG tube. The number of turns in the exciter and pickup coil were varied to select a combination that yielded coil inductances close to the measured values. Thus, the number of turns in the exciter and pick up coils were estimated as 190 and 250 turns, respectively.

Table 2 ASTM standard RFEC tube defects [15]

Parameters	Defect A	Defect B	Defect C
Type	Pinhole	Flat milled flaw	2D groove
Dimensions (mm)	2.3 (diameter)	8.6 × 4.3	15.88
Depth (mm)	2.3 (100%)	1.15 (50%)	0.46 (20%)

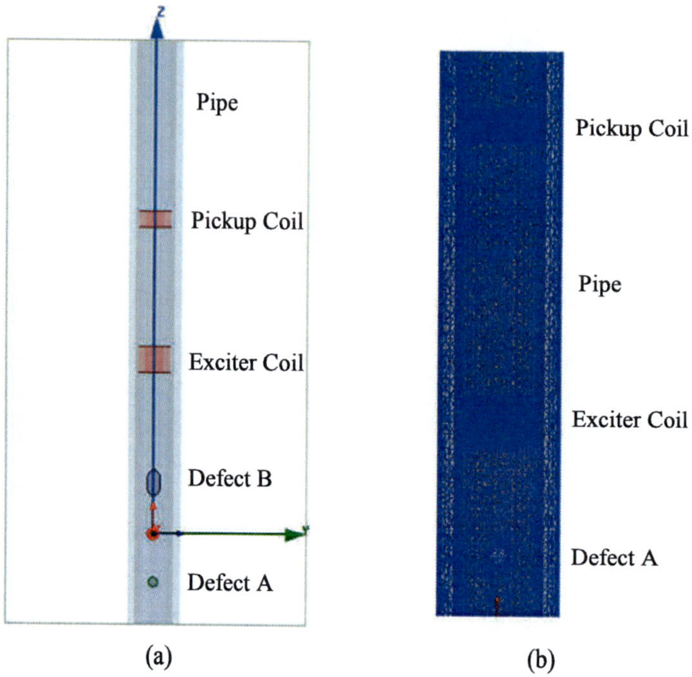

Fig. 1 Numerical model of SG tube. **(a)** RFEC probe and defect positioning, **(b)** user defined mesh plot

2.2 *Numerical Simulations*

The defect was moved across the exciter and pickup in steps of 1 mm for a span of 110 mm. The electrical conductivity σ of the tube was varied from 1.5 to 2.5 MS/m in steps of 0.5 MS/m and the relative magnetic permeability μ_r was varied over 30–40 in steps of 5. The range for the SG tube material property was chosen based on the alloy composition. Numerical simulations were carried out for 9 combinations of tube material properties, and voltage induced in the RFEC coil was calculated for 800 and 1100 Hz excitation frequency. Due to the non-uniform lift-off in measured signals, the experimental data was de-trended for comparison with numerical results.

Figure 2 shows the impedance plane plot of the voltage induced in the pickup coil of the RFEC probe for defect B and SG tube of varying $\mu_r = 30$ to 40 and $\sigma = 1.5$ to 2.5 MS/m. Experimental data is superimposed on the simulation data for comparison. With increase in μ_r and σ, the RFEC signal rotates towards the second quadrant at both excitation frequencies. It should be noted that the measurements were gathered at 20 mm/s probe velocity. As the permeability increased, the amplitude of the RFEC signals reduced smaller and the signal rotated towards the second quadrant. As measurement data was noisier at 800 Hz than at 1100 Hz, comparison between numerical simulations and measurements was done at 1100 Hz. It can be

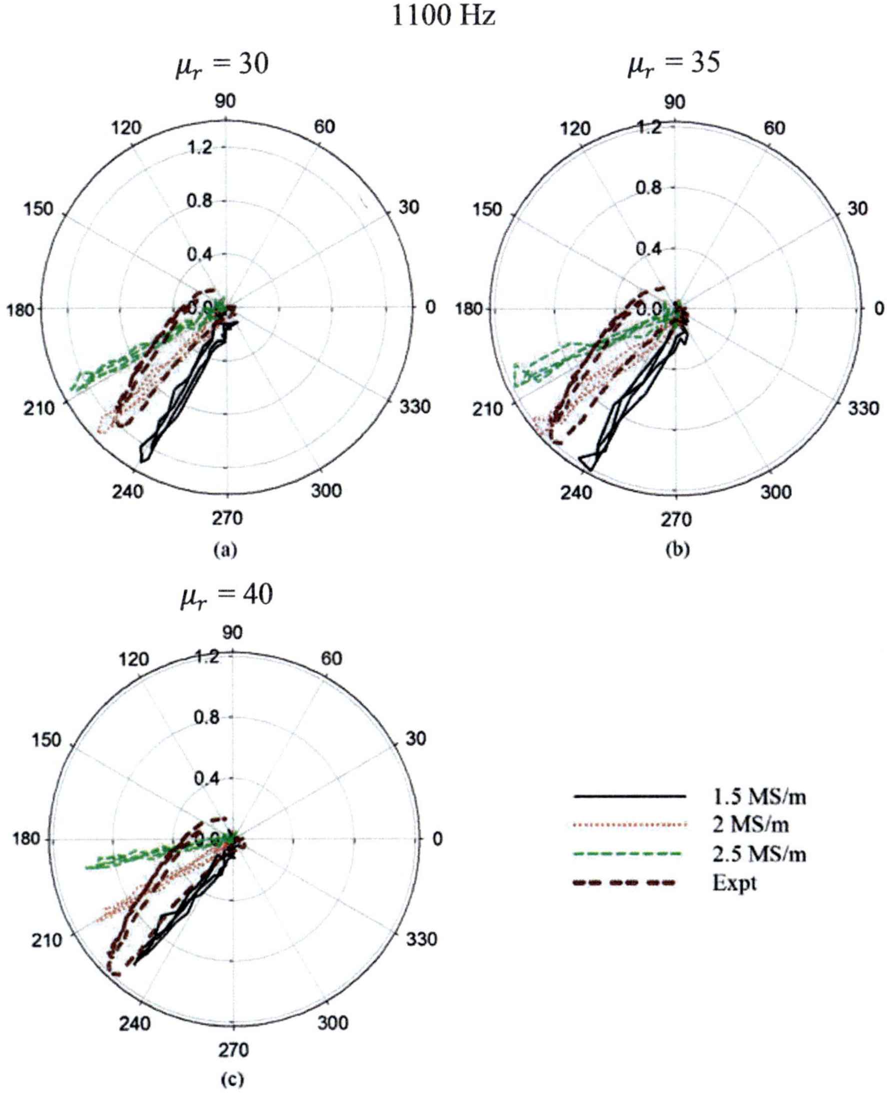

Fig. 2 Polar plots for defect B at 1100 Hz for **(a)** $\mu_r = 30$, **(b)** $\mu_r = 35$, **(c)** $\mu_r = 40$ and varying electrical conductivity

observed that the agreement between numerical and measurement data is good for $\sigma = 2$ MS/m and $\mu_r = 30$ at 1100 Hz.

Figure 3 shows the amplitude of the RFEC signals for defect A at 1100 Hz for $\sigma = 1.5$ to 2.5 MS/m and $\mu_r = 30$. The amplitude of the experimental signals is in good agreement with the numerical signals. From Figs. 2 and 3, the electrical conductivity (σ) and relative magnetic permeability (μ_r) of the SG tube were concluded as 2 MS/m and 30, respectively.

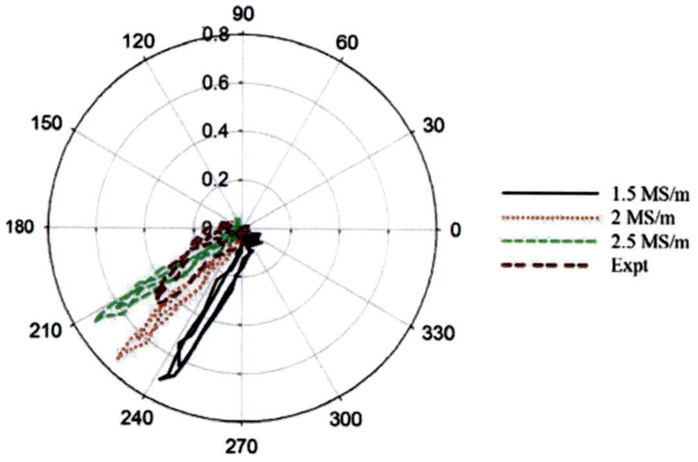

Fig. 3 Polar plot for defect A at 1100 Hz for $\mu_r = 30$

3 Influence of Magnetic Coupling

SG has a number of seamless tubes packed closely with a pitch of 32.2 mm (centre–centre distance) where each tube is surrounded by 6 adjacent ferromagnetic tubes. The extent of electromagnetic interactions between the surrounding tubes was numerically investigated considering a SG tube with and without six neighbouring tubes. There is a probability of detecting noise due to electromagnetic interferences between the adjacent tubes due to which defects in the tube could be missed. There is also the possibility of detecting defects from adjacent tubes which gives false alarm leading to incorrect location and quantification of defects. Numerical simulations were carried out to analyse these two cases. Numerical simulations were carried out for the OD groove (defect C) in the ASTM standard as the mesh density is relatively low compared to the 3D defects (A, B) in Table 2.

3.1 Selection of Optimum Frequency

In order to study magnetic field coupling between tubes, simulations were carried out for ASTM standard defect C for the chosen tube properties. Simulations were carried out for varying excitation frequency from 100 to 4000 Hz in steps of 100 Hz to identify the optimum frequency for which the induced coil voltage was the maximum. Figure 4 shows the voltage induced in the pickup coil for defect C and the difference between the peak amplitude of the defect and the defect free regions [1] are plotted. Simulation results show the maximum voltage at 2500 Hz for defect C (20% defect depth). Simulation results are also shown for defect free tube for which the maximum induced voltage was calculated at 1400 Hz.

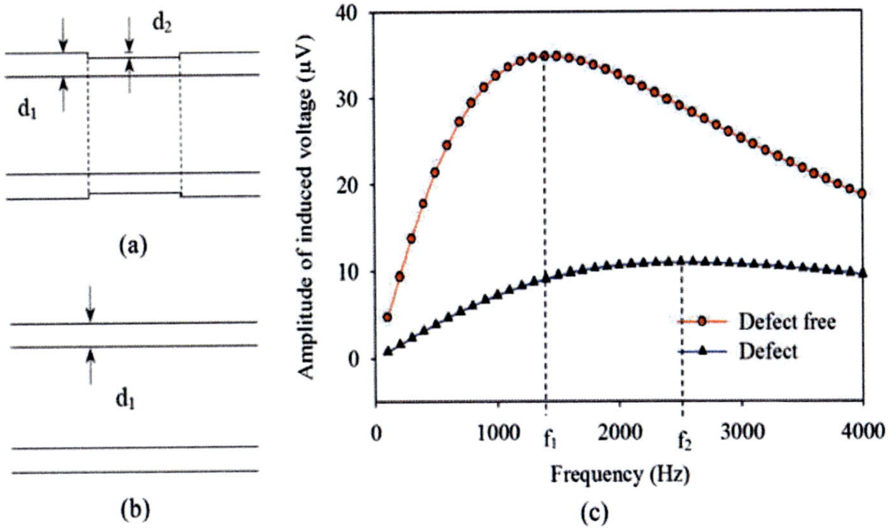

Fig. 4 SG tube cross section **(a)** with and **(b)** without OD groove, i.e. defect C, and **(c)** optimum frequency plot for defect C and defect free case for $\mu_r = 30$ and $\sigma = 2$ MS/m

3.2 Modelling Adjacent SG Tubes

To understand magnetic field coupling between the neighbouring ferromagnetic SG tubes, simulations were carried out for defect C in the centre tube and defect C in the one of the six adjacent tubes. Voltage induced in the pickup coil was calculated for the RFEC probe in the centre SG tube for both cases. Simulations were carried out at 2500 Hz for defect depths of 20% (0.46 mm), 30% (0.69 mm), and 40% (0.92 mm) of the wall thickness. Simulation was also carried out at 1400 Hz. Figure 5a shows the numerical model of seven SG tubes. Magnetic field strength induced in the adjacent SG tubes was calculated to be $\leq$15% of the maximum field strength induced in the centre tube with the RFEC probe. Skin depth mesh was assigned to the centre tube with the defect, and the coil setup and length-based mesh was assigned to the adjacent defect free tubes restricting the mesh element size to 3 mm. The mesh generated in the model is shown in Fig. 5b. The number of mesh elements was 4,54,984 for one position of the defect.

Figure 6 shows the simulation results for defect C of varying depths in the centre tube at 1400 and 2500 Hz. Simulations results are also shown for single SG tube with defect C, i.e. without the six neighbouring SG tubes. Single pipe in Fig. 7 denotes the simulation results for single SG tube and 6 pipes denote the results in the presence of six adjacent tubes. It can be observed that induced coil voltage for defect C increases with increase in material loss at both excitation frequencies. It is also evident that the coil voltage is similar for single and 6 pipe models except for a phase change of 1.2° at 2500 Hz which is negligibly small.

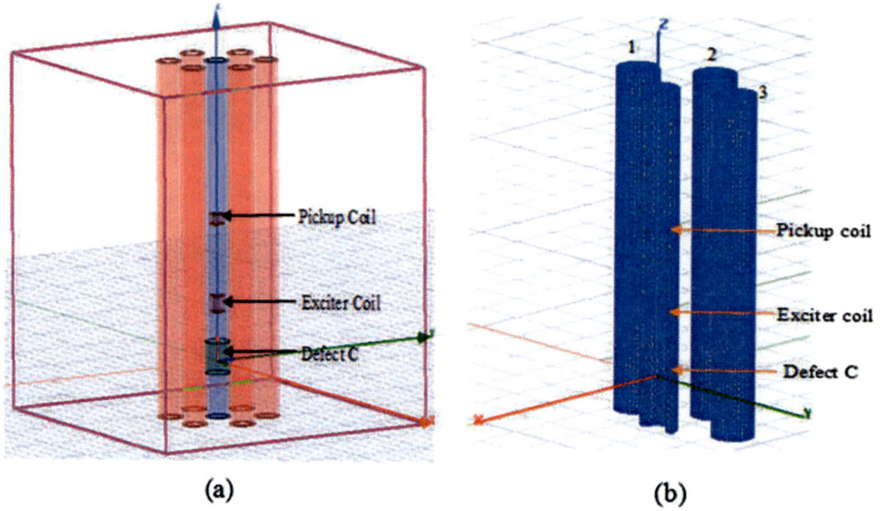

Fig. 5 Numerical model of SG tube with neighbouring tubes. **(a)** 3D model with RFEC probe in the centre tube, **(b)** mesh plot for the defect position shown in Fig. 5**a**

Figure 7 shows the simulation results for 20% defect C in one of the six adjacent SG tubes and RFEC probe in the centre SG tube. It is obvious that the pickup coil does not show change in the induced voltage at 1400 and 2500 Hz excitation frequencies. The noise induced in the pickup coil is negligible when the defect is located in the adjacent tube compared to the signal induced when the defect and coil are positioned in the same tube. Thus, it is concluded that the presence of defect in the adjacent tube will not interfere with RFEC measurements gathered in the nearby tube for the inter-tube spacing and SG tube properties investigated in this work.

4 Conclusion

A 3D model of the SG tube with RFEC probe was presented. As electrical property of ferromagnetic tube is likely to vary with time and exposure to magnetic field, the electrical properties of the ferromagnetic SG tube were estimated using the 3D tube model with machined defects. The estimated tube electrical properties were used to calculate RFEC signal induced in the pickup coil for three defects. Subsequently, 3D numerical simulations were carried out to investigate the influence of the adjacent ferromagnetic tubes on RFEC inspection. Simulations for defect free tube arrangement showed weak induced magnetic field strength in the adjacent tubes and no measurable difference was observed in the pickup signal.

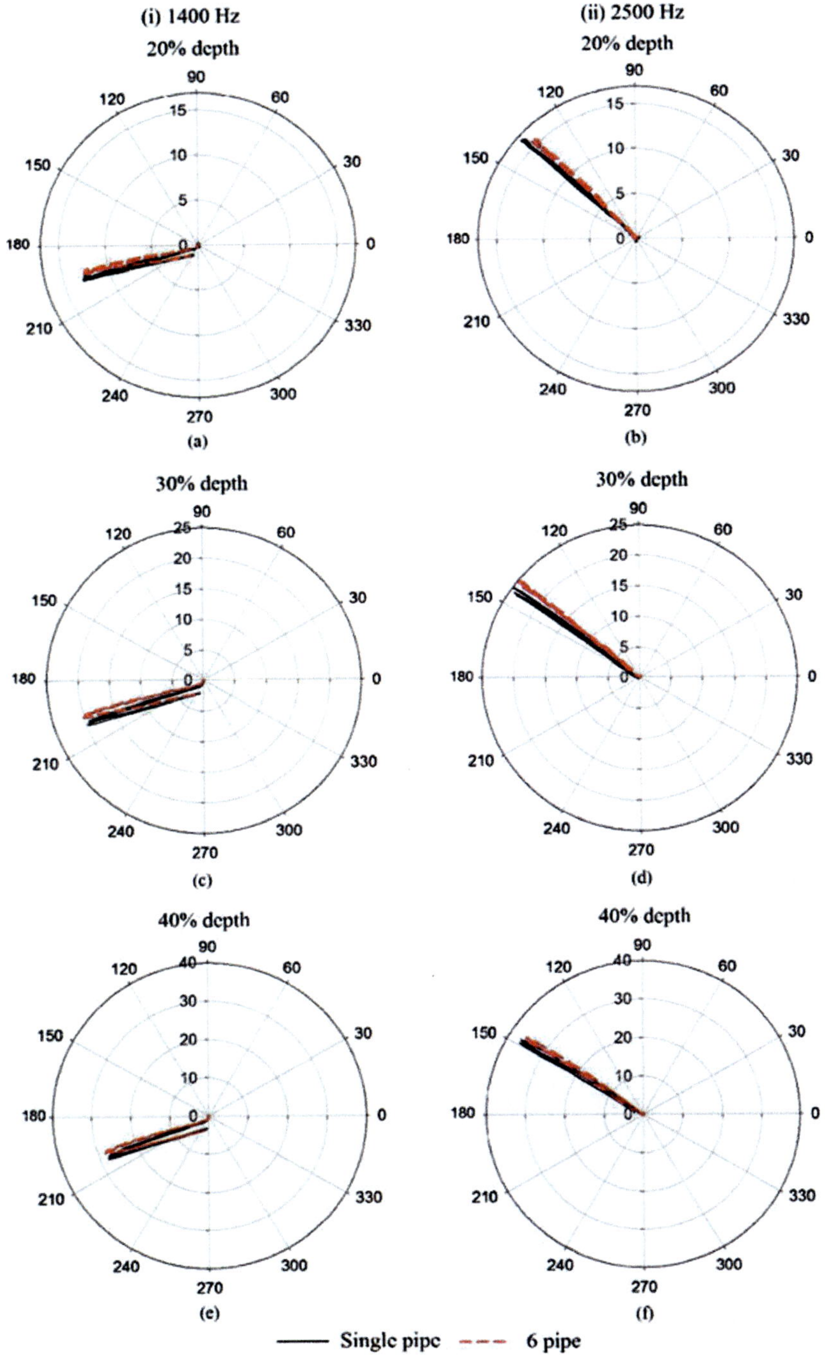

Fig. 6 Impedance plane plot of RFEC coil voltage for defect C in centre pipe, **(a)** 20% depth at 1400 Hz, **(b)** 20% depth at 2500 Hz, **(c)** 30% depth at 1400 Hz, **(d)** 30% depth at 2500 Hz, **(e)** 40% depth at 1400 Hz, **(f)** 40% depth at 2500 Hz

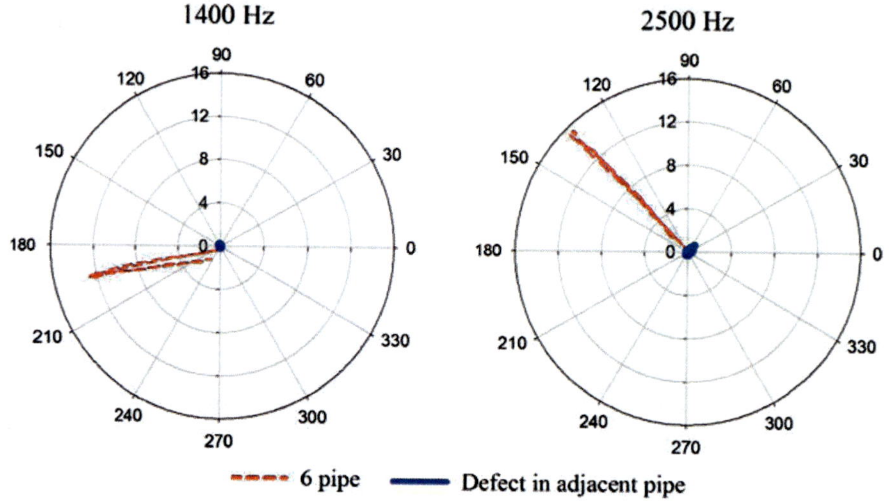

Fig. 7 Polar plot for defect C (20% depth) at **(a)** 1400 Hz, **(b)** 2500 Hz

References

1. Thirunavukkarasu S (2013) Remote field eddy current based approaches for high sensitive detection of defects in ferromagnetic steam generator tubes
2. Blitz J (1998) Electrical and magnetic methods of nondestructive testing, vol 31, no 1
3. M. R. P. (2017) Design of an eddy current probe for sub-surface defect detection in electrically thick conducting specimens
4. Janoušek L, Marek T, Gombárska D, Čápová K (2006) Eddy current non-destructive testing of magnetic tubes. J Electr Eng 57(8 SUPPL):126–129
5. Huang S, Wang S (2016) New technologies in electromagnetic non-destructive testing
6. Fischer G, Kocher J, The science behind eddy current and remote field testing, pp 1–6
7. Tao A, Zhang W, Wang Z, Luo Q (2014) Design on forward modeling of RFEC inspection for cracks. In: Proceedings of 2014 international conference information science, electronics and electrical engineering. ISEEE 2014, vol 1, pp 579–584
8. Sullivan S, Atherton DL, Schmidt TR (1989) Comparison of conventional, through-wall and remote field eddy current techniques. NDT Int 22(4):203–208
9. Falque R, Vidal-Calleja T, Dissanayake G, Miro JV (2017) From the skin-depth equation to the inverse RFEC sensor model. In: 2016 14th international conference on control, automation, robotics. Vision, ICARCV 2016
10. Nath Lord SW, Sun YS. Theoretical and experimental studies of the remote field eddy current effect
11. Luo Q, Shi Y, Wang Z, Zhang W, Ma D (2017) Method for removing secondary peaks in remote field eddy current testing of pipes. J Nondestruct Eval 36(1)
12. Luo Q, Shi Y, Wang Z, Zhang W, Li Y (2017) A study of applying pulsed remote field eddy current in ferromagnetic pipes testing. Sensors (Switzerland) 17(5)
13. Ansys, Maxwell Help, April 2018
14. Todorov E (2012) Measurement of electromagnetic properties of heat exchanger tubes. NDT E Int 48:70–76
15. ASTM SE-2096 Standard practice for in-situ examination of ferromagnetic heat-exchanger tubes using RFT.pdf

Asymmetrical Analysis of Breast Thermal Images for Detection of Breast Cancer

N. Kavya, N. Sriraam, N. Usha, D. Sharath, Prabha Ravi, Bharathi Hiremath, B. Venkatraman, and M. Menaka

1 Introduction

Medical infrared imaging technique is applied in many medical domains like oncology (breast, skin), vascular disorders, neurology, surgery, etc. Breast thermal imaging have unique characteristics like non-invasive, inexpensive, its ability to detect the lesions in early stage, used for dense breasts diagnosis, and it is effective for all age groups. The physics behind the breast thermal imaging is that all objects emit infrared radiation from their surface when their temperature is above absolute zero. According to Stefan–Boltzmann law, the total radiation energy emitted by an object per unit surface is directly proportional to the fourth power of its absolute temperature and is given by Eq. (1) [1].

$$M = \in \sigma T^4 \mathrm{W/m^2} \tag{1}$$

where T is absolute temperature in kelvin, σ is Stefan–Boltzmann constant (5.67 × 10–8 W/(m^2-K^4), and $\in$ is the emissivity. The infrared radiation emitted by human skin can be converted into temperature using thermal imaging, and since human skin has high emissivity, thermal imaging can be easily applied. The breast infrared imaging can detect the precancerous and malignant tumours. Precancerous lesions will encourage the surrounding cells to form new blood vessels, and the process is

N. Kavya (✉) · N. Sriraam · N. Usha · D. Sharath · P. Ravi
Center for Imaging Technologies, M.S. Ramaiah Institute of Technology, Bengaluru, India
e-mail: kavyan28@gmail.com

B. Hiremath
Department of Surgery, M.S. Ramaiah Medical College and Hospital, Bengaluru, India

B. Venkatraman · M. Menaka
Safety, Quality and Resources Management Group, IGCAR, Kalpakkam, India

C. K. Mukhopadhyay and R. Mulaveesala (eds.), *Advances in Non-destructive Evaluation*, Lecture Notes in Mechanical Engineering,
https://doi.org/10.1007/978-981-16-0186-6_25

called angiogenesis. The cancerous or precancerous cells will release angiogenesis factor (AF) from the cells which forms the new blood vessels. This increases the blood supply in a particular area and raises the temperature, and also we can notice the vascular asymmetry [2]. An extremely sensitive thermal camera and a temperature controlled room are used to capture the temperature distribution of the human body.

The asymmetry between the breasts could be a risk indicator of breast cancer. The bilateral asymmetry analysis is one of the evidenced procedures in breast thermogram analysis for breast cancer detection. The temperature distribution of healthy breasts will be symmetrical in bilateral breasts. Proposed work is concentrated on the bilateral asymmetry analysis of both normal and abnormal breasts. The analysis was done by extracting statistical features and by obtaining the maximum temperature.

The paper is organized as follows. In Sect. 2, the related works were discussed. Section 3 explained the methodology used. The results were discussed in Sect. 4, and Sect. 5 concluded the proposed work.

2 Related Work

Breast cancer is the second leading cancer among the women worldwide. Early detection of cancer will increase the survival rate, and it will also reduce the cost of treatment. Thermal imaging is one of the imaging techniques which can be used as screening tool for all age grouped people with symptoms or without any symptoms. Thermal imaging will indicate the cancer by showing an asymmetric temperature profile, change in vascular patterns, high temperature of the particular region, etc. [3]. Asymmetric analysis is one of the well-recognized procedures for breast thermal image analysis. Minimum average temperature and maximum temperature difference values can be used for the measure of symmetry. Using this analysis, one can differentiate benign (fibrodenoma and cyst) and malignant (ductal carcinoma) conditions [4]. The asymmetric heat patterns of thermal image can be analysed by extracting the statistical parameters from the bilateral breasts. Cumulative histogram will qualitatively differentiate the normal and abnormal breasts [5].

The thermal image can be automatically segmented using Canny edge detection and Hough transform. Histogram of both right and left breasts was compared to get the asymmetry between the breasts. The symmetry analysis can also be done by using k-means clustering algorithm [6]. The raw thermal image contrast can be enhanced to get maximum accuracy. Contrast-Limited Adaptive Histogram Equalization (CLAHE) method will enhance the image by dividing the images into tiles. First-order and second-order statistical features can be extracted from the enhanced image [7]. Moments of intensity images were extracted as features for asymmetry analysis, and bilateral ratio was calculated [8]. Block matching and 3D filtering method were used for removing noise. Second-order features like energy, entropy, contrast and differential variance were extracted for asymmetry analysis [9]. Front views of breast thermograms have different shapes in the upper and lower regions of the breast, so shape features will help in the measure of asymmetry. Polynomial curve fitting procedure separates the right and left breasts [10]. Symmetry between the bilateral breasts can be analysed by using inframammary curves, and these curves are obtained using Horizontal Projection Profile (HPP). Bifurcation point was calculated using these curves; hence, the right and left breasts were separated [11].

Normalization of temperature matrices can be done by calculating the minimum temperature and maximum temperature of each breast. DWT was applied on these normalized matrices to extract energy and statistical features [12]. Intensity-based features and temperature features were extracted to get different sets of features. Normal and abnormal images were classified using Support Vector Machine with Radial basis function kernel [13]. The lowest breast boundary can be identified using horizontal edge. Using this edge, the lower unwanted regions were removed, and right and left breast were separated using centroid. Texture features were extracted from Gabor filter, and SVM-based textural asymmetry was computed for normal and abnormal cases [14]. GLCM features were extracted from the manually segmented right and left breasts. SVM classifier was used for the automatic classification of normal and abnormal breasts [15]. The IR images of silicone breast with heat source were taken using infrared camera to note the highest temperature on the breast surface. This method was based on the principle of heat transfer, and 3D scanning was required [16]. Along with the front views, lateral views also help in contralateral comparison of breasts. The lateral view includes the lymph nodes which show the early symptoms of breast cancer. Automated segmentation of both the views can be done using edge maps created with anisotropic diffusion filter and active contour techniques [17]. The transforms were used such as wavelet, curvlet and contourlet, and statistical analysis of features is done [18]. Infrared Breast Thermography (IBT) is highly sensitive technique in abnormality detection using asymmetry analysis.

3 Methodology

In the normal condition, both left and right breasts are symmetrical in nature. The presence of tumour is indicated by the existence of asymmetry among bilateral breasts in terms of temperature distribution and statistical features. The raw images were pre-processed and segmented before feature extraction. Computer-Aided Diagnosis (CAD) helps in improving image quality and assists in image interpretation. Figure 1 shows the steps involved in asymmetrical analysis of breast thermogram.

3.1 Data Collection and Protocol Used

The thermal images were collected from Ramaiah Memorial Hospital, Bangalore using FLIR T650sc thermal camera. While acquiring the thermal images, certain protocols were followed strictly to reduce the errors in image capturing process. The screening room was maintained between the temperatures of 19 and 24 °C. The

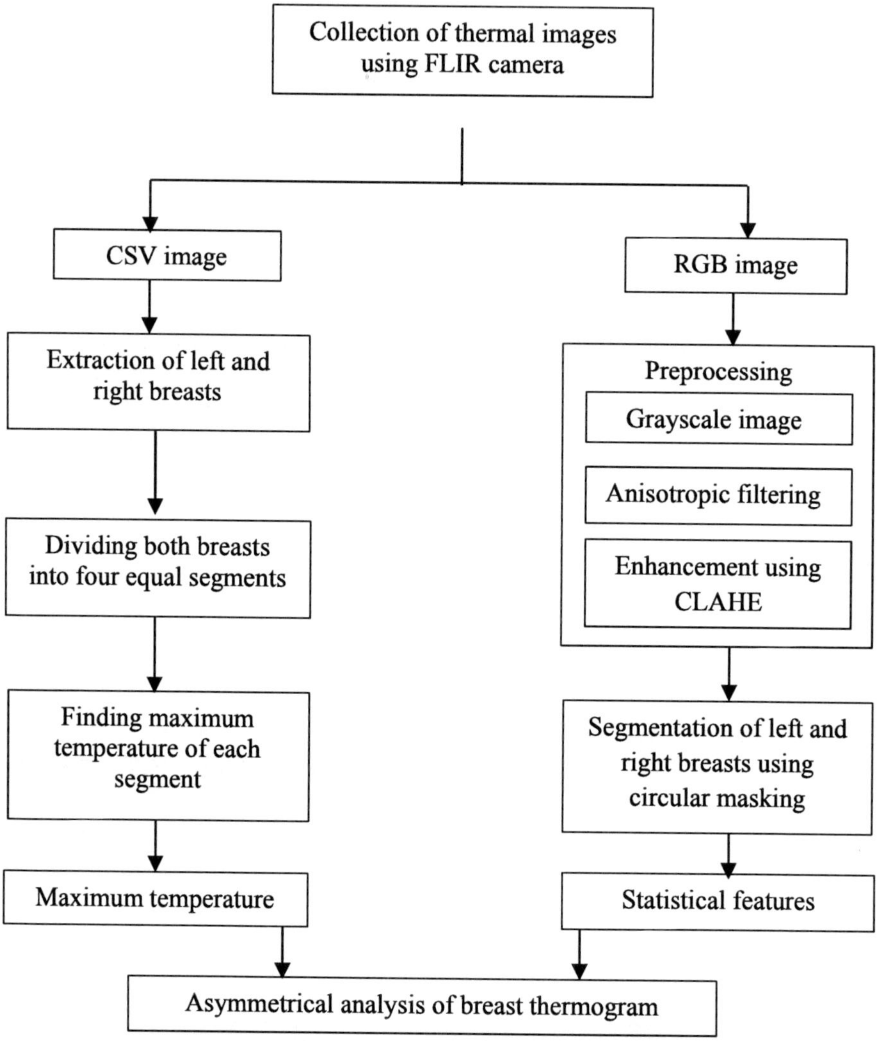

Fig. 1 Block diagram for the asymmetrical analysis of breast thermogram

subjects were instructed not to apply any lotions or cosmetics before screening, and they were asked to complete the consent and breast health history form. After the subject's consent, they were requested to disrobe from the waist up and to sit for about 15–20 min in the temperature controlled room. Then, the breast thermal images were captured in five different orientations, i.e. front, left oblique, right oblique, left lateral and right lateral.

3.2 Steps Involved to Obtain Statistical Features of Breast Thermograms

The heat information of the thermal image can be obtained using intensity values of the pixels. Hence, the statistical features contribute more in asymmetrical analysis of thermal images. The statistical features were extracted using pseudo coloured image (RGB) and processed using MATLAB. The image pre-processing and segmentation are the steps involved prior to the feature extraction.

Image pre-processing

The unwanted labels on images were removed using FLIR software. The raw images were converted to greyscale and filtered using anisotropic diffusion filter which reduces the image noise without eliminating any significant information such as edges and lines. After filtering, Contrast-Limited Adaptive Histogram Equalization (CLAHE) technique was applied to improve the contrast of image. CLAHE method computes many histograms corresponding to each image section; thus, it enhances the edge information in each region of an image.

Image Segmentation

After pre-processing, the images were segmented using circular masking. In this technique, the circle parameters such as centre and radius were initialized and masked the image with the circle. Then, the unwanted under breast regions were removed using Otsu thresholding.

Statistical feature extraction

After image segmentation, the statistical features such as mean, variance, skewness, kurtosis, entropy and energy were extracted as shown in Eqs. (2), (3), (4), (5), (6) and (7), respectively. All these features were computed from image histogram based on pixel intensity values. The feature values were not depend on the co-occurence of neighbour pixels.

$$Mean\mu = \sum_n f_n p(f_n) \tag{2}$$

$$Variance\sigma^2 = \sum_n (f_n - \mu)^2 p(f_n) \tag{3}$$

$$Skewness = \frac{1}{\sigma^3} \sum_n (f_n - \mu)^3 p(f_n) \tag{4}$$

$$Kurtosis = \frac{1}{\sigma^4} \sum_n (f_n - \mu)^4 p(f_n) - 3 \tag{5}$$

$$Entropy = -\sum_n p(f_n)^2 .logp(f_n) \tag{6}$$

$$Energy = \sum_n p^2(f_n) \tag{7}$$

where f_n is the image intensity, and $p(f_n)$ is the probability value of intensity in the image [2, 19].

3.3 Steps Involved to Obtain Temperature Values of Breast Thermograms

The thermal images were exported to Comma Separated Values (CSV) file format, and left and right breasts were separated using FLIR software. The CSV file contains temperature values which help for processing the thermograms.

Using MATLAB, both left and right breasts were divided into four equal segments and were named as RB1, RB2, RB3 and RB4 for right breast segments, similarly, LB1, LB2, LB3 and LB4 for left breast segments. The temperature values of all segments of both breasts were computed and compared. The highest temperature value indicates the presence of tumour and also gives the location of the tumour.

4 Results and Analysis of Asymmetricity in Breast Thermograms

In the proposed work, the qualitative and quantitative analysis was done using computer-aided approach to show the asymmetricity in breast thermograms.

Figure 2 shows the results of image segmentation. Figure 2a shows the original thermal image. Labels on the images give the error. Hence, the labels were removed using software, and Fig. 2b shows the label removed image. Figure 2c shows the anisotropic filtered image. Figure 2d is the contrast enhanced image to improve the image quality. Figure 2e, f is the right and left segmented images, respectively, using circular masking.

The presence of asymmetry between left and right breasts was obtained by calculating bilateral ratio of statistical features derived from left and right breasts. The statistical features such as mean, variance, skewness, kurtosis, energy and entropy of left and right breasts were extracted. Bilateral ratio value closer to 1 indicates less asymmetry and higher correlation of features. The bilateral ratio values of normal breast thermogram and abnormal breast thermogram were tabulated in Tables 1 and 2, respectively.

$$\text{Bilateralratioclosenessto1} = \left| \frac{\text{featurevaluefromleftsegment}}{\text{featurevaluefromrightsegment}} - 1 \right| \tag{8}$$

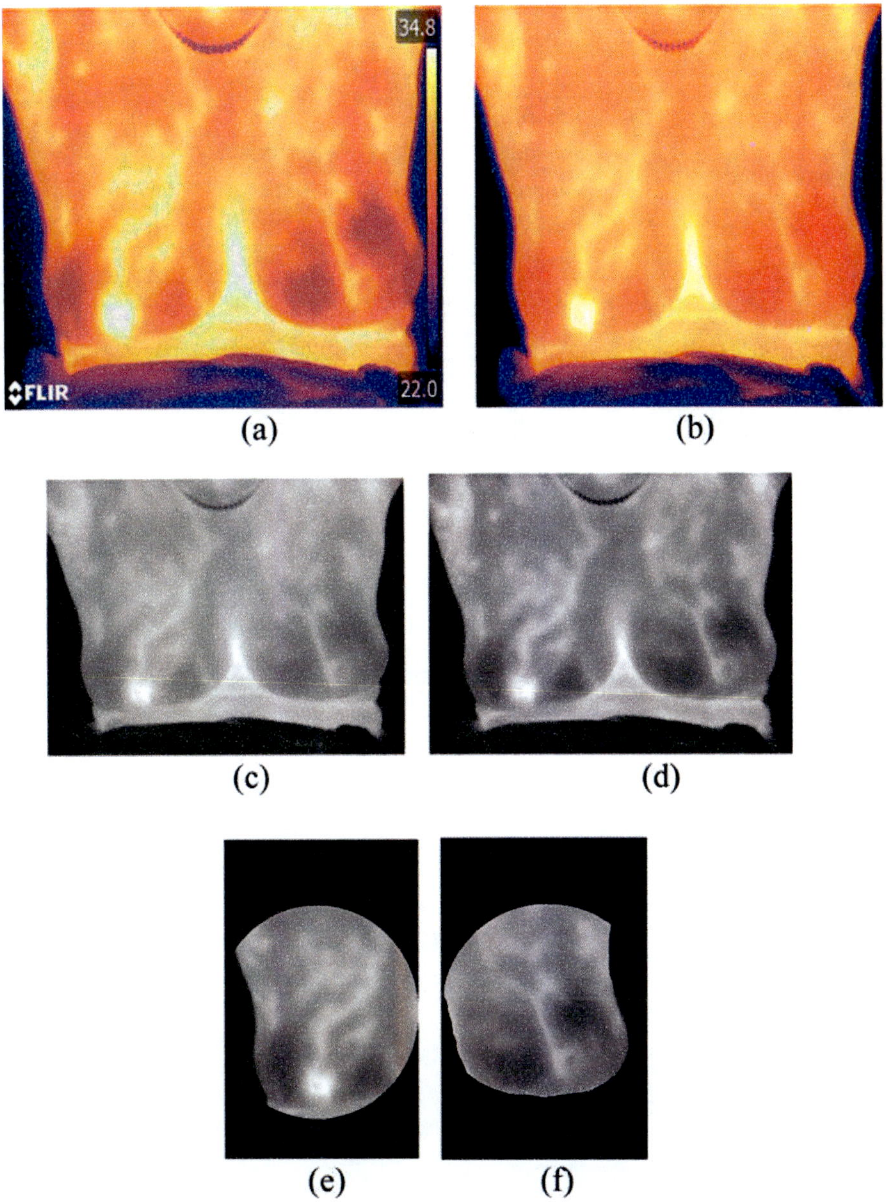

Fig. 2 **a** Original thermal image. **b** Label removed image. **c** Anisotropic filtered image. **d** Contrast enhanced image. **e, f** Segmented right and left breasts, respectively

Table 1 Evaluation of bilateral ratio for normal subject

Features	Left breast	Right breast	Bilateral ratio
Mean	2.7321	2.796289	1.02
Variance	4.848907	4.961657	1.03
Skewness	0.532792	0.59338	1.11
Kurtosis	1.438134	1.574749	1.07
Energy	0.423192	0.420205	0.97
Entropy	1.108283	1.14954	1.04

Table 2 Evaluation of bilateral ratio for abnormal subject

Features	Left breast	Right breast	Bilateral ratio
Mean	1.9882	3.4441	1.7
Variance	4.0932	6.494	1.6
Skewness	0.2348	0.4795	2.04
Kurtosis	1.0048	1.9612	1.9
Energy	0.4736	0.3159	0.65
Entropy	1.0784	1.9499	1.9

In Table 1, the bilateral ratio values calculated for normal subject were closer to 1 compared to bilateral ratio values of abnormal breast shown in Table 2 which indicates the presence of asymmetry in abnormal breast thermogram.

Figure 3 was plotted using the values of bilateral ratio of normal and abnormal images using Eq. 8 [2]. The values of closeness metric of normal image are minimum compared to the closeness metric values of abnormal image. This helps in the evaluation of asymmetricity.

Histograms were plotted for normal and abnormal breast images which indicate the asymmetry of thermal images in both breasts. The *X*-axis of the histograms indicates the intensity values of image, and the *Y*-axis represents the number of

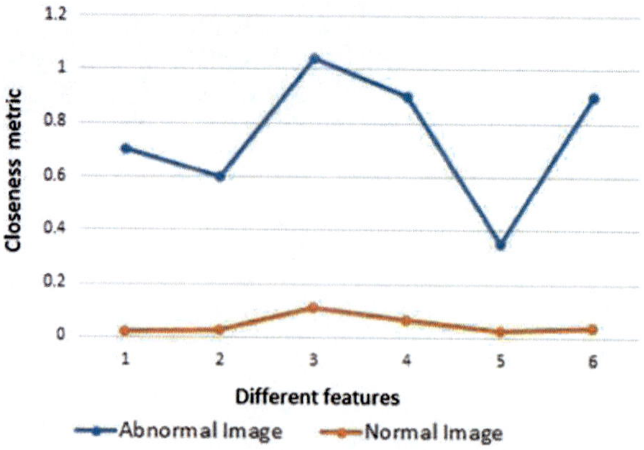

Fig. 3 Asymmetrical evaluation of breast thermograms using bilateral ratio

pixels having that intensity values. Figure 4a is the normal breast thermogram, and Fig. 4b is abnormal breast thermogram. Figure 4c, d is the histograms of right and

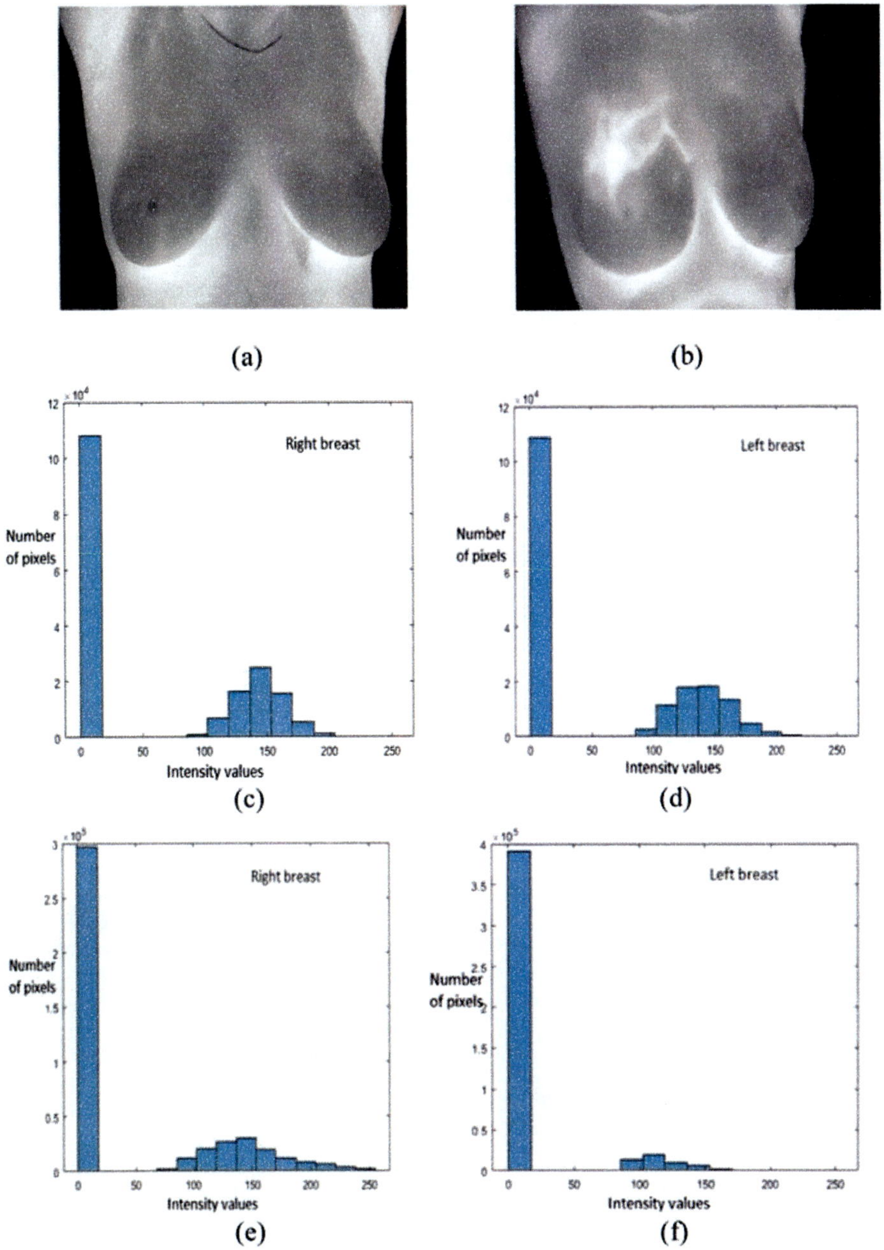

Fig. 4 Histograms of right and left breasts for normal and abnormal breast thermograms

left breasts of normal image, respectively. Figure 4e, f is the histograms of right and left breasts of abnormal image, respectively.

The left and right breasts of normal subject give the symmetrical histogram due to the thermal symmetry in both breasts. In abnormal condition, the histogram of right and left breasts shows the asymmetrical pattern. Hence, plotting histogram helps in asymmetrical analysis of breast thermograms.

Figure 5a is the CSV image, and Fig. 5b, c is manually segmented right and left breasts, respectively. Figure 5d is the four segments of right breast, and Fig. 5e is the four segments of left breast. The temperature values were computed from CSV segments as shown in Table 3. In the above image segments, the abnormality is

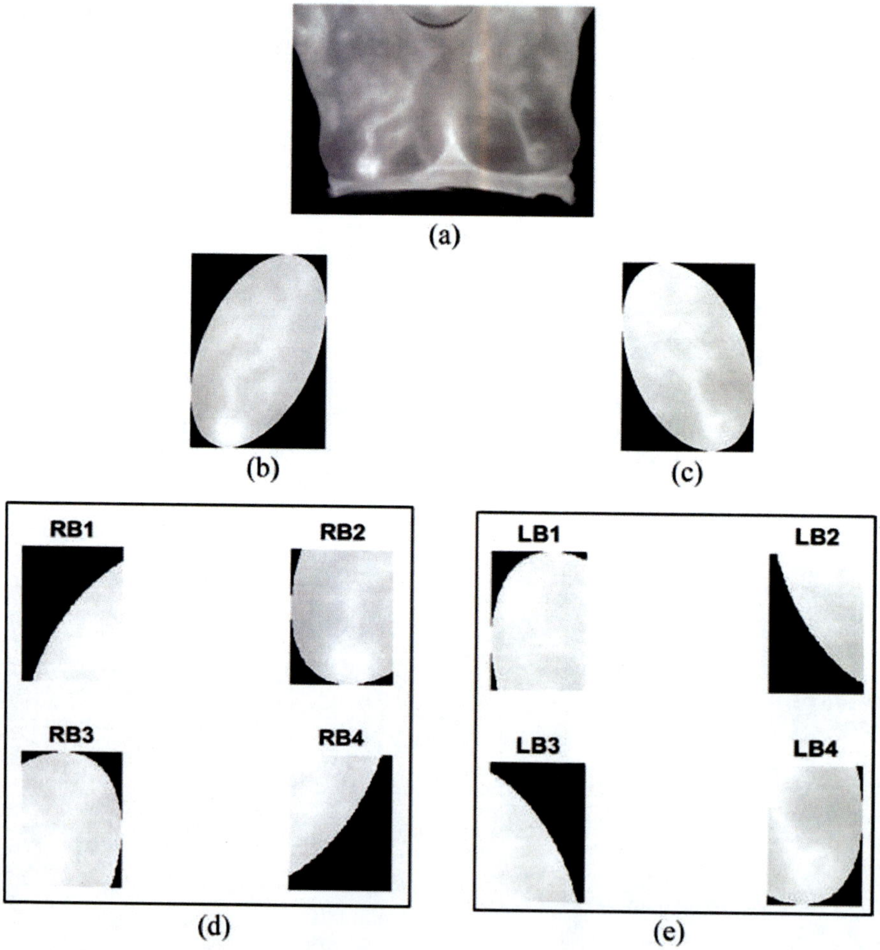

Fig. 5 Segments of breast thermogram for temperature measurement. **a** CSV converted image. **b** Segmented right breast. **c** Segmented left breast. **d** Four segments of right breast (RB1, RB2, RB3 and RB4). **e** Four segments of left breast (LB1, LB2, LB3 and LB4)

Table 3 Temperature values of right and left breast's segments

Right breast (°C)		Left breast (°C)	
RB1	33.62	LB1	33.72
RB2	**35.56**	LB2	32.44
RB3	33.93	LB3	32.70
RB4	33.89	LB4	32.93

present in the right breast which is indicated by the maximum temperature value of right segment RB2.

Compared to other segments, the RB2 segment has highest value which indicates the abnormality and location of tumour in the breast. The temperature difference between right and left breasts helps in the analysis of asymmetricity.

5 Conclusion

This paper implemented a computer-aided approach for analysis of asymmetrcity in breast thermograms. The temperature distribution profile and statistical features were analysed. From the experimental results, it can be observed that the different intensity distribution can be quantified using statistical features. The histogram essentially describes the image texture. The feature extraction is an efficient approach to find the asymmetry between bilateral breasts. The change in the heat pattern of particular surface region compared to the surrounding area is one of the major sign for diagnosing the breast abnormality. The difference in the surface temperature between abnormal and normal breasts helps in evaluating asymmetry of breast thermograms effectively. The asymmetrical analysis approach will assist the diagnostics as a second opinion.

Acknowledgements The authors would like to express their special thanks to IGCAR, Kalpakkam for funding this project (Project ID: IGC/HSEG/RSD/CP-01/2018) and Department of Surgical and Department of Radiology of Ramaiah Memorial Hospital, Bangalore for their support.

References

1. Holst GC (2000) Common sense approach to thermal ımaging. SPIE Press Monograph, vol PM86, p 48
2. Diakides NA, Bronzino JD (2007) Medical ınfrared ımaging. CRC press, Tylor & Francis Group
3. Gonzalez-Hernandez J-L, Recinella AN, Kandlikar SG, Dabydeen D, Medeiros L, Phatak P (2018) Technology, application and potential of dynamic breast thermography for the detection of breast cancer. Int J Heat Mass Transf 131:558–573
4. Sarigoz T, Ertan T, Topuz O, Sevim Y, Cihan Y (2018) Role of digital infrared thermal imaging in the diagnosis of breast mass: a pilot study. Infrared Phys Technol 91:214–219

5. Tang X, Ding H (2005) Asymmetry analysis of breast thermograms with morphological image segmentation. IEEE engineering in medicine and biology 27th annual conference
6. Chen B, Ma Z (2008) Automated image segmentation and asymmetry analysis for breast using infrared images. In: 2008 international workshop on education technology and training & 2008 international workshop on geoscience and remote sensing. Shanghai, pp 410–413 (2008)
7. Kavya N, Usha N, Sriraam N, Sharath D, Ravi P (2018) Breast cancer detection using non invasive imaging and cyber physical system. In: 2018 3rd international conference on circuits, control, communication and computing (I4C)
8. Kuruganti PT, Qi H (2002) Asymmetry analysis in breast cancer detection using thermal infrared images. In: Proceedings of the second joint 24th annual conference and the annual fall meeting of the Biomedical Engineering Society. Engineering in Medicine and Biology
9. Prabha S, Sujatha CM, Ramakrishnan S (2014) Asymmetry analysis of breast thermograms using BM3D technique and statistical texture features. In: International conference on informatics, electronics & vision (ICIEV)
10. Sathish D, Kamath S, Prasad K, Kadavigere R, Martis RJ (2016) Asymmetry analysis of breast thermograms using automated segmentation and texture features. SIViP 11(4):745–752
11. Devi RR, Anandhamala GS (2019) Analysis of breast thermograms using asymmetry in infra-mammary curves. J Med Syst 43:1–9
12. Sathish D, Kamath S, Prasad K, Kadavigere R (2017) Role of normalization of breast thermogram images and automatic classification of breast cancer. Vis Comput
13. Rani Gogoi U, Majumdar G, Kanti Bhowmik M, Kumar Ghosh A (2019) Evaluating the efficiency of infrared breast thermography for early breast cancer risk prediction in asymptomatic population. Infrared Phys Technol
14. Khan AA, Arora AS (2018) Breast cancer detection through gabor filter based texture features using thermograms images. In: 2018 first international conference on secure cyber computing and communication (ICSCCC)
15. Karim CN, Mohamed O, Ryad T (2018) A new approach for breast abnormality detection based on thermography. Med Technol J 2(3):245–254
16. Igali D, Mukhmetov O, Zhao Y, Fok SC, Teh SL (2018) An experimental framework for validation of thermal modeling for breast cancer detection. IOP Conf Ser Mater Sci Eng 408:012031
17. Josephine Selle J, Shenbagavalli A, Sriraam N, Venkatraman B, Jayashree M, Menaka M (2018) Automated recognition of ROIs for breast thermograms of lateral view—a pilot study. Quant InfraRed Thermogr J 1–20
18. Jeyanathan JS, Selle J (2018) Transform based classification of breast thermograms using multilayer perceptron back propagation neural network. Int J Pure Appl Math 118(20):1955–1961
19. Rasyid MBA, Yunidar AF, Munadi K (2018) Histogram statistics and GLCM features of breast thermograms for early cancer detection. In: 2018 International ECTI northern section conference on electrical, electronics, computer and telecommunications engineering (ECTI-NCON)

Numerical Analysis of Partial Discharge Source Localization Using Time of Arrival Measurements and Nonlinear Least Squares Search

Krishna C. Ghanakota, Sarathi Ramanujam, and Kavitha Arunachalam

1 Introduction

Partial discharge (PD) is a localized discharge in the insulation of high-voltage (HV) electrical equipments. PDs occur due to the ionization processes that occur in the voids of insulations. PD activity monitoring is essential as these discharges produce physical and chemical aging causing erosion of the insulation material. The IEC-60270 and IEC-62478 standards, respectively, define the conventional methods (such as current and voltage measurement sensing) and non-conventional methods (such as optical, chemical, acoustical, and electromagnetic (EM) measurement sensing) for PD activity monitoring in electrical equipment. According to these standards, a range of acoustic and EM measurements are used to perform PD detection and localization. Acoustic emissions for PDs lie in the frequency range of 0.1–250 kHz, whereas the frequency range for the EM emissions is in the ultra-high frequency (UHF) range of 0.5–3 GHz [1]. The acoustic emission method has a good sensitivity and accuracy for localization but suffers from limited range and severe attenuation. UHF technique is a non-intrusive technique and is widely being used for capturing EM emissions from PD, as EM wave is insensitive to pressure and temperature variations, propagates through all dielectric medium and are suited for continuous real-time monitoring.

K. C. Ghanakota (✉) · K. Arunachalam
Department of Engineering Design, Indian Institute of Technology Madras, Chennai, Tamil Nadu 600036, India
e-mail: ed18d012@smail.iitm.ac.in

K. Arunachalam
e-mail: akavitha@iitm.ac.in

S. Ramanujam
Department of Electrical Engineering, Indian Institute of Technology Madras, Chennai, Tamil Nadu 600036, India
e-mail: rsarathi@iitm.ac.in

C. K. Mukhopadhyay and R. Mulaveesala (eds.), *Advances in Non-destructive Evaluation*, Lecture Notes in Mechanical Engineering,
https://doi.org/10.1007/978-981-16-0186-6_26

A review of the literature shows various techniques for localizing PDs. In [2], a localization approach based on the main PD signal and the two sub-band signals using correlation-based feature selection (CFS) was detailed. In [3], the last distinguishable edge point in the energy density distribution of the UHF signal obtained via canny algorithm was used to estimate the position of the PD source. In [4], the a priori information of EM wave propagation using eigen function was used to identify the source. However, this approach is limited to complex electrical devices, such as transformers and switchgears, having dimensions of the order of meters. In [5], the distance to the defect was estimated by the reconstruction of a reference signal from the measured signals and the cable transfer function. In [6], an approach based on wavelet packet transform and machine learning ensemble methods was used to locate the PDs. In [7], a time difference of arrival (TDOA) approach was used in conjunction with multivariate wavelet transform to denoise the UHF signals to locate the PD source. In [8], a self-similarity recurrence quantification analysis (RQA) parameter was used to estimate the TDOA of UHF signals and further localize the PD source. The observed deviation in [8] from the true location of the PD source was 8.7 mm, which corresponds to an error of 0.31% in the localization algorithm. In this work, an array of UHF sensors is used to detect the location of the defect in the insulation. The location bearing information in the form of Time of Arrivals is extracted from the UHF signals. Results of the source localization algorithm based on a nonlinear least squares search (NLS) are compared with the true locations of the PD source in the numerical model.

2 Modeling and Numerical Simulations

The spectral content of the PD signal generally lies in the UHF range of 0.5–3 GHz with the dominant frequency occurring around 1 GHz [1]. To facilitate the analysis of the localization problem, both UHF sensors and PD source were modeled as dipoles with resonant frequency at 1 GHz. Figure 1 shows a dipole modeled in HFSS (Ansys Inc.) version 19 R2. The length of the dipole was designed to be 13.49 cm with a wire radius of 0.225 cm and a port gap length of 0.225 cm [9]. A radiation boundary was used to truncate the solution domain to a finite region by padding an absorbing boundary in all directions at a distance of 75 mm from the surface of the dipole. Adaptive meshing was used at the solution frequency of 1 GHz to divide the model into fine tetrahedral elements, and wave equation was iteratively solved for solution convergence using a frequency domain solver. The radiation patterns in the XZ and YZ planes are shown in Fig. 2a, b, respectively, at 1 GHz. The reflection coefficient, S_{11} of the resonant dipole is shown in Fig. 3.

An array of dipoles with dimensions as above was used as UHF sensors to detect the location of a PD source also modeled as a dipole antenna of same dimensions. The four sensors were located at (0, 0) cm, (50, 0) cm, (50, 50) cm and (0, 50) cm, respectively. The unknown source whose location which is to be determined was located at (25, 35) cm. Figure 4 shows the location of the sensors and the source in

Fig. 1 Dipole antenna modeled in HFSS (Ansys Inc.)

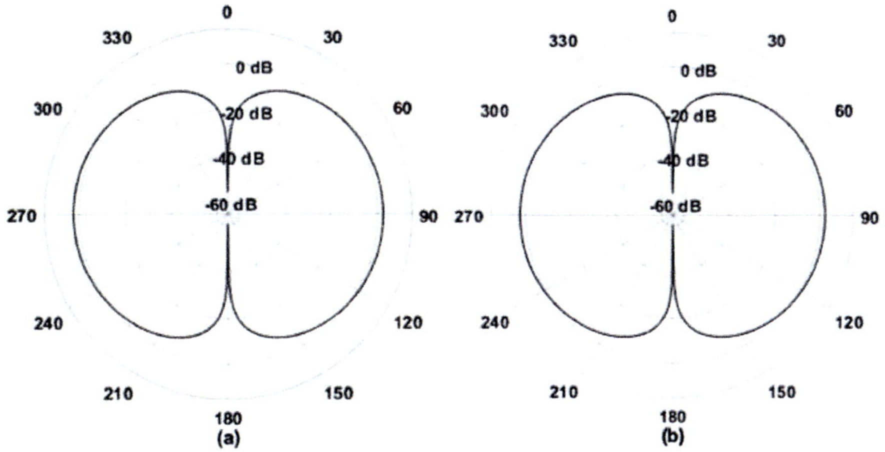

Fig. 2 Gain plot of the resonant dipole antenna at 1 GHz. **a** XZ plane. **b** YZ plane

the numerical model. The model in Fig. 4 was solved in HFSS using a transient solver based on the discontinuous Galerkin method (DGTD). An active port was used for the PD source with the excitation signal being a broadband pulse in the UHF range as shown in Fig. 5a, whereas passive ports were used for the sensors. Wave equation was iteratively solved in the solution domain until convergence. Figure 5a shows the excitation pulse and the reflected signal at the active port of the PD source. The received signal at the various passive ports of the sensors is shown in Fig. 5b. As sensors 1 and 2 are located at an identical distance from the PD source, it can be observed from Fig. 5b that sensor signals 1 and 2 overlap with each other.

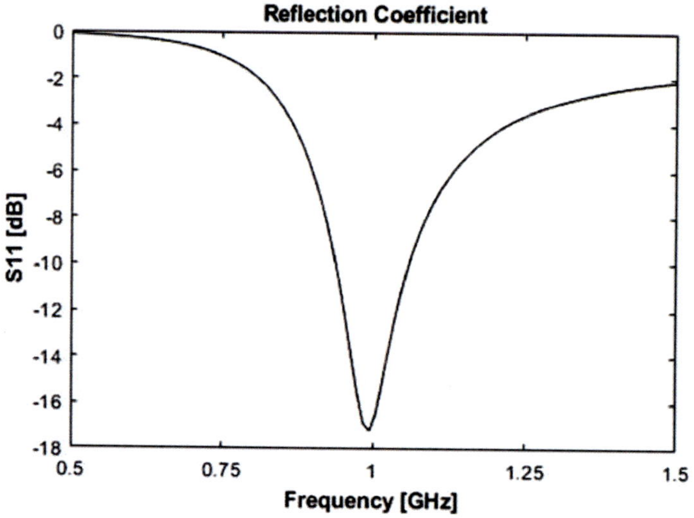

Fig. 3 Reflection coefficient of the resonant dipole antenna

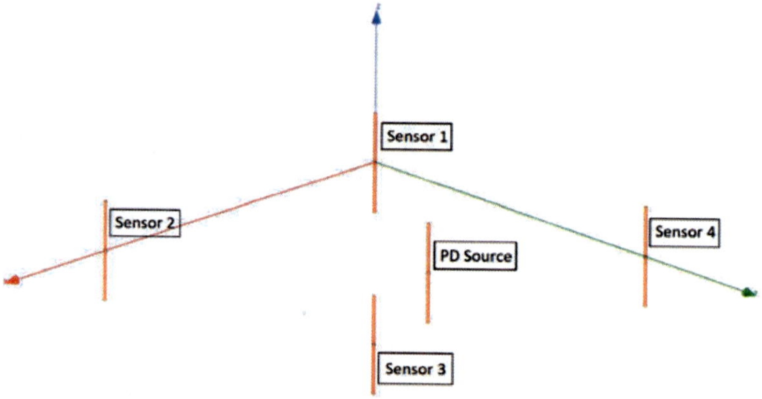

Fig. 4 Geometry of the location of PD source and UHF sensors

3 Signal Analysis and Source Localization

Time of Arrival (TOA) is one-way propagation time of the signal traveling between the source and the receiver [10]. Multiplying the TOAs by the propagation speed of EM wave in free space $c = 3 \times 10^8 \text{m/s}$, gives the distances between the source and the receivers. In the absence of measurement errors, each TOA corresponds to a circle centered at the receiver on which the source must lie. Geometrically, three or more circles deduced from the noise free TOAs will result in a unique intersection, which is the source position. The TOA was calculated using the differences between the time taken by the signal to reach 10% of the maximum value of the received signal

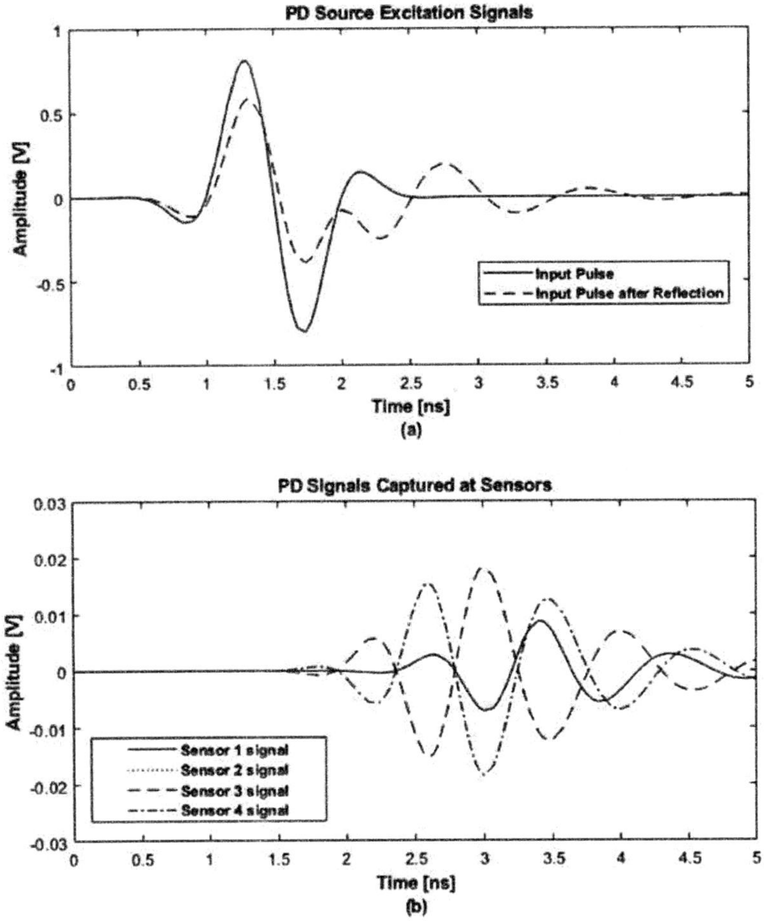

Fig. 5 Simulated UHF signals. **a** PD source, **b** PD signals received by the UHF sensors

and the excitation signal. Let $\boldsymbol{X} = [x, y]^T$ be the unknown PD source position and $\boldsymbol{X}_l = [x_l\, y_l]^T$ be the known coordinates of the lth sensor, where $l = 1, 2, 3, 4$.

The distance between the source and the lth sensor, denoted by d_l is given by,

$$d_l = \|\boldsymbol{X} - \boldsymbol{X}_l\|_2 = \sqrt{(x - x_l)^2 + (y - y_l)^2}, \quad l = 1, 2, 3, 4. \tag{1a}$$

The relation between the TOA at the lth sensor $\{t_l\}$ and d_l is given by,

$$t_l = \frac{d_l}{c}, \quad l = 1, 2, 3, 4. \tag{1b}$$

The range, r obtained by multiplying t_l by c is subjected to error, n and is given by,

$$r = d_l + n = \sqrt{(x - x_l)^2 + (y - y_l)^2} + n, \quad l = 1, 2, 3, 4. \tag{1c}$$

In vector form, the TOA model can be expressed as,

$$\boldsymbol{r} = \boldsymbol{f}(\boldsymbol{X}) + \boldsymbol{n}, \tag{1d}$$

$$\boldsymbol{r} = [r_1 r_2 r_3 r_4]^T \tag{1e}$$

$$\boldsymbol{n} = [n_1 n_2 n_3 n_4]^T \tag{1f}$$

$$\boldsymbol{f}(\boldsymbol{X}) = \boldsymbol{d} = \begin{bmatrix} \sqrt{(x - x_1)^2 + (y - y_1)^2} \\ \sqrt{(x - x_2)^2 + (y - y_2)^2} \\ \sqrt{(x - x_3)^2 + (y - y_3)^2} \\ \sqrt{(x - x_4)^2 + (y - y_4)^2} \end{bmatrix} \tag{1g}$$

The NLS cost function for the TOA model denoted by $\boldsymbol{J}(\boldsymbol{X})$ is given in Eq. (2a) and the position estimate $\widehat{\boldsymbol{X}}$ which minimizes $\boldsymbol{J}(\boldsymbol{X})$ is expressed in Eq. (2b).

$$\boldsymbol{J}(\boldsymbol{X}) = \sum_{l=1}^{4} \left(r_l - \sqrt{(x - x_l)^2 + (y - y_l)^2} \right)^2 \tag{2a}$$

$$\widehat{\boldsymbol{X}} = \min \boldsymbol{J}(\boldsymbol{X}) \tag{2b}$$

The NLS cost function, $\boldsymbol{J}(\boldsymbol{X})$ is a multi-modal 2D surface with multiple local minima apart from a global minimum. An iterative algorithm based on an initial position estimate can be considered to minimize the NLS cost function, $\boldsymbol{J}(\boldsymbol{X})$. Three local search techniques namely, Newton-Rapshon, Gauss–Newton and steepest descent were compared. In Newton–Raphson method, the iterative update rule is given as,

$$\widehat{\boldsymbol{X}}^{\boldsymbol{k}+1} = \widehat{\boldsymbol{X}}^{\boldsymbol{k}} - \boldsymbol{H}^{-1}(\boldsymbol{J}(\widehat{\boldsymbol{X}}^{\boldsymbol{k}}))\nabla(\boldsymbol{J}(\widehat{\boldsymbol{X}}^{\boldsymbol{k}})) \tag{3a}$$

$$\boldsymbol{H}(\boldsymbol{J}(\boldsymbol{X})) = \begin{bmatrix} \frac{\partial^2 J(X)}{\partial x^2} & \frac{\partial^2 J(X)}{\partial x \partial y} \\ \frac{\partial^2 J(X)}{\partial y \partial x} & \frac{\partial^2 J(X)}{\partial y^2} \end{bmatrix} \tag{3b}$$

$$\nabla(\boldsymbol{J}(\boldsymbol{X})) = \begin{bmatrix} \frac{\partial J(X)}{\partial x} \\ \frac{\partial J(X)}{\partial y} \end{bmatrix} \tag{3c}$$

where H and ∇ are the Hessian matrix and the gradient vector computed at the kth iteration estimate. The iterative update rule for Gauss–Newton method is given in Eqs. (4a–4b), where $\boldsymbol{G}\left(\boldsymbol{f}\left(\widehat{\boldsymbol{X}}^{\boldsymbol{k}}\right)\right)$ is the Jacobian matrix of $\boldsymbol{f}\left(\widehat{\boldsymbol{X}}^{\boldsymbol{k}}\right)$ computed at $\widehat{\boldsymbol{X}}^{\boldsymbol{k}}$.

$$\widehat{X}^{k+1} = \widehat{X}^{k} + \left(G^{T}\left(f\left(\widehat{X}^{k}\right)\right)G\left(f\left(\widehat{X}^{k}\right)\right)\right)^{-1} G^{T}\left(f\left(\widehat{X}^{k}\right)\right)\left(r - f\left(\widehat{X}^{k}\right)\right) \quad (4a)$$

$$G(f(X)) = \begin{bmatrix} \frac{\partial\sqrt{(x-x_1)^2+(y-y_1)^2}}{\partial x} & \frac{\partial\sqrt{(x-x_1)^2+(y-y_1)^2}}{\partial y} \\ \frac{\partial\sqrt{(x-x_2)^2+(y-y_2)^2}}{\partial x} & \frac{\partial\sqrt{(x-x_2)^2+(y-y_2)^2}}{\partial y} \\ \frac{\partial\sqrt{(x-x_3)^2+(y-y_3)^2}}{\partial x} & \frac{\partial\sqrt{(x-x_3)^2+(y-y_3)^2}}{\partial y} \\ \frac{\partial\sqrt{(x-x_4)^2+(y-y_4)^2}}{\partial x} & \frac{\partial\sqrt{(x-x_4)^2+(y-y_4)^2}}{\partial y} \end{bmatrix} \quad (4b)$$

The iterative update rule for the steepest descent method is given in Eq. (5), where μ is a positive constant, which controls the convergence rate and stability.

$$\widehat{X}^{k+1} = \widehat{X}^{k} - \mu\nabla\left(J\left(\widehat{X}^{k}\right)\right). \quad (5)$$

Using an initial position estimate as (30, 30) cm for the iterative algorithms, it can be observed from Fig. 6a, b, that the convergence rate of the steepest descent method is slower owing to a small value being used for the learning rate ($\mu = 0.1$). However, all the three iterative search techniques converge to the true location of the PD source

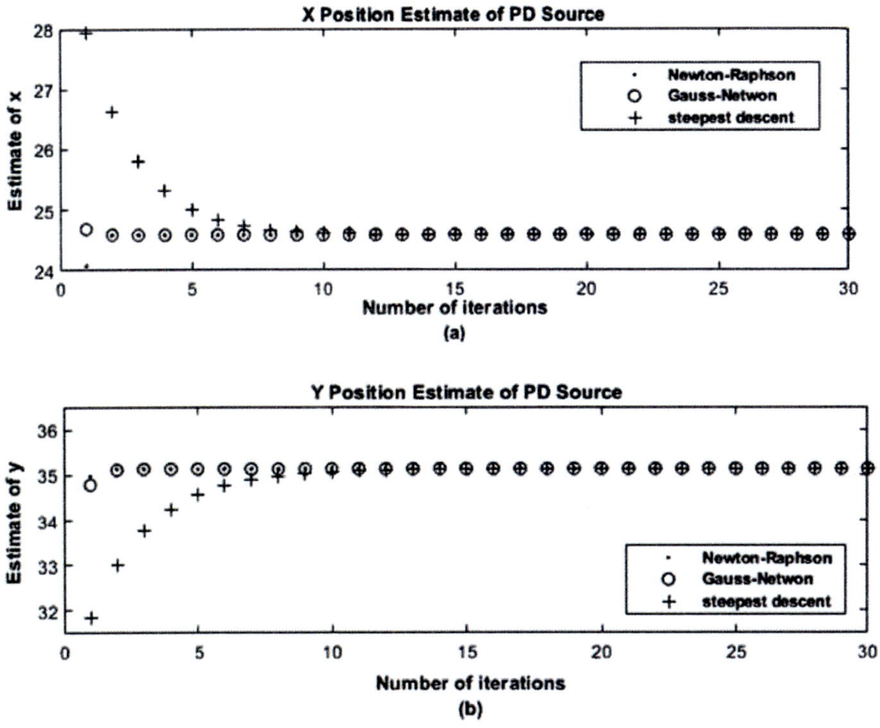

Fig. 6 Position estimation of the PD source. **a** Estimate of x. **b** Estimate of y

Table 1 Influence of pulse width on location estimation

S. No.	Pulse bandwidth (GHz)	True location (cm)	Estimated location (cm)	Euclidean error (cm)
1	0.5	(25, 35)	(24.51, 35.44)	0.66
2	1.0	(25, 35)	(24.59, 35.13)	0.42

with negligible error, with the estimated location of the PD source as (24.59, 35.13) cm. The effect of increasing the excitation pulse bandwidth on the accuracy of the localization algorithm is shown in Table 1. It can be observed from Table 1 that as the PD bandwidth was increased from 0.5 to 1 GHz, the euclidean error in the position estimate decreased from 0.66 to 0.42 cm. The error in the estimated distance is 0.3% for the proposed algorithm which is comparable to the error reported in [8] for PD source localization.

4 Conclusion

A source localization algorithm based on Time of Arrival measurements (TOA) and nonlinear least squares search (NLS) was presented. Numerical simulations show the results of the source localization algorithm are within 0.3% of the true location of the PD source with negligible error. Experimental validation of the algorithms will be evaluated using identical set of ultra wide band antennas and a laboratory scale test cell to generate PDs occurring in high-voltage power frequency transmission line system.

Acknowledgements The authors acknowledge the support of Central Power Research Institute, Bangalore, India. The first author also acknowledges the support of Robert Bosch Engineering and Business Solutions Private Limited, India.

References

1. Sarathi R, Koperundevi G (2008) UHF technique for identification of partial discharge in a composite insulation under AC and DC voltages. IEEE Trans Dielectr Electr Insul 15(6):1724–1730. https://doi.org/10.1109/TDEI.2008.4712677
2. Iorkyase ET, Tachtatzis C, Glover IA, Atkinson RC (2019) RF-based location of partial discharge sources using received signal features. High Voltage 4(1):28–32. doi: https://doi.org/10.1049/hve.2018.5027
3. Li X, Wang X, Yang A, Rong M (2019) Partial discharge source localization in GIS Based on image edge detection and support vector machine. IEEE Trans Power Deliv 34(4):1795–1802. https://doi.org/10.1109/TPWRD.2019.2925034
4. D'Antona G, Perfetto L (2019) Partial discharge localization in insulated switchgears by eigen-function expansion method. IEEE Trans Instrum Meas 68(5):1294–1301. https://doi.org/10.1109/TIM.2019.2904806

5. Mahdipour M, Akbari A, Werle P, Borsi H (2019) Partial discharge localization on power cables using on-line transfer function. IEEE Trans Power Deliv 34(4):1490–1498. https://doi.org/10.1109/TPWRD.2019.2908124
6. Iorkyase ET et al (2019) Improving RF-based partial discharge localization via machine learning ensemble method. IEEE Trans Power Deliv 34(4):1478–1489. https://doi.org/10.1109/TPWRD.2019.2907154
7. Dukanac D (2018) Application of UHF method for partial discharge source location in power transformers. IEEE Trans Dielectr Electr Insul 25(6):2266–2278. https://doi.org/10.1109/TDEI.2018.006996
8. Desai BMA, Sarathi R (2018) Identification and localisation of incipient discharges in transformer insulation adopting UHF technique. IEEE Trans Dielectr Electr Insul 25(5):1924–1931. https://doi.org/10.1109/TDEI.2018.007294
9. Balanis C (2012) Antenna theory, analysis, and design, 3rd edn. Wiley, New York
10. Zekavat R, Michael Buehrer R (2019) Source localization: algorithms and analysis. In: Handbook of position location: theory, practice, and advances, IEEE, pp 59–106. doi: https://doi.org/10.1002/9781119434610.ch3

Characteristics of Acoustic Emissions Generated During Steel Rebar Corrosion in Reinforced Concrete

P. Anjali and R. Vidya Sagar

1 Introduction

Chloride induced corrosion of embedded steel reinforcement in concrete is one of the primary reasons for the deterioration of RCC structures all over the world. As a result of this, the service life of the structures is drastically reduced. Most of the cases, this deterioration is detected only when the corrosion has progressed to a stage when the cracking and spalling of the concrete cover occurs. The repair and retrofitting of such structures are not viable from an economic point of view.

Different non-destructive testing (NDT) methods have been developed for the detection of steel reinforcement corrosion in concrete [1]. These methods are broadly classified into six major categories.

1. Visual inspection
2. Electrochemical methods
3. Elastic wave methods
4. Optical sensing methods
5. Electromagnetic methods
6. Infrared thermograpy.

Visual inspection is the regular surface inspection of the concrete structures. Corrosion could be identified as cracking, spalling, or reddish-brown rust formation on the surface. The disadvantage with this method is that surface effects could be identified only after corrosion has progressed to a sufficient stage. Corrosion of embedded steel in concrete could be very hard to detect. Electrochemical methods like Half Cell Potential measurement and Polarization Resistance methods are few of the other methods. Half Cell Potential measurement enables to know about the

P. Anjali (✉) · R. Vidya Sagar
Department of Civil Engineering, Indian Institute of Science, Bangalore 560012, India
e-mail: anjalip1@iisc.ac.in

C. K. Mukhopadhyay and R. Mulaveesala (eds.), *Advances in Non-destructive Evaluation*, Lecture Notes in Mechanical Engineering,
https://doi.org/10.1007/978-981-16-0186-6_27

probability of corrosion whereas Polarization Resistance method gives an idea about the rate of corrosion. Half Cell Potential method depends on the condition of the RC structure during testing; both moisture content and chloride concentration can influence the readings and give erroneous results. Polarization Resistance method assumes uniform corrosion along the reinforcement bar. But in most of the structures, pitting corrosion is more common than uniform corrosion. Other methods like electromagnetic methods, infrared thermography, and optical sensing method gives only a qualitative assessment of corrosion. It is known that acoustic emission (AE) technique is one of the NDT methods for the evaluation of reinforcement corrosion in concrete [1–4]. AE technique can detect both the corrosion of steel reinforcement as well as the cracking of concrete due to expansion of the corrosion products. Two periods of AE activity have been identified by Kawasaki et al. one due to the onset of corrosion in rebar and the other one due to the nucleation of corrosion-induced cracks in concrete [2]. Comparative study of AE and electrochemical techniques was conducted by Patil et al. It was observed that cumulative signal strength (CSS) can be used as a promising parameter of AE technique to monitor progress of damage in RC due to corrosion [5]. A relationship between gravimetric mass loss and cumulative signal strength value has been developed to quantitatively account for the steel rebar corrosion in concrete [6]. The effectiveness of AE in detecting and characterizing the initiation of the corrosion process was investigated based on results from small-scale, pre-cracked RC specimens [7]. Similar utility of AE in monitoring the corrosion process was reported by other researchers [8–12]. This study presented the usefulness of time driven parameters to monitor the corrosion process continuously. It was reported that the knee in the CSS curve that is associated with the nucleation of cracks cannot be clearly identified owing to the scale of the graph. Thus, AE method can be used for both quantitative and qualitative assessment of corrosion of steel bar embedded in concrete.

2 Aim of the Study

It was already reported by a lot of researchers that AE technique could be effectively used to monitor steel rebar corrosion in reinforced concrete. But comparative study of the variation in the aggressive environment and corresponding change in the AE data recorded were not studied. In this experimental study, the variation in the AE generated due to corrosion of steel rebar in concrete due to variation in the corrosive environment is studied.

During the corrosion of steel rebar embedded in concrete, AE hits from two sources are generated: one is due to corrosion of steel rebar and the other due to micro and macro-cracking of concrete. With the increase in concentration of NaCl solution, the intensity of steel corrosion increases. The signal generated due to steel rebar corrosion cannot be distinguished from those due to cracking of concrete as both of their strength is almost of the same order. And therefore, the knee in the CSS curve cannot be clearly distinguished owing to the scale of the graph. Hence, we have to resort to other methods of analysis to distinguish the AE hit data recorded.

3 AE Parameters for Corrosion Monitoring

AE are the transient elastic waves generated by the release of energy within a solid. Burst type AE is typically detected in concrete, characterized by rapid rise to peak and slower decay. Continuous emissions are encountered without any decay under phenomena like plastic deformation. Signal strength is defined as the area under the voltage signal of AE over the duration of the waveforms. It is measured in pico-volt second.

In this study, the characteristics of AE signals are estimated by two parameters:

$$\text{RA value} = \frac{\text{Rise Time}}{\text{Amplitude}} \tag{1}$$

$$\text{Average frequency} = \frac{\text{Counts}}{\text{Duration}} \tag{2}$$

Tensile type cracks are characterized by AE signals with high average frequency and low RA value. A shear type crack is identified by low average frequency and high RA value. This criterion is used to classify data in corrosion monitoring using AE.

4 Experimental Setup

Four concrete cylinders of dimension 150 mm diameter and 300 mm height were used in the experimental study and subjected to accelerated corrosion. The cylinders were cast with a steel rebar embedded concentrically inside it. A height of 50 mm was kept protruding outside for electrical connection. All the steel rebars were connected to the positive terminal of 5V DC source (anode). Copper wire was used as cathode and connected to negative terminal of the battery to supply constant voltage. Two cylinders were casted using 8 mm diameter steel reinforcement and the two others with 12 mm. The concentration of NaCl solution was as 3% and 5% in the present study. The following table shows the details about the specimens used in the experimental study (Tables 1 and 2):

All the specimens were subjected to accelerated corrosion and generated AE were recorded continuously. The acquiring of data was stopped after every seven days to change the NaCl solution to ensure the concentration remains same throughout the experiment. The study on cumulative signal strength and RA versus AF with the progression of corrosion was studied (Figs. 1 and 2).

Table 1 Details of the cylinder specimens used in the study

Cylinder	Dimensions Diameter × Height (mm)	Diameter of steel bar (mm)	Concentration of NaCl solution (%)
C1	150 × 300	12	3
C2	150 × 300	8	3
C3	150 × 300	12	5
C4	150 × 300	8	5

Table 2 The concrete mix used for casting cylinders

Cement (kg/m^3)	520
Sand (kg/m^3)	894
Coarse aggregates (10 mm size) (kg/m^3)	825
w/c ratio (by weight)	0.4

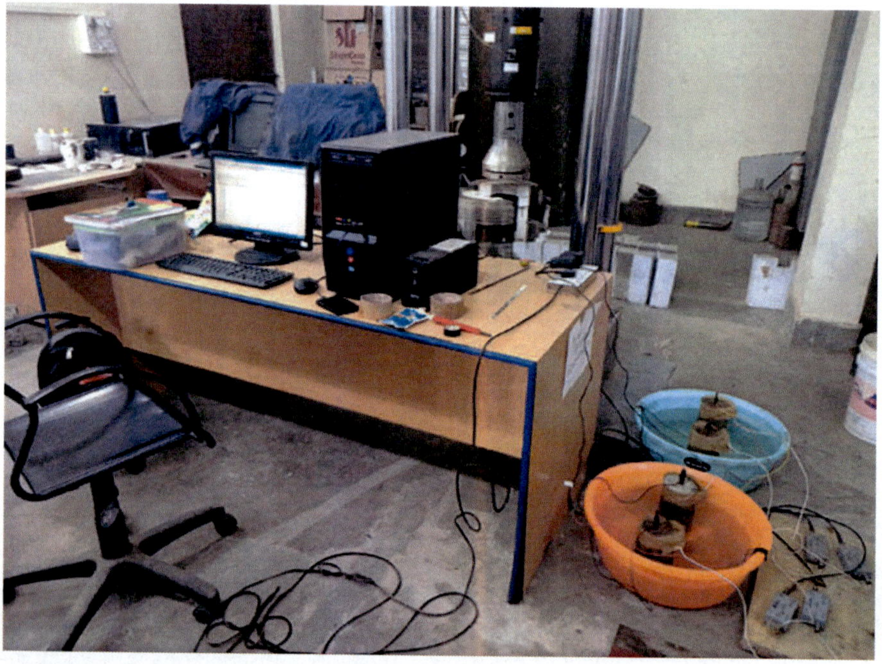

Fig. 1 Accelerated corrosion test setup using AE monitoring system in Structures Laboratory, Department of Civil Engineering, Indian Institute of Science, Bangalore, India

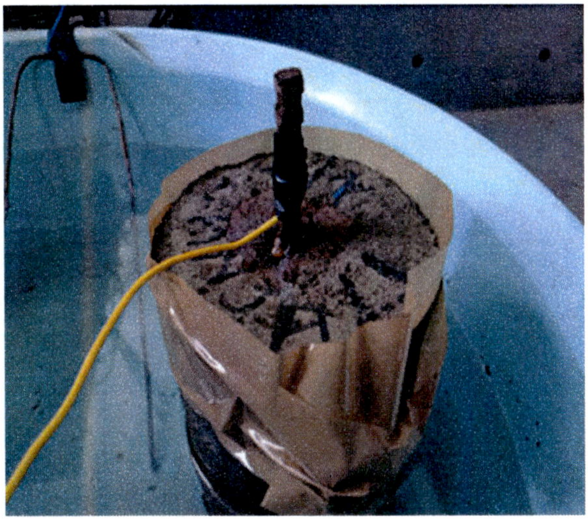

Fig. 2 Visible signs of corrosion in the test specimen

5 Results and Discussion

Figures 3 and 4 show the variation in signal strength and cumulative signal strength, respectively, with time for cylinder C1. In case of cylinder C1, from the signal strength variation with time, the hits generated due to the corrosion of steel reinforcement is clearly distinguishable from the hits generated due to the cracking of concrete. The same can also be understood from the steep rises in the cumulative signal strength curve. First major cracking occurred after 3 days of accelerated corrosion.

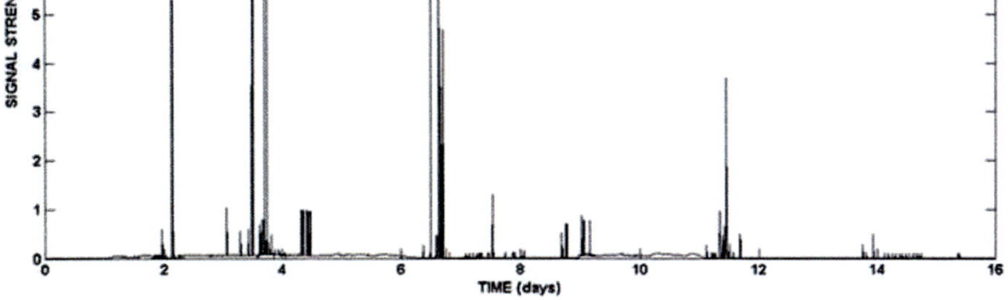

Fig. 3 Signal strength variation with time for cylinder C1

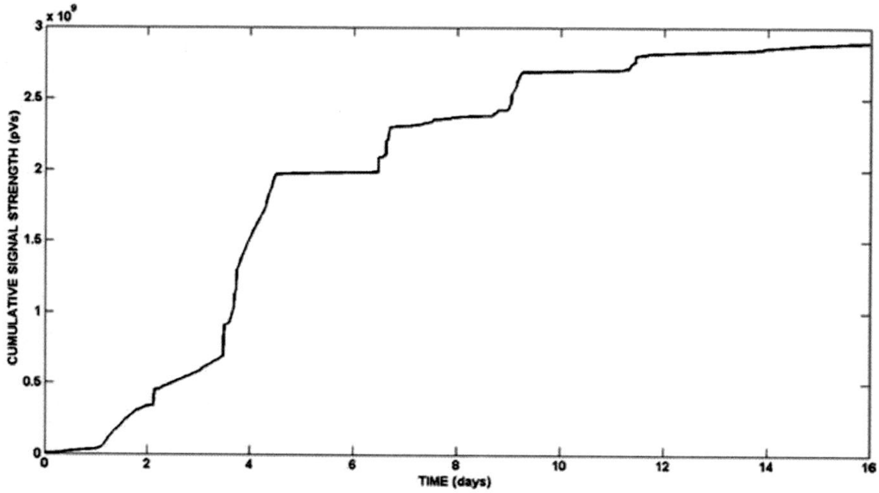

Fig. 4 Cumulative signal strength variation with time for cylinder C1

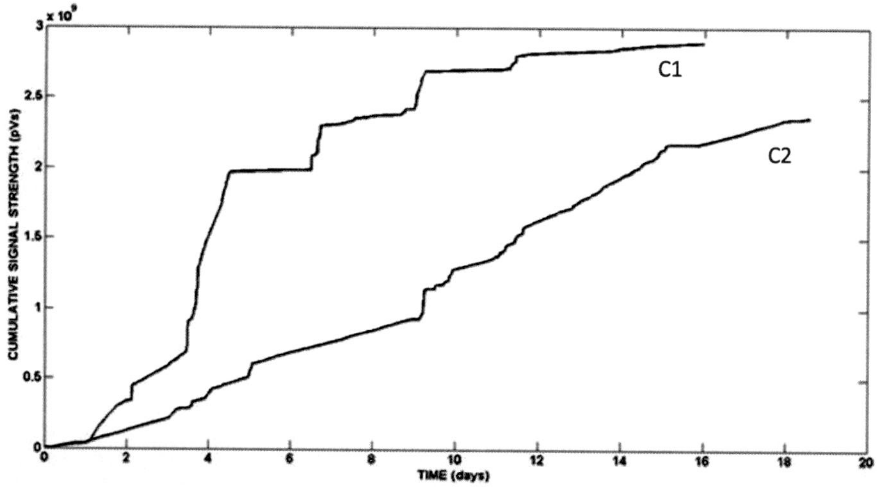

Fig. 5 CSS variation with time for cylinder C1 and cylinder C2

Figure 5 shows the variation of cumulative signal strength value for the cylinders C1 and C2. Because of smaller diameter of the bar in cylinder C2, the corrosion resulted in AE signal strength of lower value as accumulation of corrosion products were not enough to cause cracking as in cylinder C1. The CSS value for cylinder C2 was observed to be less than that for cylinder C1 owing to the same reason.

When the concentration of NaCl solution was increased from 3 to 5% in cylinders C3 and C4, the intensity of steel corrosion increased shown in Fig. 6. As a result, the AE signal strength generated due to steel corrosion was comparable to that of

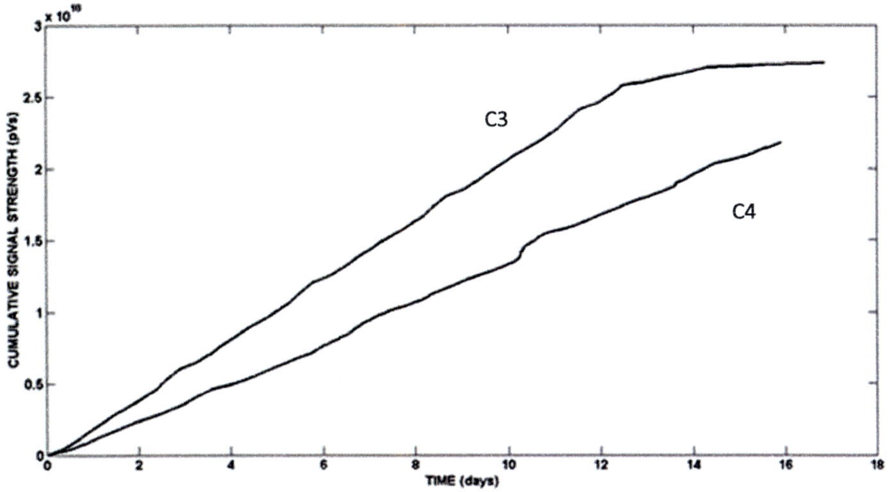

Fig. 6 CSS variation with time for cylinder C3 and cylinder C4

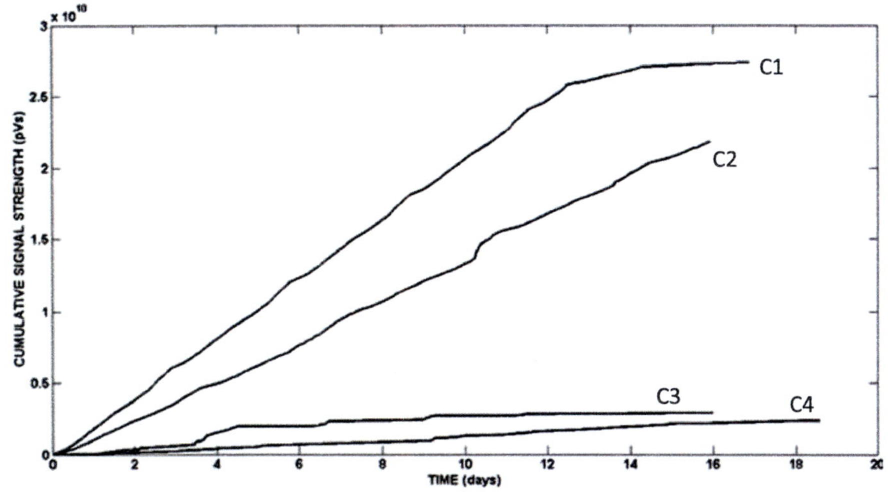

Fig. 7 CSS variation with time for test specimen cylinder-C1, cylinder-C2, cylinder-C3 and cylinder-C4

concrete cracking. Therefore, cracking of concrete was not clearly distinguishable from the knee in the CSS curve. Also, the CSS value for cylinder C4 was observed to be lesser than cylinder C3. The reason could be explained as because of smaller perimeter and area of cross-section of 8 mm bar, the hits generated were lower in number in C3 than in C4 (Fig. 7).

A clear trend in the variation was observed with the progression of corrosion. Initially, when the corrosion was started, de-passivation of steel rebar occurs resulting

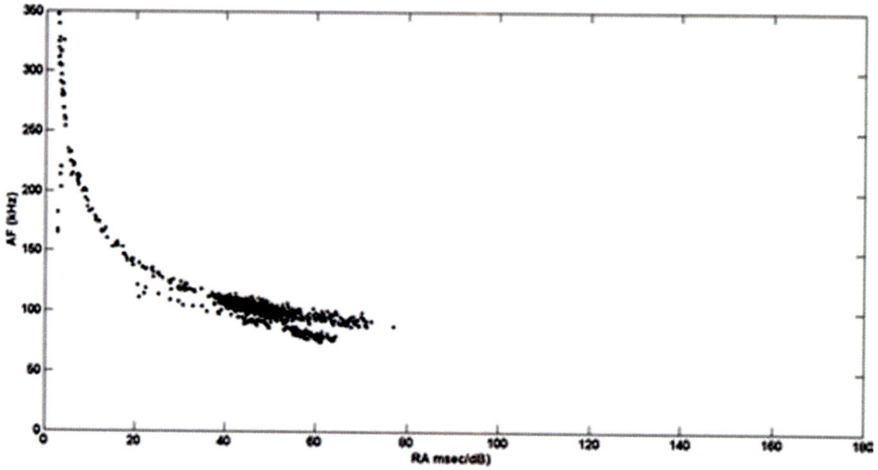

Fig. 8 RA versus AF plot for cylinder C1 (first 3 days of accelerated corrosion)

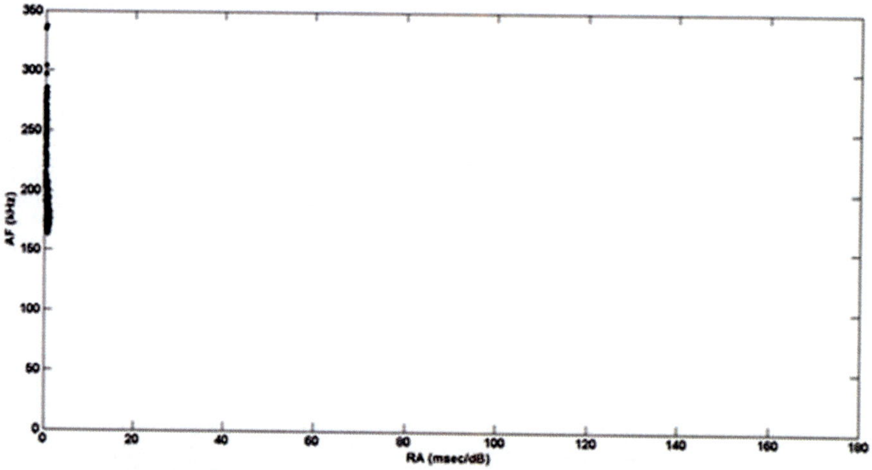

Fig. 9 RA versus AF plot for cylinder C1 (3 to 5 days of accelerated corrosion)

in release of AE with high AF and low RA value. But with the progression of steel corrosion, AF value started decreasing and RA value started increasing (Fig. 8). As more and more corrosion occurs, corrosion products accumulate on the surface of the steel bar which further affects the type of stress waves released. The corrosion rate progressively decreases and resulting in lower AF value. And RA value progressively increases as shown in Fig. 9. The tensile cracking of concrete is characterized by stress waves with high AF value and low RA value. The hits generated due to this are distinctively observed in Fig. 9 (Figs. 10, 11, 12 and 13).

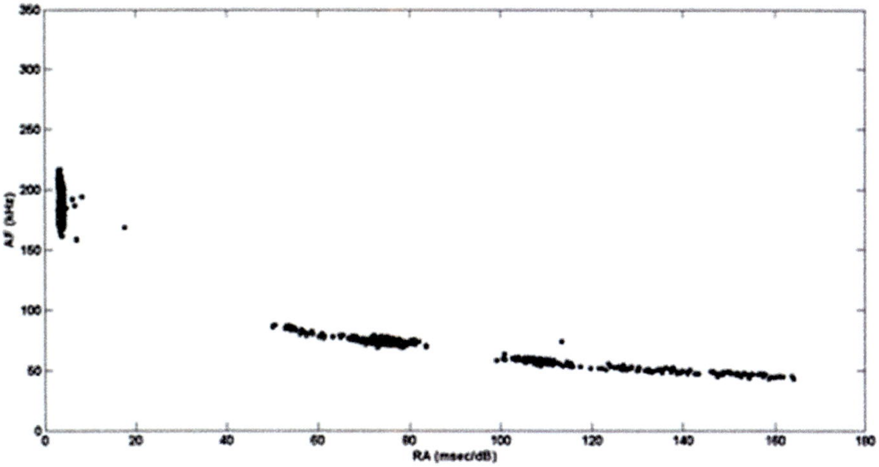

Fig. 10 RA versus AF plot for cylinder C1 (5 to 16 days of accelerated corrosion)

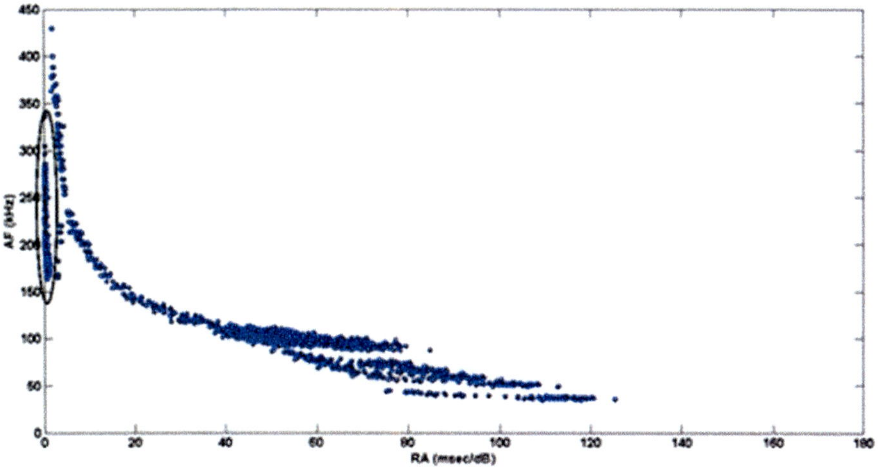

Fig. 11 RA versus AF plot for cylinder C2

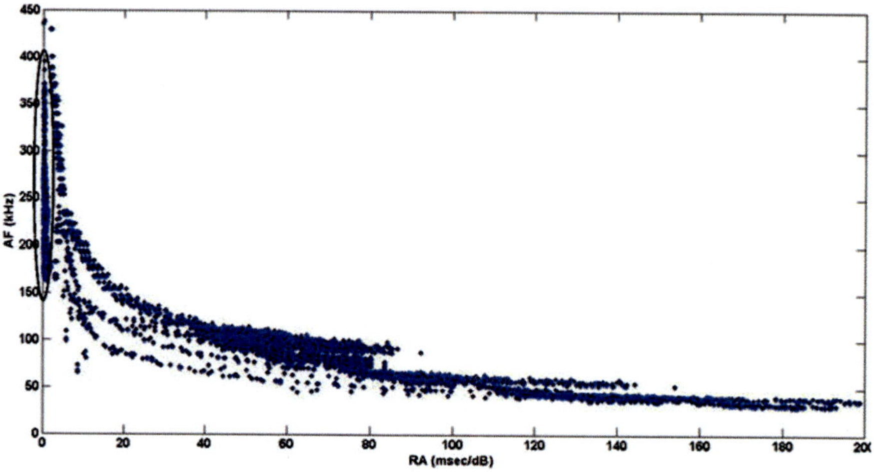

Fig. 12 RA versus AF plot for cylinder C3

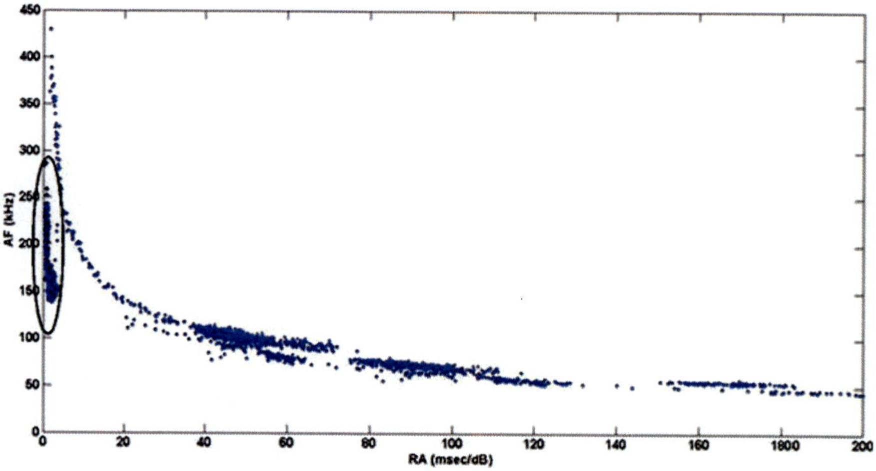

Fig. 13 RA versus AF plot for cylinder C4

6 Conclusions

Based on the above experimental observations, the given below major conclusions are drawn:

1. Corrosion of steel reinforcement in concrete can be monitored using AE technique. But this is very sensitive to the corrosive environment in which the steel reinforcement is degrading. In the above experimental work, the variation in the acoustic emission to varied corrosive environment is studied by changing the concentration of NaCl solution. The CSS curve showed a clear change in trend with increase in the percentage of NaCl solution.
2. The RA value increases while the AF value decreases with the progression of corrosion. The plots of AF versus RA for different specimens were plotted. The hits generated due to concrete cracking and steel corrosion can be clearly distinguished from these plots.

References

1. Zaki A, Chai HK, Aggelis DG, Alver N (2015) Non-destructive evaluation for corrosion monitoring in concrete: a review and capability of acoustic emission technique. Sensors 15:19069–19101
2. Kawasaki Y, Wakuda T, Kobarai T, Ohtsu M (2013) Corrosion mechanisms in reinforced concrete by acoustic emission. Constr Build Mater 48:1240–1247
3. Kalyanasundaram, Mukhopadhyay CK, Subba Rao SV. Practical acoustic emission. Indian Society for Non-Destructive Testing
4. Song H-W, Saraswathy V (2007) Corrosion monitoring of reinforced concrete structures—a review. Int J Electrochem Sci 2:1–28
5. Patil S, Karkare B, Goyal S (2014) Acoustic emission vis-à-vis electrochemical techniques for corrosion monitoring of reinforced concrete element. Construct Build Mater 68:326–332
6. Patil S, Goyal S, Karkare B (2015) Acoustic emission-based mathematical procedure for quantification of rebar corrosion in reinforced concrete. Curr Sci 109:943–948
7. Di Benedetti M, Loreto G, Matta F, Nanni A (2013) Acoustic emission monitoring of reinforced concrete under accelerated corrosion. J Mater Civ Eng 25:1022–1029
8. Ohtsu M, Tomoda Y (2011) Phenomenological model of corrosion process in reinforced concrete identified by acoustic emission. Concr Res Lett 2:280–285
9. El Maaddawy TA, Soudki KA (2003) Effectiveness of impressed current technique to simulate corrosion of steel reinforcement in concrete. J Mater Civ Eng 15:41–47
10. Yoon D-J, Jason Weiss W, Shah SP (2000) Assessing damage in corroded reinforced concrete using acoustic emission. J Eng Mechan 126:273–283
11. Idrissi H, Limam A (2014) Reliability of acoustic emission technique to assess corrosion of reinforced concrete. J Acoust Emission 18:307–313
12. Idrissi H, Limam A (2003) Study and characterization by acoustic emission and electrochemical measurements of concrete deterioration caused by reinforcement steel corrosion. NDT and E Int 36:563–569

Implementation of Acoustic Emission Testing to Study the Type of Cracking in Reinforced Concrete Beams

Swarnangshu Ghosh, Rithik Agrawal, and R. Vidya Sagar

1 Introduction

Reinforced concrete (RC) beams are one of the most important structural members used in modern construction. The RC beams form integral parts of buildings and various civil engineering structures. But many of the buildings around us have exceeded more than 45 years after their construction, and it is required to study the structural integrity of the same. Hence, it is of importance to assess the structural integrity of these buildings. It is known that the acoustic emission (AE) testing is a nondestructive evaluation method which can be used to assess the damage status and the type of cracking occurring in RC structures in real-time. AE refers to the transient elastic waves generated by the release of energy within a material during the fracture process in a solid [1]. Several investigators implemented AE testing to assess the damage and to study the fracture process in RC structures. Gaussian mixture modeling (GMM) and AE testing were implemented to study the crack classification in RC structures [2]. AE testing was also used for monitoring the structural health of steel–concrete (SC) composite shear walls [3]. The same methodology was implemented to show that RA-AF analysis can be used to differentiate between tensile and shear crack development in RC beams [4]. The damage in the RC beams is classified into four zones symbolizing the formation of microcracks, visible cracks, steel yielding, and concrete crushing. The AE parameters such as amplitude, rise time, count, duration and average frequency (=counts/duration), and RA (=rise time/peak amplitude) are quantified in each damage levels and a parametric analysis is performed between average frequency and RA value [5]. But studies related to

S. Ghosh (✉) · R. Agrawal
Department of Civil Engineering, National Institute of Technology, Surathkal 575025, India
e-mail: swarnangshughosh.171cv146@nitk.edu.in

R. Vidya Sagar
Department of Civil Engineering, Indian Institute of Science, Bangalore 560012, India

C. K. Mukhopadhyay and R. Mulaveesala (eds.), *Advances in Non-destructive Evaluation*, Lecture Notes in Mechanical Engineering,
https://doi.org/10.1007/978-981-16-0186-6_28

the sensitivity of AE toward the shear reinforcement in RC structures are reported minimum. In this study, the strain on the surface of the RC beam was measured at different time intervals during the experiment in the laboratory and is compared with the percentage of AE generated during tensile cracking and shear cracking. Also, the measured strain at mid-span of the RC beam may indicate the critical loads in an RC beam. In this experimental study, an attempt is made to connect that strain readings in different time intervals are in accordance with the GMM results. The aim is to study the sensitivity of the AE toward the shear reinforcement in RC beams.

2 Implementing Gaussian Mixture Modeling to Acoustic Emissions Generated During the Fracture Process in RC Beams

A Gaussian mixture model is a type of probability model which is used to group a large dataset into numerous clusters which also follow Gaussian distribution. Gaussian mixture modeling is a widely used probability model. Generally, it is used in statistical studies. It is known that there are two types of Gaussian distributions namely (1) univariate Gaussian distribution and (2) multivariate Gaussian distribution.

Univariate Gaussian distribution is given in Eq. (1)

$$N(x|\mu, \sigma) = \frac{1}{\sqrt{2\pi\sigma^2}} e^{\frac{-(x-\mu)^2}{2\sigma^2}} \tag{1}$$

where μ, σ and x are mean, standard deviation, and random variable, respectively. Multivariate Gaussian distribution is given in Eq. (2)

$$N(x|\mu, \Sigma) = \frac{1}{\sqrt{2\pi|\Sigma|}} e^{\frac{-(x-\mu)^T \sum^{-1}(x-\mu)}{2}} \tag{2}$$

where $\mu, \sum$ are mean and covariance matrix, respectively. GMM uses expectation maximization (EM) algorithm to group into clusters. EM algorithm is an iterative optimization technique. The EM algorithm has two steps namely (1) estimation step and (2) maximization step. In this study, the recorded AE dataset can be broadly divided into two clusters, one with a higher RA value and the other with a higher AF value. Linear superimposition of Gaussians is given in Eq. (3)

$$p(x) = \sum_{k=1}^{K} \pi_k N(x|\mu_k, \Sigma_k) \tag{3}$$

where K refers to the number of Gaussians and π_k refers to the weightage for each Gaussian distribution. In this experimental study, $K = 2$, as there are two clusters namely (1) AE related to tensile cracks and (2) AE related to shear cracks. The boundary conditions of this GMM are

$$0 \le \pi_k \le 1, \sum_{k=1}^{K} \pi_K = 1 \tag{4}$$

π_k refers to the weightage of each Gaussian distribution, i.e., weightage of the shear cluster and weightage of the tensile cluster in this case. The weightage of either of these Gaussians cannot be more than 1.0. The sum of these weightages also cannot exceed 1.0. A computer program to implement the Gaussian mixture modeling of AE was written in MATLAB. First, the RA and AF data points were plotted. Then, the mean and covariance of these data points were found out. The "ezsurf" function available in MATLAB was used to plot the 3D graph. The minimum and maximum values of RA and AF were kept as the parameters for this function. The ezcontour function was used to plot the contour graphs. Moving average function was also used to smoothen the data points in order to get the best-fit graphs. More details on the application of GMM of AE are given in [2].

3 Experimental Program

In the present experimental study, two RC beams were cast and tested using an MTS machine of 1200 kN capacity. These two RC beam specimens were cast with concrete having 28 days compressive strength of 21 MPa. The geometric details of the both RC beams are shown in Fig. 1. The spacing of the steel stirrups in the beam ff1 was 160 mm. No stirrups were present in the beam sf2. The cross-sectional area of tensile steel reinforcement in each RC beam specimen is 276.32 mm^2. Both the specimens were subjected to monotonically increasing loading under displacement control method. The loading rate for ff1 was 0.0035 mm/s and that of sf2 was 0.005 mm/s. The reinforcement details for both the beams are shown in Fig. 1. The surface strain at mid-section and near to a section present close to the left side and right-side supports were measured manually with a Demec gauge at different time intervals during testing.

3.1 AE Monitoring Test Setup

In this experiment, an eight-channel AE data acquisition system was used. 2D Planar *X-Y* mode was used to mount AE sensors on the RC beam specimen. The sensors used for the AE recording were resonant type, differential AE sensor, R6D (57 kHz) with a preamplifier gain of 40 dB (PAC, NJ, USA). The sensor has a sensitivity and frequency response over a range of 35–100 kHz. Silicon grease was used as a couplant to attach the AE sensor at the specified location and was tied firmly using

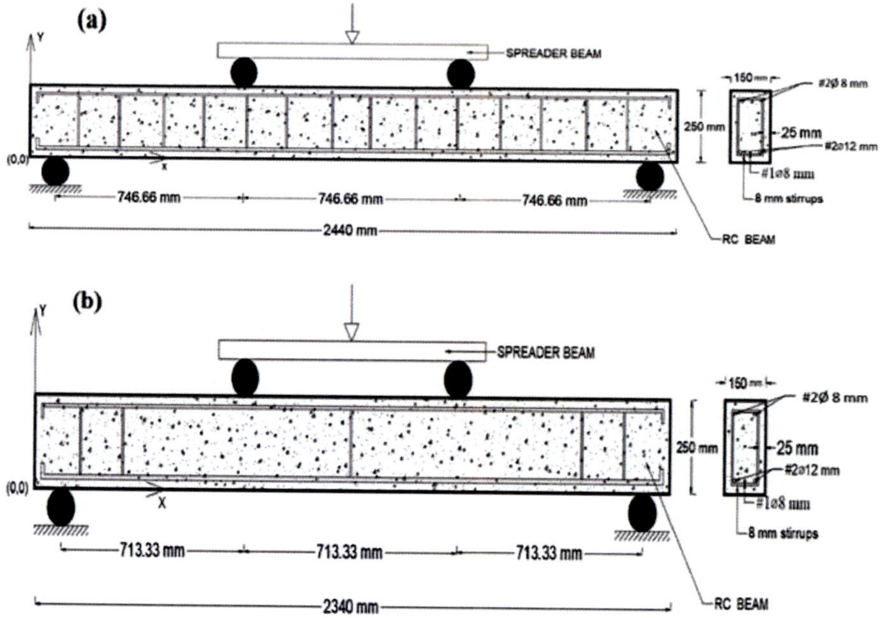

Fig. 1 Schematic representation of geometry and reinforcement details in RC beams **a** with shear stirrups **b** with no shear stirrups

tapes. Sensors were placed on the face of the beam considering the extreme left corner of the beam as the origin. Five AE sensors were placed on the RC beam in 2D planar with the coordinates as given in Table 1. The AE sensors locations on the tested RC beams are shown in Fig. 2. The experimental setup is shown in Fig. 3.

4 Results and Discussion

From Fig. 3a, it is observed that the beam with shear reinforcement (ff1) took more time to reach peak load than the RC beam with NO shear reinforcement (sf2). There is a sudden drop in load at failure in sf2, thereby indicating brittle failure. On the contrary, RC beam ff1 showed ductile nature before failure, i.e., steel started yielding and prevented the beam from sudden failure. From Fig. 3b, it is observed that specimen ff1 sustained load for a longer duration than RC beam sf2. The amount of energy released from the active microcracks is more, and it can be observed from Fig. 3b. There is an increase in the number of recorded hits in both the specimens as time progresses, but in sf2, it is a little more abrupt. From Fig. 3c, it can be observed that AE energy released from ff1 is relatively more than RC beam sf2 as ff1 underwent ductile failure thereby generating more microcracks before failure. It can be observed from both the plots (Fig. 3b, c) that there is a sudden rise in the

Table 1 AE sensor location coordinates

Specimen	AE sensor location coordinates [2D planar] (mm)									
	Ch-4		Ch-5		Ch-6		Ch-7		Ch-8	
	X	*Y*	*X*	*Y*	*X*	*Y*	*X*	*Y*	*X*	*Y*
ff1	1200	200	685	50	1755	50	150	200	2290	200
sf2	1220	50	685	175	1755	175	150	75	2290	75

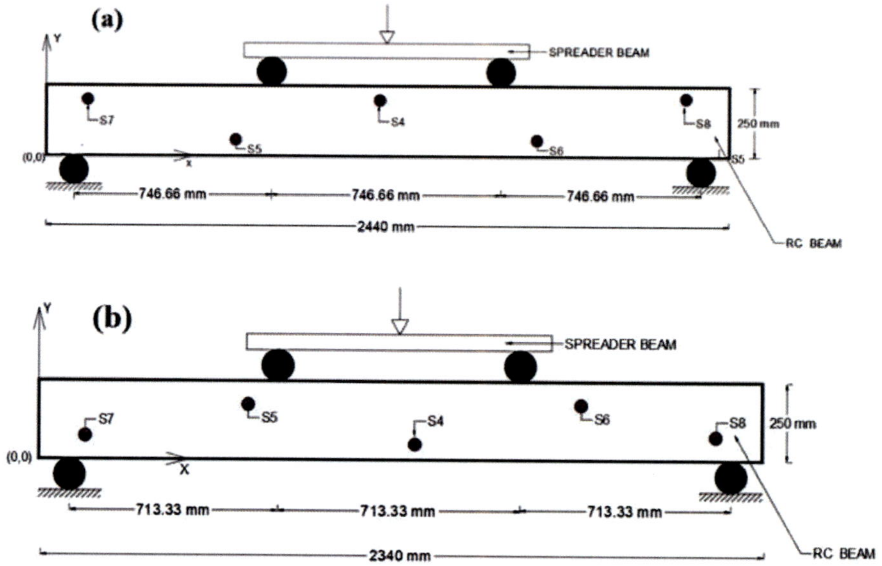

Fig. 2 Schematic representation of AE sensors located on the RC beam **a** with shear reinforcement **b** without shear reinforcement

Fig. 3 Experimental setup in the Structures Laboratory, Department of Civil Engineering, Indian Institute of Science, Bangalore, India

cumulative energy which is the outcome of initiation, coalescence, and propagation of microcracks (Fig. 4).

It is known that the area under the load versus time curve represents the energy consumed by the specimen during the fracture process. It can be seen that the beam with shear reinforcement showed the higher value of energy consumed than the beam without shear reinforcement. It can also be noticed that the total duration of

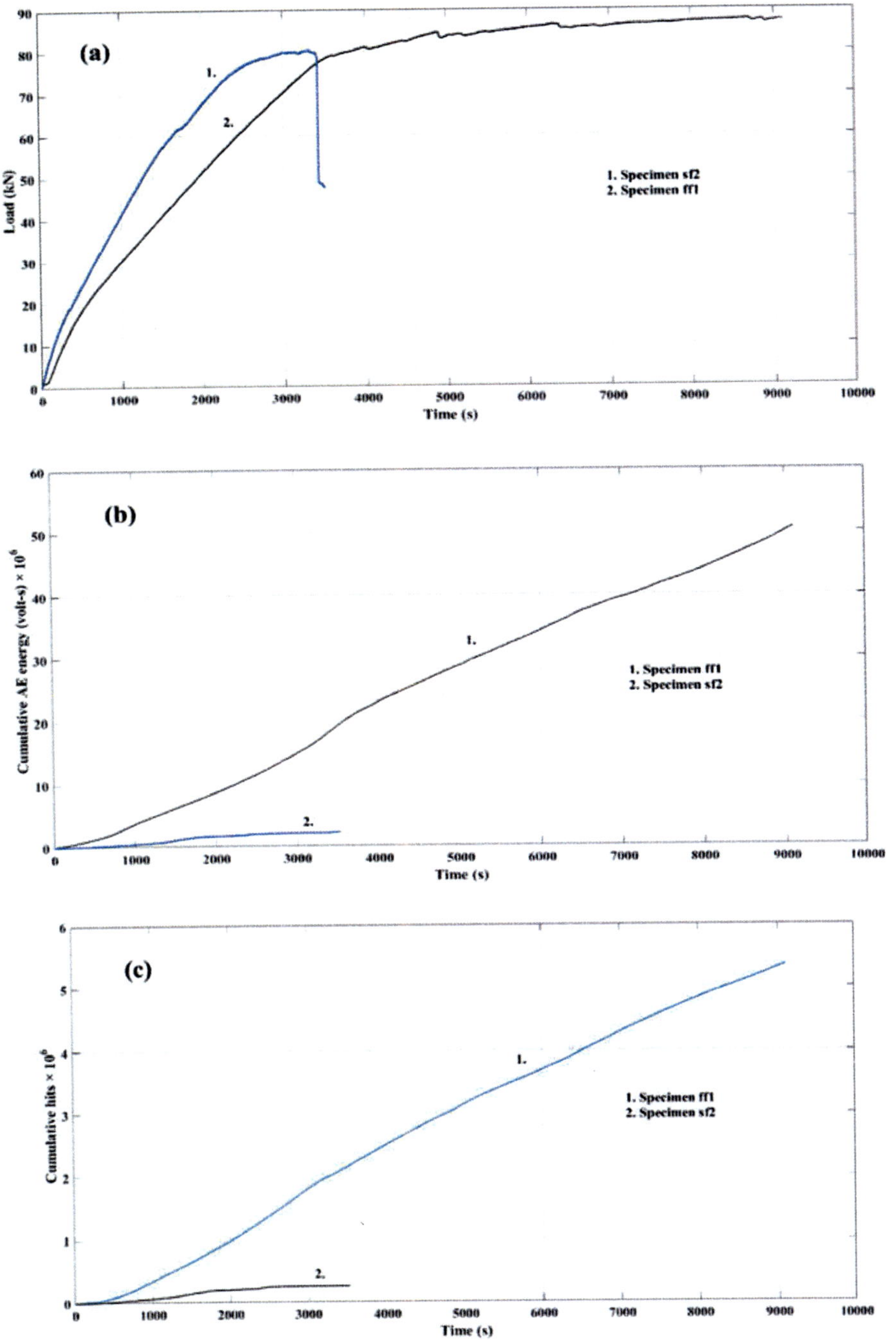

Fig. 4 Variation of **a** load **b** AE energy **c** AE hits with time in RC beams with shear reinforcement and without shear reinforcement

the experiment for the shear reinforced beam is higher than the beam with no shear reinforcement.

4.1 GMM Analysis of AE Generated During the Fracture Process in RC Beams

It can be observed from the GMM graphs shown in Fig. 5 that ff1 showed slow shear growth than in sf2. From Table 2, it can be observed that AE related to tensile cracks are more than shear cracks in beam ff1. It is also observed that the percentage of shear cracks increased in a steady manner till the collapse in ff1. From Table 3, it is observed that there is a sudden increase in released AE due to shear cracking in beam sf2 near to the failure stage. Also, from Figs. 5 and 6, it can be stated that the GMM analysis of AE is useful for the classification of cracks in RC structures. The results also let us conclude that before the last interval, the number of AE hits related to tensile cracking is more in number, and they have higher AF value and lower RA value. But in the last interval, AE related to shear cracks were released almost in the same number as that of tensile cracks for ff1, whereas in sf2, we can observe that in last interval, AE related to shear cracking generated is more in number than that of tensile cracks.

In Fig. 5, (ia) refers to the contour plot of the RA and AF data points and (ib) represents the 3D representation of the contour plot.

The contour and 3D plots for interval-3 are not meaningful, and the reason could be because of external noise during the generated AE.

4.2 Variation of Surface Strain in RC Beams with Shear Stirrups and Without Shear Stirrups

It can be observed from Fig. 7 that the surface strain at mid-span in RC beam with shear reinforcement (ff1) showed a gradual increase with respect to the applied load and became almost constant after 0.003. On the contrary, the beam without shear reinforcement (sf2) showed a sudden increase in the strain around 40 kN which can be observed from Fig. 8. The reason could be the absence of the steel stirrups in sf2. The maximum surface strain in the RC beam at the outermost fiber (above neutral axis) as can be observed from both Figs. 7 and 8 is almost near to 0.0035 which is the maximum allowable strain as per IS: 456:2000.

In this experimental study, crack classification in the RC beam was carried out with an approach based on GMM in connection with the strain data. The results from the GMM analysis and measured strain in ff1 showed that with the increase in intervals there is a gradual increase in the percentage of AE related to shear cracks (Table 2), and in the last interval, percentage of AE related to shear cracks is almost same in

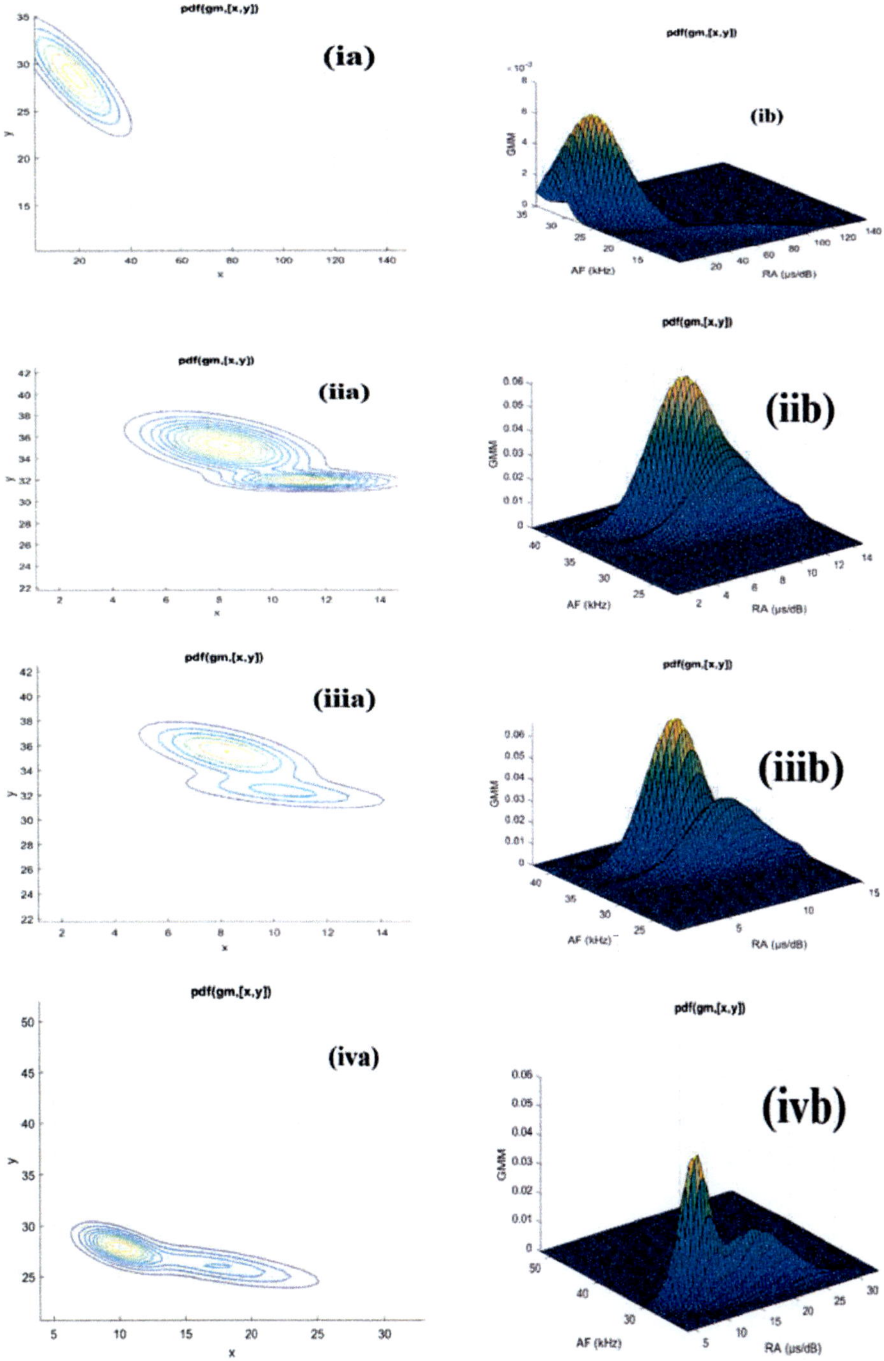

Fig. 5 GMM clusters variation in five intervals related to specimen ff1

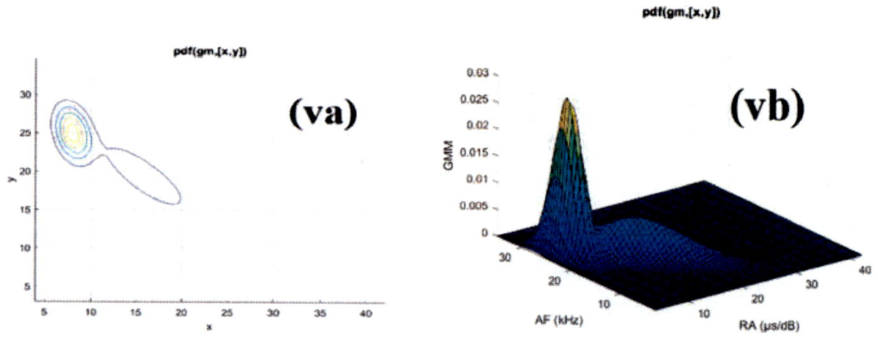

Fig. 5 (continued)

Table 2 Time intervals considered for the GMM algorithm and the AE recorded during tensile-type cracking and shear cracking for specimen ff1

S. No.	Time interval (s)	Load interval (kN)	AE hits (%)			Surface strain at mid-span
			Tensile	Shear	Mixed	
1	0–574.5	0–20.35	90.1	9.6	0.3	0.0005
2	574.5–1486	20.35–40.60	74.4	23.4	2.2	0.0013
3	1486–2277	40.60–56.86	63.6	34.3	2.1	0.0018
4	2277–3161	56.86–73.25	53.6	42.6	3.8	0.0023
5	3161–4000	73.25–81.22	51.5	46.3	2.2	0.0025

Table 3 Time intervals considered for the GMM algorithm and the AE recorded during tensile-type cracking and shear cracking for specimen sf2

S. No.	Load interval (kN)	Time interval (s)	AE hits (%)			Surface strain at a section near to the left support
			Tensile	Shear	Mixed	
1	0–15.9	0–273.3	95.6	4.3	0.1	0.0141928
2	15.9–39.9	273.3–966	80.8	15.1	4.1	0.0144440
3	39.9–55.9	966–1466	70.1	27.3	2.6	0.0160297
4	55.9–71.9	1466–2204	54.4	44.7	0.9	0.0221684
5	**71.9–79.9**	**2294–3395**	**36.9**	**62.9**	**0.2**	**0.0326246**

number as tensile cracks. The same can also be noticed from Fig. 7. On the other hand, there was a gradual increase in the percentage of AE related to shear cracks initially, but later toward the last interval, there is a sudden increase in the percentage of AE related to shear cracks (Table 3). This abrupt change in the percentage of AE related to shear cracks can also be noticed from Fig. 8. It is known that bending is accompanied by bending shear. Shear stress thus produced leads to diagonal tension in the RC beam. As concrete is weak in tension, large diagonal tensile stresses can

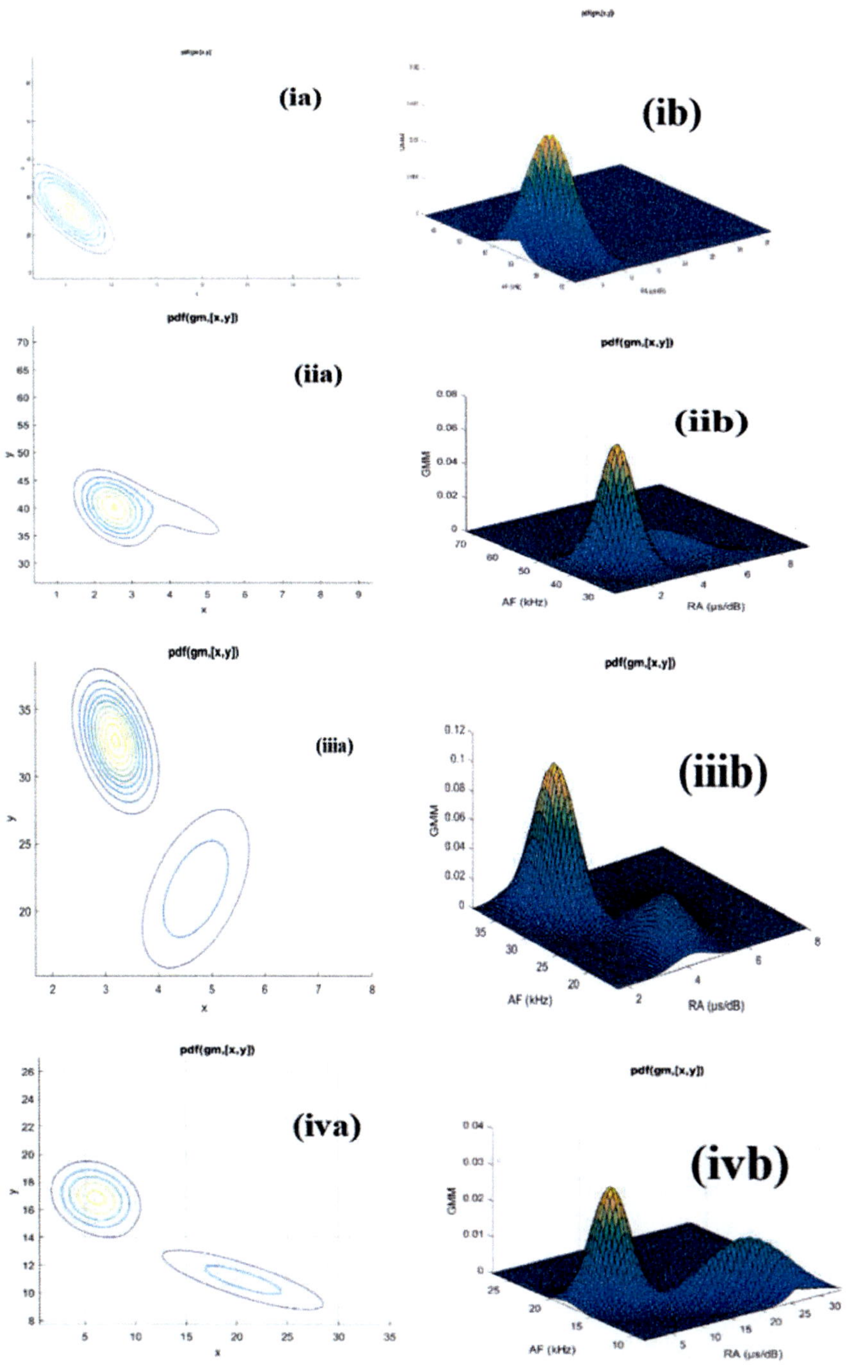

Fig. 6 GMM clusters variation at different time intervals in RC beam specimen SF-2

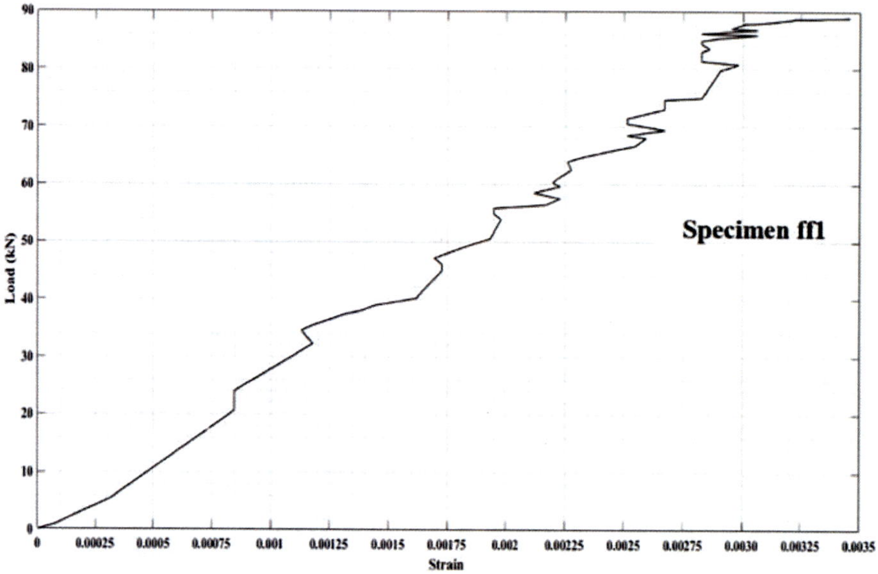

Fig. 7 Variation of surface strain at mid-span in RC beam ff1

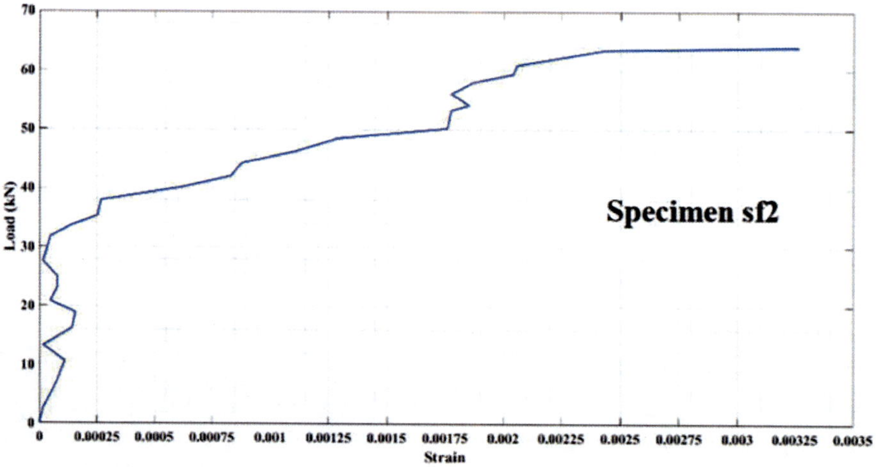

Fig. 8 Surface strain variation with load in specimen sf2 near to left support

produce cracking and even failure of the concrete member as shown in Fig. 9a, b. To prevent this diagonal failure, shear reinforcement needs to be provided. Since RC beams with shear reinforcement prevent the propagation of diagonal cracks (because of stirrups), tension cracks propagate in the majority, and the beam fails in tension.

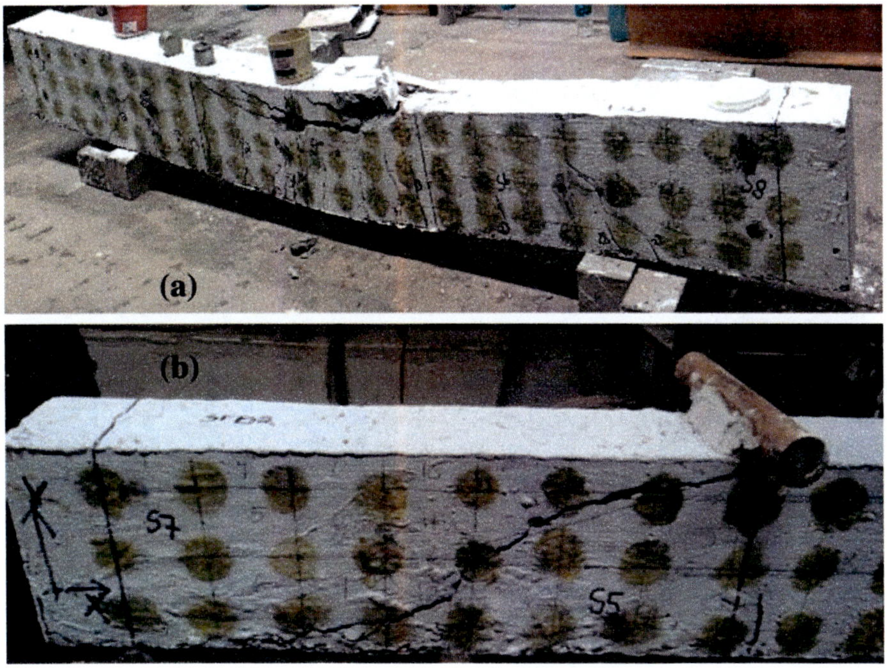

Fig. 9 RC beams after the test. **a** Development tensile crack at mid-span in specimen ff1 **b** diagonal shear crack developed in specimen sf2

$$\tan 2\theta = \left| \frac{\sigma_x - \sigma_y}{2\tau} \right| \tag{5}$$

σ_x is the normal external stress in the x-direction (Fig. 1b). σ_y is the normal external stress in y-direction, and τ is the shear stress. θ is the angle between maximum principal stress and the horizontal measured in the counter-clockwise direction. If there is no externally applied shear stress, $\tau = 0$. Hence, $\tan 2\theta = \alpha$, and hence, $\theta = 45°$. Therefore, diagonal cracks as shown in Fig. 9b are developed at angle 45°.

5 Conclusions

Based on the above experimental observations, the following conclusion is drawn.

1. It can be observed from load versus time plots that there is sudden failure in case of the beam without shear reinforcement, whereas beam with shear reinforcement showed no sudden failure.
2. It can be observed from the GMM analysis of AE plots for both the specimens that there is an initial dominance of AE related to tensile cracks. But in the beam lacking shear reinforcement, toward the final loading stages, there was a sudden

increase in the percentage of AE related to shear cracks, whereas in the beam with shear reinforcement, there was a gradual increase in AE till collapse.

3. It can be observed that in the RC beam without shear reinforcement, there was a sudden increase in the strain with a small increase in the load, depicting the large generation of cracks. But in the case of the beam with shear reinforcement, there was a gradual increase in the strain with the increase in load.

In future studies, the observations obtained from acoustic emission and thermography can be compared.

References

1. Gross CU, Ohtsu M (2008) Acoustic emission testing, 1st edn. Springer, Berlin, Heidelberg
2. Farhidzadeh A, Singla P, Salamone S (2013) A probabilistic approach for damage identification and crack mode classification in reinforced concrete structures. J Intel Mat Syst Str 24(14):1722–1735
3. Farhidzadeh A, Epackachi S, Salamone S, Whittaker AS (2015) Bayesian decision and mixture models for AE monitoring of steel-concrete composite shear walls. Smart Mater Struct 24(11):115028
4. Shahidan S, Pulin RHYS, Bunnori NM, Holford KM (2013) Damage classification in reinforced concrete beam by acoustic emission signal analysis. Constr Build Mater 45:78–86
5. Prem PR, Murthy AR (2017) Acoustic emission monitoring of reinforced concrete beams subjected to four-point-bending. Appl Acoust 117:28–38

Vacuum and Helium Leak Testing Techniques Used for Very Large Size Vacuum Chambers

Venkat N. Ramani, M. Satheesh Kumar, and S. Sachuthan

1 Introduction

Helium leak detection using mass spectrometric technique is one of the major leak testing methods adapted in NDT [1]. The application of this technique to laboratory and small vacuum systems has been amply exploited [2, 3]. In this technique of leak detection, there are mainly two broad categories. One category involves leak detection methods for pressurized vessels and the second category involves leak detection methods for vacuum vessels. In the methods for pressurized vessels, the vessel is filled fully or partially with helium to the test pressure, and the leakage of helium is detected using a sniffer probe. This leak detection method is known as detector probe technique or inside-out technique. To measure the integrated value of the helium leak rate from the system, a hood or a closed envelope is formed where the leakage of helium gets collected, and the helium concentration is then sensed to estimate the value. Detailed calibration procedures for the above techniques are well described in codes and standards for relevant manufacturing disciplines [4]. In the methods for vacuum vessels, the helium leak detector is connected to the vacuum system and helium is sprayed at the test points using a gun. If there is a leak, helium enters into the vacuum vessel and the leak is sensed by the helium leak detector. This leak detection method is known as tracer probe technique or outside-in technique. If the integrated helium leak rate is needed, the helium is injected in a hood that encloses the vessel. System calibrations are required to determine the leak detection sensitivity and response time in this method [4]. The advent of high

V. N. Ramani (✉) · M. Satheesh Kumar
Plasma and Vacuum Technologies, GIDC Kathwada, Ahmedabad 382430, Gujarat, India
e-mail: vnramani@plasvac.com

S. Sachuthan
Rajiv Gandhi Institute of Technology, Bangalore, Gujarat, India
e-mail: sachuthans@gmail.com

C. K. Mukhopadhyay and R. Mulaveesala (eds.), *Advances in Non-destructive Evaluation*, Lecture Notes in Mechanical Engineering,
https://doi.org/10.1007/978-981-16-0186-6_29

sensitive, wide range and precise helium leak detecting mass spectrometers has made the applications of helium leak testing method widespread in many industrial vacuum systems as well as pressure vessels. The application areas include petrochemical, pharmaceutical, chemical, nuclear and power plants, in addition to space simulation vessels and various components of R&D systems. The application of MSLD has found an immense relevance, particularly in determining and locating minute leaks from welds and joints. Mass spectrometric helium leak detection (MSLD) today is considered as the ultimate leak detection tool since it offers a very high sensitivity and short response time. The large range of leak rates, typically in the range from 10^{-10} to 10^{-4} Pa.m^3/s, detection of precise location of leak, its capability to be applied in complex and difficult environments, its usefulness to test under on-line or off-line conditions, and its high reliability combined with repeatability, have all made this technique very relevant in today's industrial situations.

In this paper, we present a few specific techniques for the application of helium leak detection in large size vacuum vessels. The need for these special specific techniques arises due to the fact that simple and direct approaches give a very low sensitivity of detection and a very large response time. To make this point clear, let us consider the following example. If the volume of vacuum vessel is about 100 m^3, and the typical helium pumping speed of mass spectrometric leak detector is 10 l/s, then the response time constant (= volume/pump speed) for leak detection is ~2.77 h. And, if a designed pumping system for the vacuum vessel (which may have typically a pumping speed of about 10,000 l/s), is used in parallel, the sensitivity will drop by a factor of 1000. Both the sensitivity and response time, as in the above case, are generally not acceptable in practical situations. In addition, the need to maintain the working condition during leak detection, makes the specific techniques described in this work, very relevant and useful.

2 Development of Techniques and Observations

2.1 Techniques Developed

Three different techniques for helium leak detection in large size vacuum vessels are described in this work. The developed techniques are specifically useful for: (i) vacuum vessels operating at low vacuum level; (ii) vacuum vessels operating at high vacuum level, demanding a high sensitivity of detection with a short response time; and (iii) vacuum vessels operating at high vacuum level, allowing a lower sensitivity of detection with a short response time.

2.1.1 Technique for Vacuum Vessels Operating at Low Vacuum Level

Large vacuum vessels operating at low vacuum level, typically >100 Pa, make it either unable for the mass spectrometric leak detector to be deployed or necessary to connect it in the gross leak rate measurement mode. The inherent configuration of gross leak rate measurement mode in a leak detector reduces the sensitivity of detection, generally by 3 orders of magnitude. To increase this sensitivity of detection or to enable the mass spectrometric helium leak detector to work in normal or high sensitivity mode, the use of a helium permeator is adapted in this technique. Towards this purpose, we developed the permeators and characterized them in the laboratory. When a permeator is placed in the path connecting the leak detector to the vacuum system, the background or process gas flow is restricted and relatively more of helium flows to the detector. This reduces the background, increases the sensitivity of detection, and eases the application of the technique over a large range of vacuum. Experiments were conducted with two specially designed permeators, and the observations were made on sensitivity and its variation over a wide vacuum range—1000 to 10^{-1} mbar. The experimental set-up established for the study of the performance of the permeator is given in Fig. 1.

The development of permeator involved making a diaphragm, uniformly thick, having a reasonable area, through which adequate helium can permeate while the other gases—like oxygen, nitrogen, or other process gases like methane, find it as a barrier. After many trials with many membranes, it was decided to use two geometries—Permeator I: cylindrical diaphragm made of PTFE, with a diameter ~20 mm, length 25 mm, and thickness ~500 μm and Permeator II: cylinder made of Quartz, with a diameter ~10 mm, length 250 mm and thickness ~100 microns. The permeators were mounted inside a stainless steel cylindrical pipe having standard KF25 end flanges. Permeator I was operated at room temperature while Permeator II was operated at 200 °C.

The experimental set-up was designed to provide a typical situation that is envisaged in process systems. The vessel used is a large rectangular vacuum chamber (5 m × 5 m × 5 m), and it was pumped using a roots pump (Hind High Vacuum make, RD500) and rotary pump (Hind High Vacuum make, CD120) combination. The ultimate vacuum achievable in the chamber was better than 1 Pa. A combination of 2 capacitance manometers (Alcatel make ASD2001 and ASD2004) measured the vacuum level over the range 10^5 to 1 Pa. With the help of variable leak valve V1, any vacuum in the range of 10^5 to 10 Pa could be steadily maintained in the chamber. The permeator was connected to the pumping line through an isolation valve V3. The other end of the permeator was connected to a mass spectrometric helium leak detector [MSLD] (Alcatel make, ASM310) through an isolation valve V6. A standard calibrated leak-1 (Hind High Vacuum make, Model HSL 101, 2.6×10^{-9} Pa m^3 s^{-1}) was used for the instrument calibration of the helium leak detector. Using vacuum lines with valves V2, V4, V5 and V6 as shown in schematic and using the rotary vacuum pump 2 (Hind High Vacuum make, Model ED6) along with Pirani Gauge 2 (Hind High Vacuum make, Model DHPG-222), appropriate vacuum levels were created for connecting the standard leak and permeator to MSLD. Another calibrated

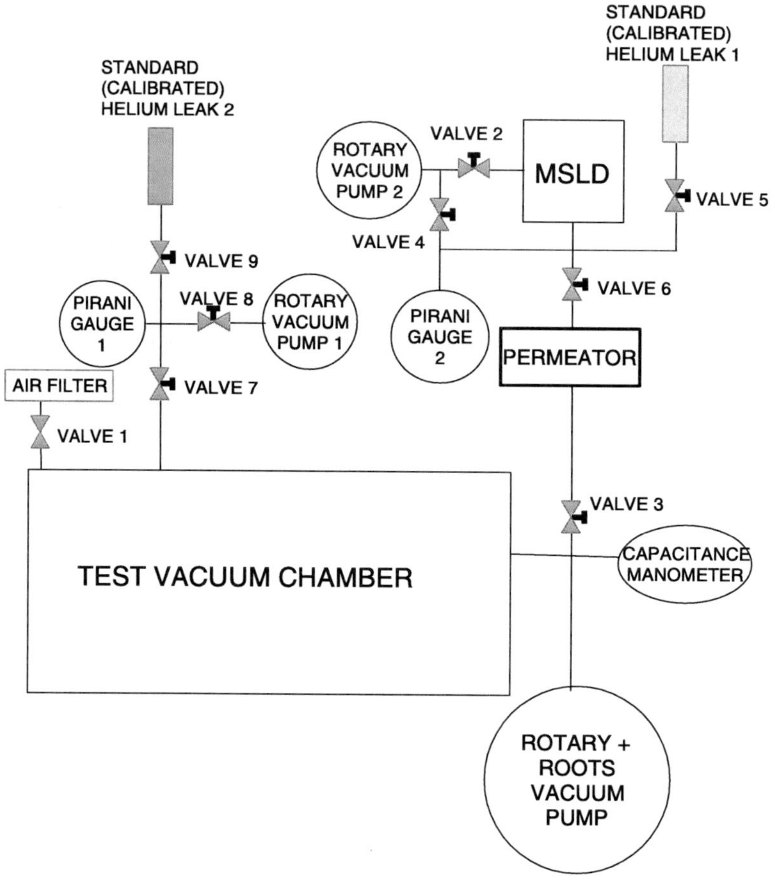

Fig. 1 Schematic of experimental set-up for the study of permeator deployed helium leak detection in a low vacuum system

leak—standard leak-2 was connected to the chamber. At low vacuum (>above 10 Pa) and high helium background values ($>10^{-7}$ Pa m^3 s^{-1}), the standard leak-2a used was of a higher value (Alcatel make, Model FV3, 1.0×10^{-5} Pa m^3 s^{-1}). At high vacuum (below 10 Pa) and low helium background values ($<10^{-7}$ Pa m^3 s^{-1}), the standard leak-2b used was of a lower value (Hind High vacuum make, HSL101, 1.0×10^{-7} Pa m^3 s^{-1}). These standard leaks were used to determine the leak detection response of the configured system. Using vacuum lines with valves V7, V8 and V9 as shown in schematic and using the rotary vacuum pump 1 (Hind High Vacuum make, Model ED6) along with Pirani Gauge 1 (Hind High Vacuum make, Model DHPG-222), appropriate vacuum levels were created for connecting the standard leak-2 to chamber without affecting the vacuum level in chamber.

The instrument calibration was first performed on MSLD using the standard leak-1, and the sensitivity of the instrument was noted to be better than 1.0×10^{-10} Pa m^3 s^{-1}. Permeator I was introduced in the vacuum circuit at the location shown in the schematic. With this permeator, it was found that the MSLD could be operated at around 1 kPa. The background level in MSLD was ~1.0×10^{-7} Pa m^3 s^{-1}. When the standard leak-2a was introduced into the chamber, the observed signal was ~5.0×10^{-7} Pa m^3 s^{-1} and the sensitivity was evaluated as 0.04. The experiment was repeated at vacuum levels 100 and 10 Pa. The observed sensitivities were found to be 0.06 and 0.09, respectively.

The experiment was repeated with Permeator II. With this permeator, it was found that the MSLD could be operated even at 100 kPa. The background level in MSLD was ~1.0×10^{-9} Pa m^3 s^{-1}. When the standard leak-2b was introduced into the vessel at a pressure of 50 kPa, the observed signal was ~5.5×10^{-9} Pa m^3 s^{-1}. The sensitivity was evaluated as 0.055. The sensitivity was found to improve at lower pressures.

2.1.2 Technique for Vacuum Vessels with Low Gas Load Operating at High Vacuum Level

The experimental set-up, used for studying helium leak detection in large vacuum vessels with low gas load operating at high vacuum, typically <1 Pa, is given in Fig. 2.

In this set-up, a high vacuum is achieved in the vessel using the diffusion pump, which is backed by a roots and rotary pumps combination. The term "low gas load" is defined for the situation where the total integrated leakage rate to the vacuum system is very small causing an insignificant gas load to the backing pump combination. In such a case, after the required vacuum is achieved, the roots and rotary pumps combination is isolated by closing valve V7, and the helium leak detector is directly connected to the backing side of the high vacuum pump. The vacuum pump of the leak detector acts as the backing pump and the detector has the maximum sensitivity for helium detection.

All the components used in this set-up have been already described in the last section. The only addition is the high vacuum pump, which is a diffusion pump (Hind High Vacuum make, Speed 20,000 l/s), and the ultimate vacuum achieved was better than 0.001 Pa. The instrument calibration was first performed on MSLD using the standard leak-1, and the sensitivity of the instrument was noted to be better than 1.0×10^{-10} Pa m^3 s^{-1}. The leak detector was then connected to the pumping system at the backing of the diffusion pump. Valve 7 was then closed. The background level in MSLD was ~2.0×10^{-10} Pa m^3 s^{-1}. When the standard leak-2b was introduced into the chamber, the observed signal was ~9.9×10^{-8} Pa m^3 s^{-1}, and the sensitivity observed was ~1. The time taken for achieving the steady value was found to be ~1 min. Thus, a very high sensitivity and fast response in helium leak detection in the large vessel is achieved using this configuration. Theoretically, near steady state is expected to be achieved in 5 times the response time constant. The opening and closing of the standard leak—2b were performed 3 more times with a time interval of 30 min, and the stability of the helium leak detection method was found to be very good.

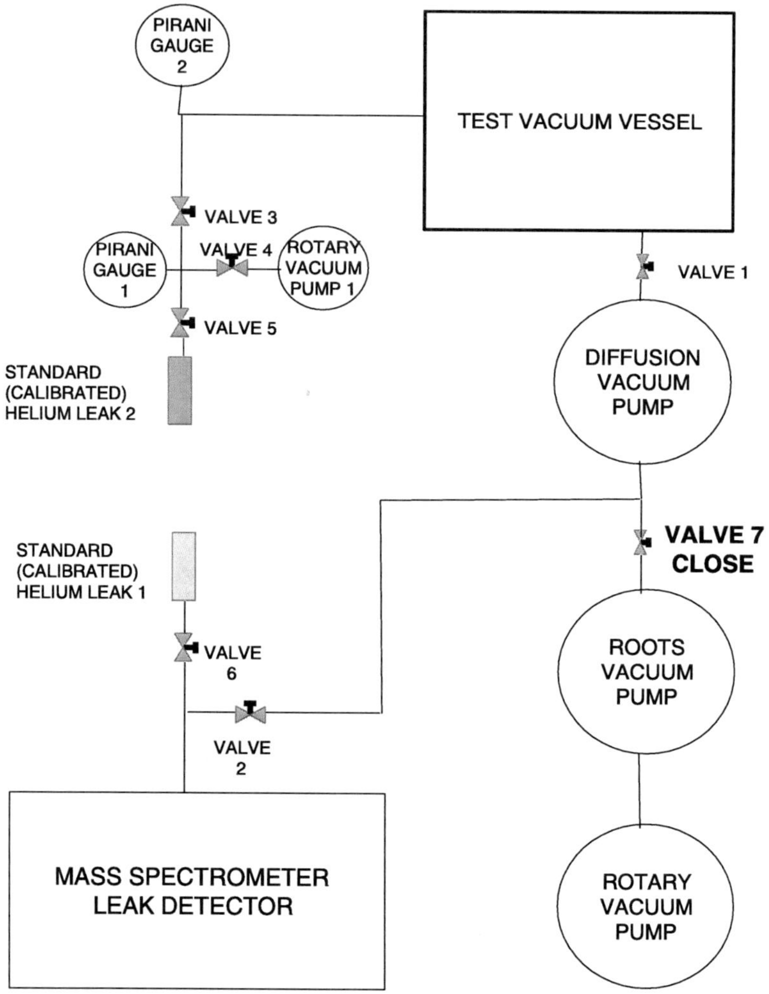

Fig. 2 Schematic of experimental set-up for the study of helium leak detection in a low gas load high vacuum system

2.1.3 Technique for Vacuum Vessels with High Gas Load Operating at High Vacuum Level

The experimental set-up, used for studying helium leak detection in large vacuum vessels with high gas load operating at high vacuum, typically <1 Pa, is given in Fig. 3. The same set-up used in previous section was used, except the condition of the Valve 7. The term “high gas load” is defined for the situation where the total integrated leakage rate to the vacuum system is significant, and hence, it causes a significant gas load to the backing pump combination. In such a case, after the required vacuum is achieved, the roots and rotary pumps combination cannot be

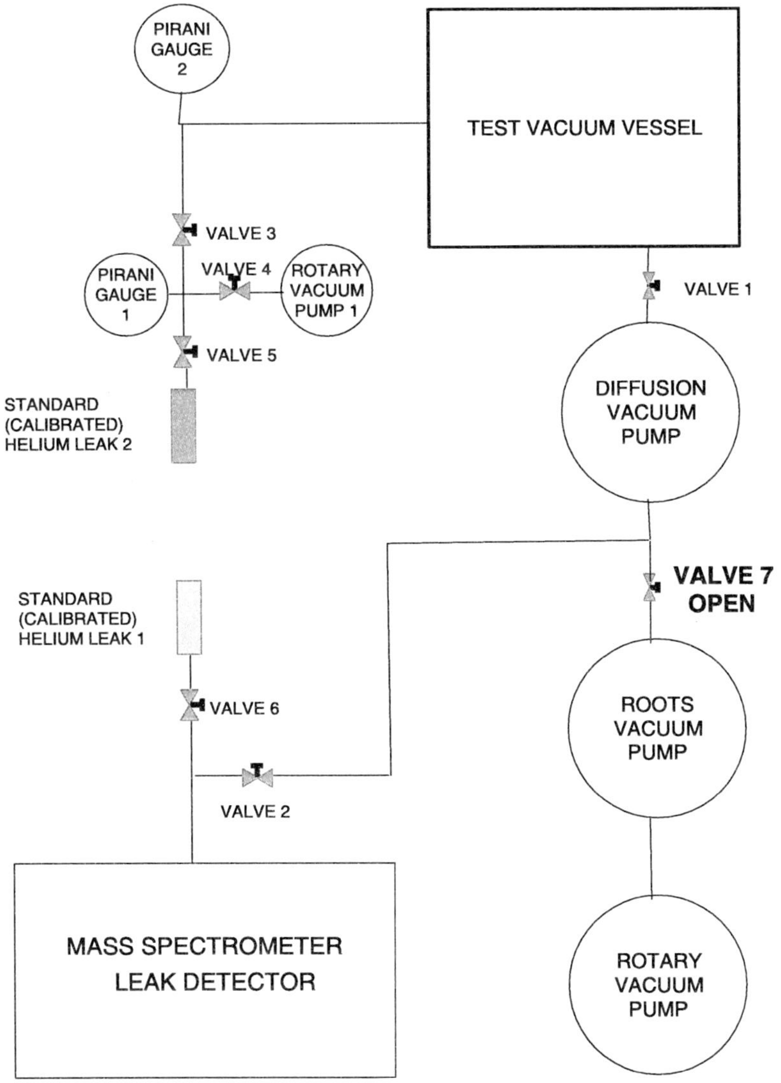

Fig. 3 Schematic of experimental set-up for the study of helium leak detection in a high gas load high vacuum system

isolated maintaining the achieved vacuum. Hence, the valve V7 is kept open and the helium leak detector is connected to the backing side of the high vacuum pump in parallel. The gas load is divided in the two paths—partly to the vacuum pump of the leak detector and the remaining to the roots and rotary pump combination. The ratio of the gas loads to the two branches is exactly as per the ratio of the pumping speeds for helium gas in the two branches. Thus, for example, if the roots and rotary pumps combination has a pumping speed nine times that of the pumping speed of helium leak detector, then only one tenth of the gas flow will go to the helium leak detector. Then, the sensitivity of the detector is reduced by 10 times. However, the response time remains very small due to the presence of high vacuum pump in the path. When the leak detector was connected to the pumping system at the backing of the diffusion pump, with Valve 7 kept open, the background level in MSLD was ~1.0×10^{-11} Pa m^3 s^{-1}.

When the standard leak-2b was introduced into the chamber, the observed signal was ~1.1×10^{-9} Pa m^3 s^{-1} and the sensitivity observed was ~0.02. The time taken for achieving the steady value was found to be ~1 min, same as in the previous case. Thus, although the sensitivity was found to get reduced, the fast response in helium leak detection in the large vessel was achieved using this configuration. The opening and closing of the standard leak—2b were performed 3 more times with a time interval of 30 min, and the stability of the helium leak detection method was found to be very good.

2.2 *Discussion*

The table below summarizes the tests conducted and results obtained on the sensitivity and measured response time for the different techniques at specific vacuum levels.

Technique deployed	Vacuum category	Gas load category	Test vacuum (Pa)	Permeator used	Reference standard leak used (Pam^{-3} s^{-1})	Sensitivity obtained	Observed response time (S)
1	Low	High	1000	PTFE	1.0E−5	0.04 ± 0.005	910 ± 15
			100	PTFE	1.0E−5	0.06 ± 0.008	890 ± 10
2	High	Low	10	PTFE	1.0E−7	0.09 ± 0.01	905 ± 6
			5000	Quartz	1.0E−5	0.055 ± 0.006	925 ± 30
3	High	High	0.001	–	1.0E−7	1 ± 0.1	60 ± 4
			0.001	–	1.0E−7	0.02 ± 0.002	60 ± 10

The technique adapted, using the permeator, demonstrated the applicability and usefulness for the helium leak detection in low vacuum vessels. The experiments show that leak rates of value 1.0×10^{-7} Pa m^{-3} s^{-1} and above can well be detected with good sensitivity and reasonable response time using the permeator as barrier between vacuum system and the mass spectrometric leak detector. The variation in the sensitivity is attributed to the transitional nature of gas flow in the low vacuum regime and due to the collisions of the helium atoms with the background gas molecules. The techniques adapted for high vacuum vessels demonstrated the high efficiency of the helium leak detection with very good response time for very large vacuum vessels.

One important aspect, relevant to the helium leak testing in very large vacuum vessel, is the temporary joints that are made for connection to the pumping system and leak detector. Permeation of helium through the large diameter temporary joints, if made from elastomer material, like Neoprene or Viton, vitiates the background value considerably and thereby prohibiting the measurement of helium leak rate through the permanent and demountable joints of the vacuum vessel. A successful test set-up for large volume vacuum vessel must ensure all the large diameter temporary joints are made of double elastomer rings or metal seals.

3 Conclusions

Specially designed permeators can enable successful application of helium leak detection technique in low vacuum large volume vessels. In high vacuum systems, the connection of the mass spectrometer leak detector to the backing side of high vacuum pump, leads to a good response and high sensitivity leak detection. The technique can be also used in high gas load systems by keeping the backing pumps in operation, although sacrificing in the sensitivity of leak detection. Prediction of the sensitivity and response time is possible to a limited extent.

Acknowledgements The authors wish to acknowledge the technical support of PlasVac team members—Mr Ranjit Kage, Mr Pritesh Modi, Mr Bhautik Patel, Mr Akshay Patel, Mr Dharmendra Makwana and Mr Mahendra Rajput, in carrying out the leak testing works at site.

References

1. Rozanov LN (2002) Vacuum technique, edited by M.H. Hablanian. Taylor & Francis, Milton Park
2. Methods of Experimental Physics (1979) In: Weissler GL, Carlson RW (eds) Vacuum physics and technology, vol 14. Academic Press, Cambridge
3. O'Hanlon JF (1980) A user's guide to vacuum technology. Wiley, Hoboken
4. ASNT (2011) In: Moore PO (ed) Non-destructive testing handbook, 3rd edn, vol 1. Leak testing. ASNT, USA

Coded Excitation for Low-Power Operation in Guided Ultrasonic Wave Non-destructive Evaluation

Shashvat Jayakrishnan, Dileep Koodalil, Nived Suresh, and Krishnan Balasubramaniam

1 Introduction

Ultrasonic NDE measurements are widely used in industry to ensure the integrity and worthiness of a structure/component. The advantage of ultrasonic methods is that they can mechanically interrogate a structure. Guided ultrasonic waves, which can travel long distances, can be employed to probe hidden and inaccessible areas which are far apart and complex structures can be inspected. However, the downside is that—especially for non-contact methods of transduction like air-coupled and electromagnetic acoustic transducers (EMATs)—the input power needs to be very high to achieve satisfactory SNR. To match the high input power needs, the current practice in the industries is to employ high-power electronics and increase the extent of averaging of the acquired signal [1, 2]. On one hand, high-power electronics tend to make the entire setup bulky and, these are intrinsically unsafe as well. Therefore, meeting the safety standards of the industry—especially for safety critical industries—becomes cumbersome. On the other hand, applying high averaging increases the pause time between consecutive data acquisitions, which thereby increases the overall time required for inspection. This is a very critical issue in large-scale production and manufacturing industries. In the case of coded excitation, SNR can be improved—without compromising the energy budget—by employing temporally long low-power sequences instead of the conventional, short spanning, high power bursts. Such sequences also reduce the time required for averaging and thereby reduce

S. Jayakrishnan (✉)
Department of Instrumentation and Control Engineering, National Institute of Technology Tiruchirappalli, Tiruchirappalli 620015, Tamil Nadu, India
e-mail: shashvatjk@gmail.com

D. Koodalil · N. Suresh · K. Balasubramaniam
Centre for Non-Destrcutive Evaluation, Indian Institute of Technology Madras, Chennai 600036, Tamil Nadu, India

C. K. Mukhopadhyay and R. Mulaveesala (eds.), *Advances in Non-destructive Evaluation*, Lecture Notes in Mechanical Engineering,
https://doi.org/10.1007/978-981-16-0186-6_30

acquisition time. Safety critical industries like aerospace, nuclear, process and manufacturing industries, where low-power inspection and fast signal acquisition is crucial, can employ this technique.

2 EMAT-Guided Waves

EMATs generate ultrasound based on the principle of Lorentz force generation. Components of an EMAT comprise of a permanent magnet and a current-carrying coil. The biasing magnetic field from the permanent magnet induces a magnetic field (B) inside the specimen and the current passing through the coil causes an eddy current (J_e) within the conducting specimen. The induced magnetic field and the electric field interact with each other to generate a Lorentz force defined as:

$$f_L = J_e \times B$$

The Lorentz force generated triggers the motion of charged particles in the specimen [3, 12]. These charged particles in motion collide with the crystal lattice and lead to wave propagation. For the inverse process, i.e., the reception, the resulting ultrasonic waves force the charged particle to move in the presence of the magnetic field. Such charged particles moving under the influence of perpendicular magnetic field experience a Lorentz force (or rather generate a potential difference), giving rise to a current similar to an eddy current dictated by the equation:

$$J = \sigma * (V \times B)$$

where V is the velocity of the charged particles and σ is the conductivity. These currents are inductively picked by the coil of the receiving EMAT.

In periodic permanent magnet (PPM) EMATs, permanent magnets are arranged in alternating polarity, periodically. This constrains the wavelength of the generated ultrasound which is equal to the periodicity of the magnet arrangement. This PPM arrangement of magnets is equivalent to combing of transducers. The shear horizontal waves generated from each magnet, in such an arrangement, constructively interfere, causing the SH-guided wave to propagate [4, 5, 11, 13]. Due to electromagnetic transduction, efficiency of EMAT is poor which significantly affects the SNR.

3 Coded Excitation

Pulse compression, which has seen wide range applications in RADAR and medical ultrasonics, is a method that has the benefit of a long pulse width transmission's high energy combined with the advantage of a short pulse width transmission's high

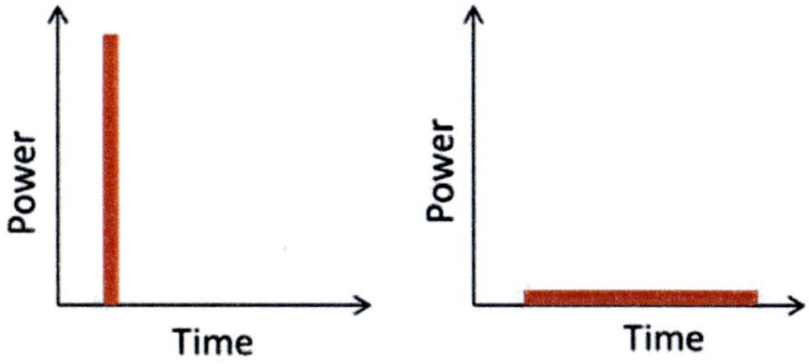

Fig. 1 Power versus time plot for conventional (left) and pulse compression (right) experiments

resolution. Cegla et al. have reported the use of this technique in the field of NDE by proposing an ultrasonic bulk wave echo measurement and inspection technique of samples [6, 7]. Lin et al. have presented their studies on waveform design for enhancing SNR from coded excitation, taking into consideration parameters such as type of input burst and signal frequency [8]. Pulse compression is usually done to enhance signals hidden in noise. It deals with exciting long transmission pulses at low power, hence ensuring no reduction in supplied energy as compared to the conventional excitation (Fig. 1). Pulse compression is predominantly of two types: (1) frequency-modulated pulse compression and (2) phase-modulated pulse compression. Frequency modulation comprises of transmitting a chirp signal which encompasses a wide band of signal frequencies. The frequency can be varied in three ways: (i) linearly variation, (ii) nonlinearly varied and (iii) time–frequency-based coded variation. In this work, we have focused on the second type of pulse compression, i.e., phase-modulated pulse compression. Phase-modulated pulse compression usually deals with encoding a regular transmission signal of continuous transmission pulses with a phase shift code. Phase shift codes can be binary, tertiary or may have more than three encoded values. In this work, our prime focus is on binary phase shift keyed (BPSK) transmissions with intermittent reception pauses a.k.a. coded excitation. Figure 2 depicts the mechanism of the coded excitation technique for ultrasonic inspections.

Code Train. In BPSK, a train of code is generated wherein each element corresponds to a 180° phase shift marked by a '−1' or 0° phase shift marked by a '1'. The original transmitted signal is modulated using this coded information, which, in our demonstration is solely pseudo-random code. These are randomly generated code streams of '−1' and '1'.

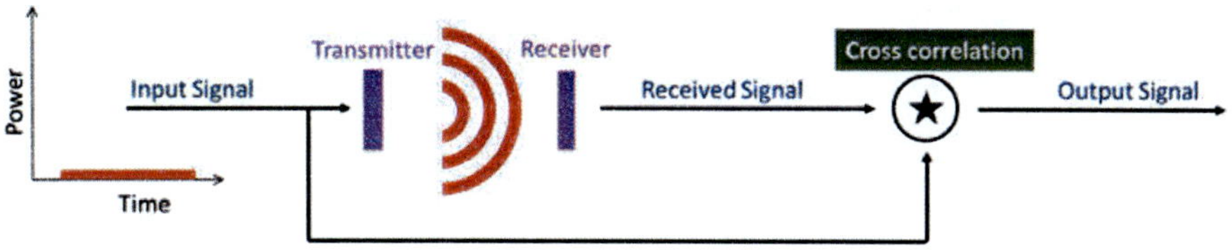

Fig. 2 Schematic showing concept of coded excitation

Gap Train & Combinational Product Sequence. Introduction of receive intervals in between coded signal transmission has been proposed by Isla et. al [6]. These receive intervals ensure pauses in transmission where the echoes from previous transmission can be retrieved. Each element in this signal can take a value corresponding to a reception marked by '0' or corresponding to a transmission marked by '1'. In addition, we define these transmission pulses as transmission entities, and reception pauses in between these transmissions as reception entities. In our technique, we have encoded our transmission with a combinational product sequence which is effectively a product of appropriately up-sampled pseudo-random codes and pseudo-random gaps, combined to form a signal with phase modulation and intermittent pauses for reception. Each entity corresponds to a part of the signal within the width characterized by the formula:

$$T_{entity} = \frac{N}{f}$$

where f is the frequency of excitation, and N is the number of cycles.

This is ideally the width of one transmission pulse which will henceforth be referred to as a transmission entity. A reception pause of equal width shall be referred to as a reception entity. The ideal reception interval should be no more than 3–5 T_{entity} in width [6]. This constraint was applied to the gap train while processing the transmission signal. Additionally, the number of transmission entities and receptions entities was kept at par for optimal SNR [9]. It must also be ensured that the total number of entities is above a certain threshold. It has been reported to ensure at minimum of 2048 total entities for a signal of 2 MHz frequency and a 5-cycle Hann windowed sine tone burst transmission entity [10]. Dependency of this threshold on frequency and other criteria has not been studied.

Final Transmission Signal. The final step in generating transmission signal is to give form to this encoded information of phase shift as well as transmit and receive intervals by convolving the combinational product sequence with a transmission burst of any choice. Bursts of several types have been explored such as pure sine, Hann windowed sine and chirp. Bursts can usually be sinusoidal or windowed/modulated derivatives of the fundamental burst signal. The final all-encompassing transmission (obtained as a result of the convolution stated above) shall be henceforth referred to as a coded excitation. A 500 kHz centered 5-cycle Hann windowed sine tone burst signal with 2^{16} entities will span for 655.36 ms and this coded excitation signal is transmitted using Keysight 33512B, an arbitrary waveform generator.

Processing Reception Signals. The reception signals obtained as a result of sourcing the above-mentioned coded excitation through a variety of transducers and specimens, are passed through a tuned amplifier 48–53 dB gain, and this amplified signal is sampled through a USB oscilloscope Picoscope 3203D. The obtained reception is then processed by first masking it with an inverse gap mask to remove remnants of magnetically coupled transmission imprints.

The inverse gap mask is obtained by inverting the gap train and convolving it with a T_{entity} wide square pulse. The modified reception signal shall henceforth be referred to as the masked received signal. This masked received signal is further cross-correlated with the coded excitation input to look for latent echoes.

$$T_{\text{transmission}} = e \times T_{\text{entity}}$$

where $T_{\text{transmission}}$ is the total transmission time, and e is the total number of entities (transmissions + receptions).

Cross-correlated Output. The cross-correlated output is seen analogous to a conventional echo acquisition result. The peaks of the cross-correlated result mark the times of flight of echoes unlike the conventional output where the zero-crossing of the observed peaks and not the peaks themselves mark the times of flight of echoes.

4 Experiment and Results

In this section, we discuss the experimental results with coded excitation for electromagnetic acoustic transducers (EMATs) based on Lorentz force transduction mechanism. Figure 3 shows the complete experimental setup. The coded excitation signal generated as discussed in the previous section is loaded onto an arbitrary waveform generator (Keysight *Trueform Series*—33512B AWG). A pitch-catch experiment is performed by feeding this transmission signal to a PPM-EMAT transmitter placed on an aluminum sample of 2 mm thickness. Another identical EMAT probe acts as a receiver which continuously receives the echoes from transmission. This reception

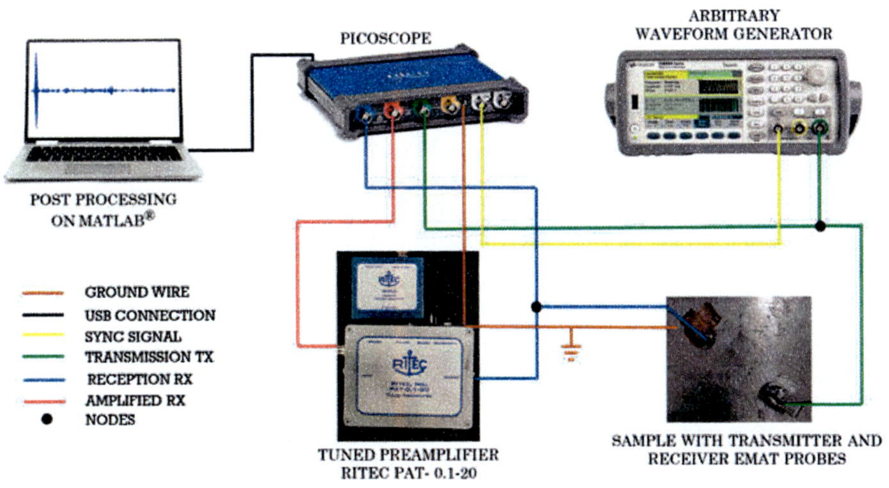

Fig. 3 Schematic of the experimental setup

Fig. 4 EMAT probes placed on aluminum sample of 2 mm thickness. Distance between the probes is 143 mm

signal is then sent through a preamplifier (RITEC PAT—0.1–20) which is tuned to selectively amplify a desired band of frequency. This frequency bandwidth is centered at the frequency of the transmitted signal. The preamp gives a flat ~40–50 dB gain. The signal from the preamp is then sent to a USB oscilloscope (PicoScope 3203D). However, the oscilloscope output, viewed by connecting to a PC, is not the final output. The oscilloscope signal is analyzed, logged and processed in MATLAB® as discussed in the previous section. The post-processing finally yields a time response which compares to a conventional experiment's A-scan result.

An experiment was conducted with the following EMAT probes configuration in pitch-catch mode as shown in Fig. 4. To select excitation frequency, a line whose slope is equal to the wavelength of the PPM-EMAT is imposed on the phase velocity dispersion curve as shown in Fig. 5. Exciting at a frequency at the point of intersection of this line with a particular mode in dispersion curve generates the mode at maximum efficiency. The plate dimensions are 910 mm x 343 mm x 2 mm. We transmit a coded excitation generated from a 5-cycle Hann windowed tone burst centered at 500 kHz frequency from one PPM-EMAT transmitter and receive the echoes from the other. From the phase velocity plot shown in Fig. 5, for 2-mm-thick aluminum plates, we conclude primarily the presence of SH0 wave mode only. Based on theoretical velocity calculations, the times of flight (ToF) of echoes for SH0 wave mode for the setup shown in Fig. 4 are 46.13, 104.19 and 117.10 μs.

The correlated echo scan for this setup is shown in Fig. 6. The transmitted coded excitation used to obtain the scan shown in Fig. 6 has a voltage peak-to-peak of $10Vpp$.

Another experiment was conducted to compare the outputs of a conventional experiment with coded excitation. For the conventional experiment, we used a transmit–receiver PowerBox H (Innerspec, USA) operated at $600Vpp$ with 32 averaging. The coded excitation was generated at $5Vpp$ with no averaging. The compared echo scans depicted in Fig. 7 clearly show that coded excitation enables low-power operation of EMATs bringing down operating voltage requirements more than 100x.

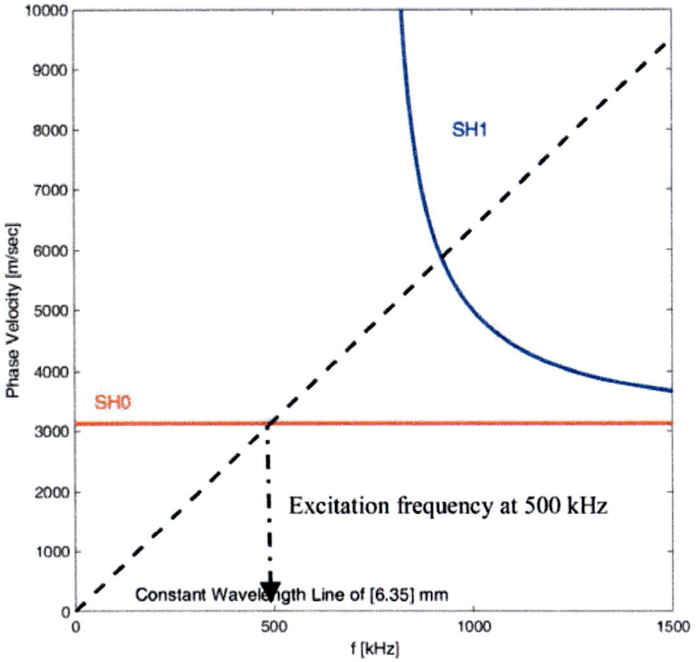

Fig. 5 Phase velocity plot for 2-mm-thick aluminum plate. Constant wavelength line of 6.35 mm (corresponding to PPM-EMAT wavelength) is overlaid

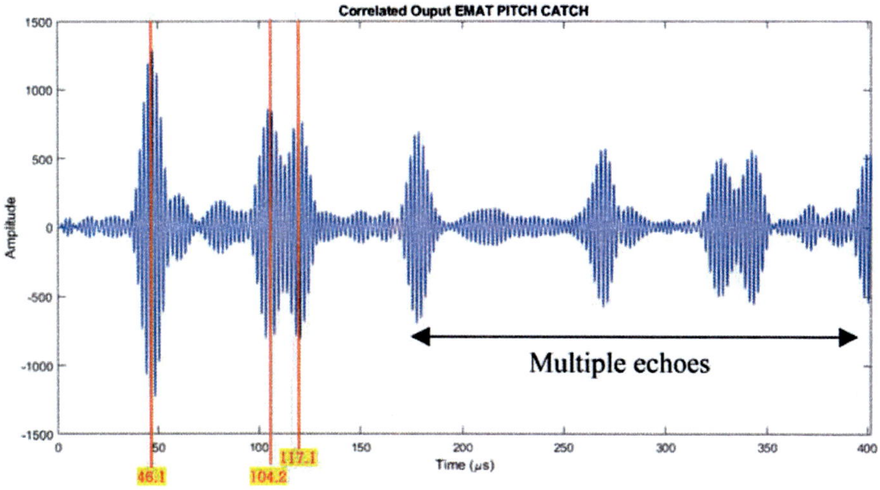

Fig. 6 Correlated output showing arrival time of echoes for SH0 mode at 46.13, 104.19 and 117.10 μs

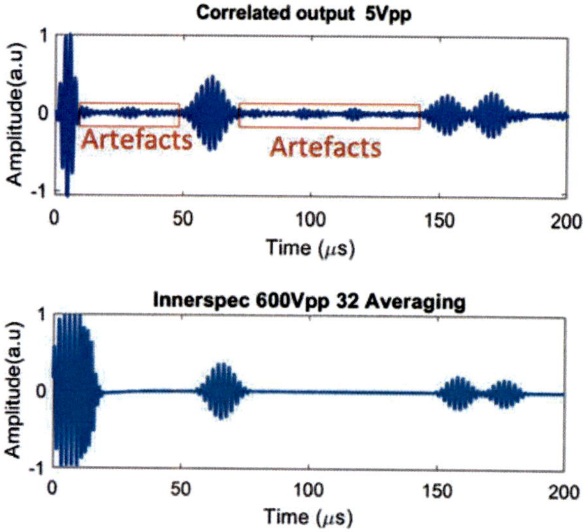

Fig. 7 Experiment comparing EMAT operation in coded excitation experiment using arbitrary waveform generator and conventional experiment using PowerBox H—Innerspec

5 Conclusions

Though electromagnetic acoustic transducers are a non-contact way of generating ultrasound, the major drawback is their poor transduction efficiency which significantly affects the SNR of the measured signal especially on ferromagnetic materials. Further, the use of high-power electronics to improve the SNR is a hindrance to the use of electromagnetic acoustic transduction in safety critical applications like petrochemical industries and aerospace industries. In this work, a pulse compression-based signal processing technique called coded excitation has been proposed to improve the SNR in guided ultrasonic wave NDE. The results obtained clearly show that coded excitation enables low-power operation of EMATs, bringing down operating voltage requirements more than 100x. This enables the successful industrial deployment of EMATs in power-limited applications to perform low-power guided ultrasonic wave inspections.

References

1. Gao H, Ali S, Lopez B, Austenitic weld inspection with EMAT phased array, pp 1–10
2. Jian X, Dixon S (2007) Enhancement of EMAT and eddy current using a ferrite back-plate. Sens Actuat Phys 136(1):132–136
3. Hirao M, Ogi H (2003) EMATs for science and industry. Springer, Boston
4. Suresh N, Balasubramaniam K (2019) Reflection study of SH0 mode with plate edge at different incident angles. In: AIP conference proceedings, vol 2102, pp 0–8
5. Koodalil D, Barnoncel D, Rajagopal P, Balasubramaniam K (2019) Interfacial adhesion (kissing bond) detection using shear horizontal (SH) waves. In: AIP conference proceedings, vol 2102, no 1, p 20042

6. Isla J, Cegla F (2017) Coded excitation for pulse-echo systems. IEEE Trans Ultrason Ferroelectr Freq Control 64(4):736–748
7. Cegla FB, Coded sequences for simultaneous acquisition of data from multiple EMAT coils
8. Hua J, Zeng L, Lin J, Huang L (2018) Excitation series design and pulse compression synthesis for high-resolution Lamb wave inspection. Struct Heal Monit 37:1–15
9. Isla JA (2017) Coded excitation for Low-Snr, p 221
10. Isla J, Celga F (2016) The use of binary quantization for the acquisition of low SNR ultrasonic signals: a study of the input dynamic range. IEEE Trans Ultrason Ferroelectr Freq Control 63(9):1474–1482
11. Koodalil D, Barnoncel D, Rajagopal P, Balasubramaniam K (2020) NDT E Int 112:102248
12. Koodalil D, Rajagopal P, Balasubramaniam K (2021) Int J Adhes Adhes 104:102761
13. Suresh N, Balasubramaniam K (2020) NDT E Int 116:102313

Mixing of Lamb Waves at Delamination Defect in a Unidirectional GFRP Laminate

Yamnesh Agrawal, Akhilendra S. Gangwar, and Dhanashri M. Joglekar

1 Introduction

Fiber-reinforced polymer (FRP) laminate composites are increasingly used in aerospace, automobile, civil industries and other applications owing to some of their desirable properties such as high strength-to-weight ratio, high modulus and good corrosion resistance than metals [4]. Delaminations are the most common defect types in FRP laminate composites resulting from high interlaminar stresses generated at the interfaces between adjacent plies. Delamination can be developed from cyclic loading, notches, manufacturing defects, impact and stress concentrations caused by changes in structural parameters and which in turn affects the safety of the component. As the safety-critical engineering components are highly susceptible to delamination defect which are difficult to visualize through naked eye, attention needs to be focused toward the study of identification and the quantification of such defects [15, 19, 20]. A good number of researchers have thrown light on vibration and wave-based techniques to detect damage, numerically as well as experimentally [14].

From the perspective of Non-Destructive Evaluation (NDE) of composite laminates, linear interaction of Lamb waves with delamination has been explored since a couple of decades. There is a correlation between time domain wave signal characteristics such as reflection, mode conversion, attenuation, speed of the wave, wave amplitude, etc. with delamination location and size. This correlation has been used toward the detection and characterization of the delamination [5, 6, 16, 17].

Y. Agrawal · A. S. Gangwar (✉) · D. M. Joglekar
Indian Institute of Technology Roorkee, Roorkee 247667, India
e-mail: agangwar@me.iitr.ac.in

C. K. Mukhopadhyay and R. Mulaveesala (eds.), *Advances in Non-destructive Evaluation*, Lecture Notes in Mechanical Engineering,
https://doi.org/10.1007/978-981-16-0186-6_31

However, in presence of operational variabilities, changes in the geometry of the model, boundary conditions and for incipient damage, analysis through linear approach offers certain constraints [7]. In that case, an interest has been triggered toward exploiting the damage-induced nonlinearity that has led to the evolution of nonlinear damage detection techniques [1, 2].

In the recent past, researchers have explored the nonlinear interaction of Lamb wave on composites with damage as a promising approach that proposes improved sensitivity, better reliability and practicability of damage detection [18]. Researches based on the application of nonlinear Lamb waves in composites have been minimal. Till date, most of the studies on application of nonlinear Lamb waves have been dedicated to metals, such as aluminum and steel. However, focusing on the understanding of nonlinear Lamb waves is essential for spotting and localizing delamination in FRP laminate composites. Moreover, researchers were mostly fascinated by classical nonlinearity raised from material nonlinearity [12], whereas very less attention was given to non-classical nonlinearity raised from CAN [9, 10, 22, 25]. The nonlinear approach to acoustic Non-Destructive Testing (NDT) is concerned with nonlinear responses of the Lamb waves, which is inherently related to the existence of nonlinear harmonics, frequency shifts and signal modulations. The onset of nonlinear harmonics of the examined signal is due to the delamination modeled by the contact pair in FRP laminated composite.

The existence of nonlinear harmonics in the form of super-harmonics or subharmonics in the frequency domain corroborates the existence of delamination and along with the temporal data, other properties of delamination can also be identified. Several researchers have proposed the nonlinear acoustics for structural health monitoring and demonstrated the generation of higher harmonics both theoretically and experimentally [13]. Conversely, the usage of higher harmonics to detect and quantify damage is limited because of their low amplitude and the accuracy of instrument to detect high frequencies.

Vibro-Acoustic Modulation (VAM) method, a nonlinear acoustic method, can be used to overcome aforementioned limitations. This method has the capacity to detect structural defects. In this method, the damaged structure is simultaneously actuated by a combination of frequencies (a super low frequency and an ultrasonic frequency signal) that leads to the generation of sidebands at frequency $f_1 \pm nf_2$ in the frequency spectrum where, 'n' is an integer, which highlights the existence of the damage [3, 24, 26]. Two different types of actuators, one of which has the ability to generate a low-frequency signal and other for the high-frequency excitation, need to be employed for the implementation of the VAM method, and the practical implementation of such method is difficult. Therefore, sufficient modifications that include cross-modulations and nonlinear frequency mixing have been reported in the literature, wherein the components to be interrogated are subjected to excitation comprising two high frequency signals with frequencies that are comparable in the magnitude.

When a dual-frequency signal is allowed to pass through a defect, it results into mixing of the frequencies [8, 11]. Although NDE for cracks using bulk waves has been studied widely, there is a scarcity of literature pertaining to such studies in case of delamination. To the best of authors' knowledge, the influence of interlaminar location and width of delamination on peculiar combination of frequencies have not yet been discussed. Therefore, in this paper, nonlinear interaction of Lamb wave using dual-frequency input signal and the effect of interlaminar location and width of delamination on higher harmonics along with a algebraic combination of input frequencies is investigated numerically and authenticated through some basic experimental results. For nonlinear waves in laminated composite plates, Contact Acoustic Nonlinearity (CAN) is introduced for modeling the interactions of sublaminates at delamination. Based on the numerical results, nonlinearity index with the presence of the delamination is discussed. The entire paper has been compiled into three sections. FE modeling along with contact pair is presented in Sect. 2. In Sect. 3, results have been discussed and Sect. 4 is devoted to summarize the conclusions.

2 Finite Element Simulation

2.1 Structural Modeling

Finite element (FE) simulations were carried out in ANSYS 19.0. A 2D model of unidirectional E-glass fiber-reinforced laminated composite were modeled. All linear orthogonal elastic properties were assigned to the structure as shown in Table 1. In these FE simulations, 8 layers of laminates are stacked over one another and each having dimensions of 450 and 0.5 mm along the thickness with layup configuration $[0]_8$ was modeled. A_0 wave mode was excited by giving in-plane transient force in opposite direction at the top and bottom nodes in the opposite direction at the actuation points, as shown in figure 1. PLANE-82 element with plane strain conditions were used in Fe simulations. After meshing of all the sub-laminates, merging of nodes at surfaces of different layers present at same location is imperative as it ensures the physical and chemical bonding between layers. To create delamination of any size in between any two layers, nodes at those surfaces should remain unmerged.

The excitation force of Gaussian window toneburst signal for the total time of 0.5 ms was used as input signal. The excitation time of 0.1 ms was kept constant rather than the number of cycles for the same spectral leakage of both excitation frequencies. Figure 2 shows time domain and the frequency domain of the excitation pulse of mix frequency signal of 70 and 100 kHz.

Table 1 Material properties of GFRP laminated composite plate studied in this work

Material	E_{11} (GPa)	E_{22} (GPa)	G_{13} (GPa)	G_{23} (GPa)	υ_{13}	υ_{23}	ρ (kg/m^3)
GFRP	44.68	6.90	2.54	2.5459	0.28	0.355	1990

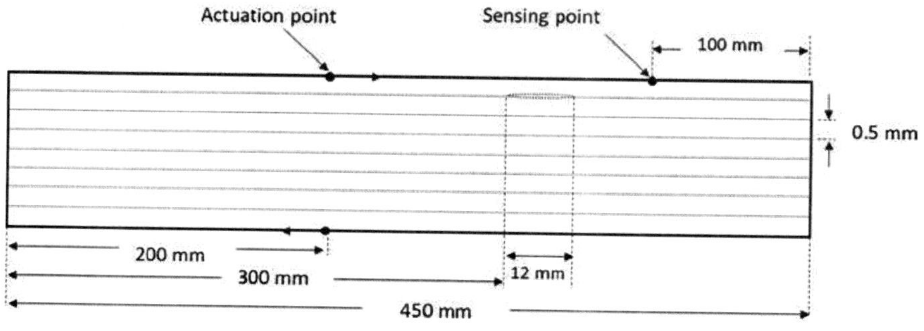

Fig. 1 Schematic 2D representation of $[0]_8$ laminate

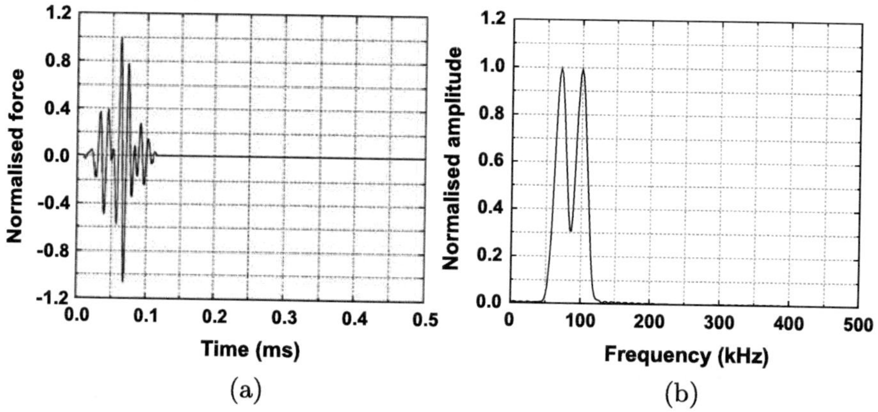

Fig. 2 The dual-frequency toneburst excitation input signal in **a** time domain, **b** frequency domain

2.2 *Element Size and Time Step*

For a wave to propagate in a FE model, mesh size should be small enough so that there are at least 20 nodes per wavelength exist. The excitation signal used in the simulations is mixed frequency signal having frequencies of 70 and 100 kHz. To capture nonlinear harmonics of mix frequency signal, the FE model should be able to capture frequency greater than or equal to 300 kHz so that all the higher harmonics and combination frequencies ($f_2 - f_1$, $2f_1$, $f_1 + f_2$, $2f_2$ etc.) get easily captured. All the higher harmonics and combination signal due to mix frequency should also be able to propagate in both sub-laminates in delamination region. From the dispersion curve obtained from Semi-Analytical Finite Element (SAFE) simulation, it can be interpreted that wavelength of a signal would be smaller for higher frequencies and smaller thickness of sub-laminate. Minimum thickness of sub-laminate can be obtained if the delamination is located in between first and second layers. For an anti-symmetric wave signal of 300 kHz in 0.5 mm thickness sub-laminate, wavelength of the signal would be 2.88 mm and so the nodal distance should be less than

0.144 mm. Since PLANE-82 element has mid-side nodes, so the element size should be less than 0.288 mm. In all numerical simulations in this study, authors have used very fine discretized mesh with element size of 0.25 mm.

Time step is an important parameter in any numerical simulations to get the desired accuracy in simulation results. Time step should be selected on trade-off between accuracy and computational cost. The element size of the meshed model is 0.25 mm and since the element type used in the simulation have mid-side nodes, the minimum nodal distance can be approximately assumed to be 0.125 mm. Maximum A_0 wave speed in the simulation is around 1250 m/s. From the wave speed and nodal distance, time step came out to be 0.1 μs.

2.3 *Modeling of Contact Pair*

In this study, modeling of contact pair is a key step to generate contact acoustic nonlinearity (CAN). Without modeling contact pair, delamination nodes would cross-over into each other without having any constraint as if both sub-laminates do not exist for each other. FE simulations without contact pair will act similarly as linear analysis. Researchers in the past have developed many algorithms in the finite element modeling to deal with such contact pair problems. When both delamination faces interact with each other, the local stiffness must change to avoid penetration. ANSYS offers many algorithms such as penalty method, MPC algorithm, Lagrange method, etc., in order to execute the contact problem. These algorithms are based on the methods of changing normal penalty stiffness and penetration tolerance with a series of iterations by updating those penalties. In this simulation, Augmented Lagrange (AL) method was used as this algorithm offers better penetration control, better satisfaction on governing equations and at the same time, it costs less in computational efforts. The governing equations for contact pressure in terms of normal contact stiffness updation, contact gap size and behavior of contact surface are elaborated in [24]. For the purpose of creating contact pair, contact elements such as CONTA172 and TARGET169 are assigned to delamination nodes. In this problem, no friction forces are assumed to be present in between the delamination faces.

2.4 *Formulation of Nonlinearity Index (NLI)*

The nonlinearity index is an important parameter to calibrate the presence of non-linearities in components. As the wave propagates in the composite laminate, due to the presence of delamination, sub-laminates interact with each other and the energy of wave transfers from the fundamental harmonic to the higher harmonics. Many researchers have developed their own methods for the mathematical calculation of CAN. Shen [21] proposed nonlinearity index that was based on the wave energy ratio, shown in Eq. 1, where A_i is the amplitude of ith harmonics in the fast Fourier transform (FFT) response.

$$NLI = \sqrt{\frac{A_2 + A_3 + A_4 + \cdots}{A_1}}. \quad (1)$$

In present analysis, as the dual-frequency wave signal propagates in the composite laminate, due to delamination, energy of the wave transfer from fundamental harmonics to the higher harmonics and the combination harmonics of those excited frequencies. So, the energy of all higher harmonics including combination frequencies is considered in the formulation of NLI as defined in Eq. 2.

$$\text{NLI} = \sqrt{\frac{A_{f_2-f_1} + A_{2f_1} + A_{f_1+f_2} + A_{2f_2} \cdots}{A_{f_1} + A_{f_2}}}. \quad (2)$$

where A_i, A_{i-j}, A_{i+j} represents the amplitude of excitation of ith, $(i - j)$th, $(i + j)$th harmonics of the excitation frequency in the frequency spectrum of the response and $i = 1, 2, 3 \ldots.$

3 Results and Discussion

In the non-destructive evaluation of structures, different properties of delamination can be found out from the nonlinearity trends obtained through numerical simulations. Some experiments were also performed whose results are consistent with the numerical approach.

3.1 *Basic Experimental Results*

The experiments were carried out for predicting the peaks at combination frequencies which are observed in the frequency spectrum of response signal as the dual-frequency Lamb wave passes through delamination. A 450 mm × 450 mm quasi-isotropic uni-directional glass fiber-reinforced plastic (GFRP) composite laminates with thickness 4 mm were fabricated from 8 layers of glass/epoxy prepregs with 12 mm delamination between first and second layers. Several samples were also cut for the ASTM tests that were carried out for characterization of the laminate, and the properties of the lamina thus obtained are given in Table 1. A 25 mm × 8 mm and 1 mm thick rectangular piezoelectric wafer (PW) transducers of material type SP-5H were adhesively bonded to both sides of the composite laminates to excite the dual-frequency fundamental anti-symmetric Lamb mode.

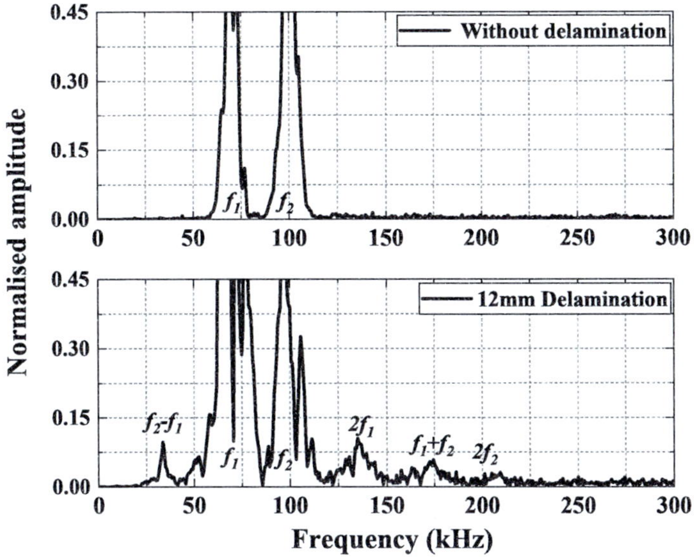

Fig. 3 FFT response of experimental results having delamination between 1 and 2 layers

The dual-frequency excitation signal was generated by a computer-controlled signal generator with a 10 V peak to-peak output range, which was further amplified up to 300 V peak-to-peak output voltage using a NF made BA4825 high-speed bipolar amplifier and then the amplified signal is applied to the PW actuator. The displacements were measured using a scanning laser Doppler vibrometer (Polytec VIB-E-400), and the data were sent to the laser controller and data acquisition system for post-processing. The same excitation signal was applied which was used in numerical simulations, as shown in Fig. 2. Figure 3 shows FFT responses of the displacement signal on intact plate and plate having delamination between first and second layers. Higher harmonics and combination frequencies can be easily differentiated for the two cases as depicted in Fig. 3.

3.2 Numerical Simulation Results

To characterize any delamination, the location and the size of delamination must be known. The longitudinal location of delamination can be easily found out by the linear analysis and hence this study focuses on two parameters, i.e., interlaminar location and delamination width which can be demonstrated by the presence of combination frequencies involving a dual-frequency excitation. The selection of excitation frequencies is based on gaps among combination frequencies, higher harmonics and the excited frequencies. All the higher harmonics and combination frequencies should have an appropriate distance in FFT response such that no two peaks merge

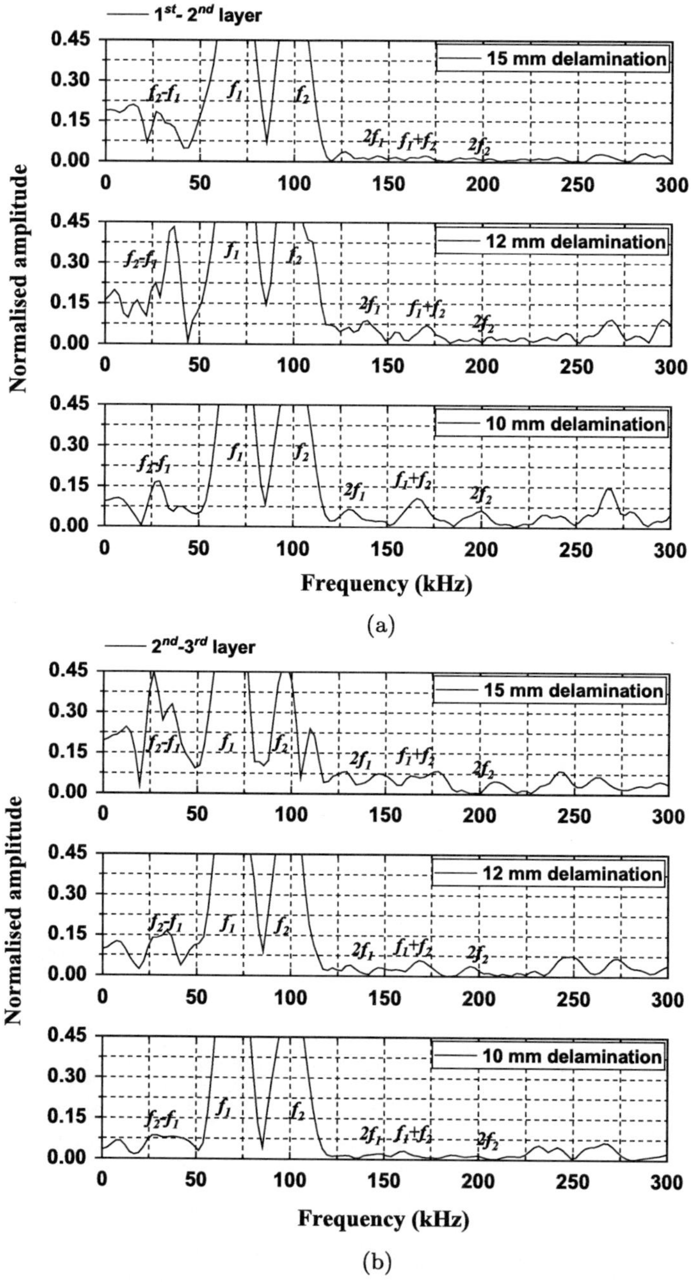

Fig. 4 FFT response for various delamination width on the plate having delamination between **a** 1 and 2 layer, **b** 2 and 3 layer

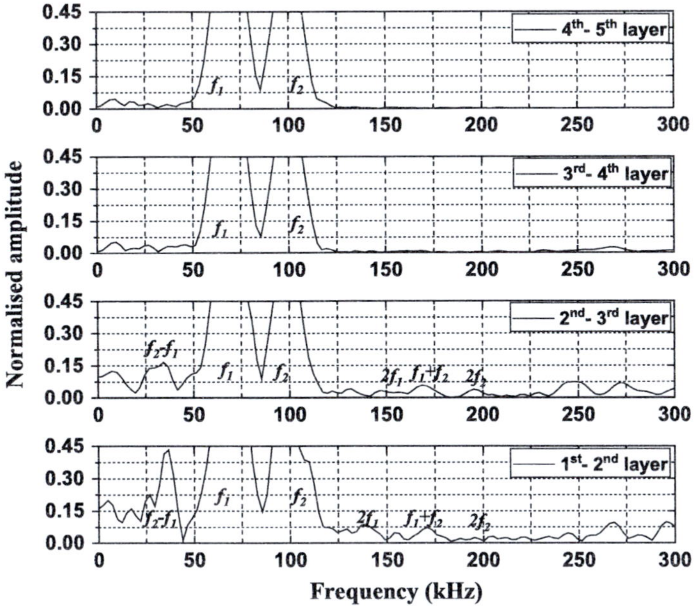

Fig. 5 FFT response for various interlaminar location on the plate having 12 mm delamination

on to each other. From Fig. 2, it is clearly noticed that two frequencies, i.e., f_1 and f_2 with a value of 70 kHz and 100 kHz, respectively, forms the excitation signal. Also, two peaks at those two frequencies are generated in the frequency domain plot. The frequency domain response reveals combination frequencies around the excitation frequencies. These combination frequencies are located at frequencies corresponding to excited frequencies, as shown in Figs. 4 and 5. All the numerical simulations were carried out on dual-frequency signal of 70 and 100 kHz with different interfacial location and delamination width. The out-of-plane displacement response in all cases was analyzed, and its FFT response was calculated for all the temporal responses. The nonlinearity index was calculated using Eq. 2 and plotted. The combination frequencies like 30 and 170 kHz were also observed apart from the higher harmonics of each tone.

The effect of sensing locations was also analyzed, and it was found that the nonlinearity is not much affected by the change of sensing location and that small effect of sensing location on nonlinearity is attributed to Lamb wave dispersion. The frequency content of all harmonics change with the change of sensing location. The effect on nonlinearity would be minimum if the sensing location is fixed w.r.t. the delamination. So, for the purpose of uniformity in the numerical results, sensing location was set to be on the other side of delamination from actuation point and 50 mm away from the starting of the delamination region. Wave actuation and sensing location are based on pitch-catch arrangement [23].

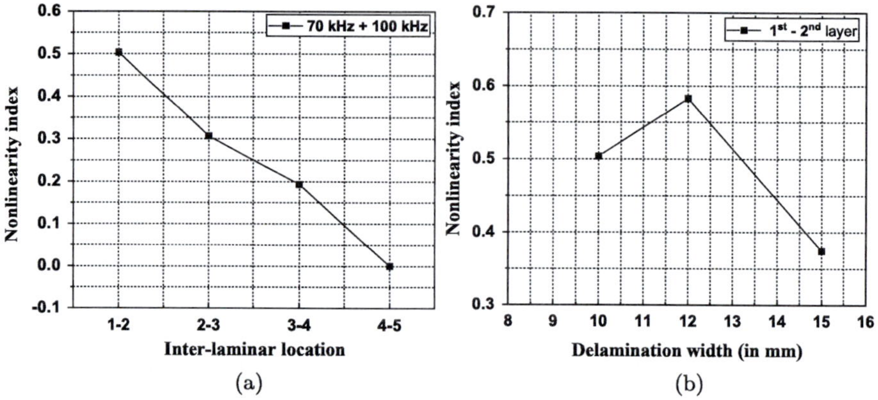

Fig. 6 Trend of nonlinearity index with different **a** interlaminar location of delamination, **b** delamination width

Figure 4a, b shows the FFT responses at various delamination width located in between first and second layers and second and third layers, respectively. Figure 5 shows FFT responses at different interlaminar location. Amplitude of higher harmonics at different combination of frequencies is clearly visible from the plot. Figure 6a shows the NLI with different interlaminar location at 12mm delamination. NLI is found to be maximum for delamination located in between first and second layer, and it decreases when interlaminar location of delamination moves from boundary to center. Trend of NLI with the delamination width is shown in Fig. 6b. It shows that NLI is maximum at 12mm delamination width and minimum at 15mm delamination.

The nonlinearity generated from dual-frequency mixing has its own advantages as lower frequencies can be more accurately captured by relatively low-precision instrument. Moreover, the amplitude of the combination frequencies is relatively more than that of higher harmonics of individual excited frequencies. So, the possibility of higher harmonics to get diminished in the noise generated by external factors would be very less. The lowest frequency in FFT responses is 30 kHz ($f_2 - f_1$). The amplitude of 30 kHz frequency is highest among all other harmonics generated by CAN.

4 Summary and Conclusions

This paper carried out a numerical analysis based on FE simulation to investigate the effect of interlaminar location and width of delamination on the amplitude of higher harmonics. Some basic experiments were also carried out to validated the presence of combination frequencies. It is concluded from FE simulations that the strength of higher harmonics and combination frequencies is related to size and the location of delamination. The nonlinearity generated from dual-frequency mixing is advantageous as sub-harmonic frequencies can be more accurately captured by relatively low-precision instrument. Overall, the numerical results show that the nonlinearity is prominent at a certain range of delamination width and location of delamination.

Acknowledgements The authors gratefully acknowledge the financial aid provided by the Science and Engineering Research Board (SERB) under the Department of Science and Technology (DST), Government of India, under the Grant No.: ECR/2017/001171.

References

1. Bovsunovsky A, Surace C (2015) Non-linearities in the vibrations of elastic structures with a closing crack: a state of the art review. Mech Syst Signal Process 62:129–148
2. Broda D, Staszewski W, Martowicz A, Uhl T, Silberschmidt V (2014) Modelling of nonlinear crack-wave interactions for damage detection based on ultrasound-a review. J Sound Vib 333(4):1097–1118
3. Donskoy D, Sutin A, Ekimov A (2001) Nonlinear acoustic interaction on contact interfaces and its use for nondestructive testing. Ndt & E Int 34(4):231–238
4. Giurgiutiu V (2015) Structural health monitoring of aerospace composites. Academic Press
5. Guo N, Cawley P (1993) The interaction of lamb waves with delaminations in composite laminates. J Acoust Soc Am 94(4):2240–2246
6. Hayashi T, Endoh S (2000) Calculation and visualization of lamb wave motion. Ultrasonics 38(1–8):770–773
7. Jhang KY (2009) Nonlinear ultrasonic techniques for nondestructive assessment of micro damage in material: a review. Int J Precis Eng Manuf 10(1):123–135
8. Joglekar D, Mitra M (2015) Analysis of nonlinear frequency mixing in 1d waveguides with a breathing crack using the spectral finite element method. Smart Mater Struct 24(11):115004
9. Joglekar D, Mitra M (2015) Nonlinear analysis of flexural wave propagation through 1d waveguides with a breathing crack. J Sound Vib 344:242–257
10. Joglekar D, Mitra M (2016) Analysis of flexural wave propagation through beams with a breathing crack using wavelet spectral finite element method. Mech Syst Signal Process 76:576–591
11. Joglekar D, Mitra M (2016) Time domain analysis of nonlinear frequency mixing in a slender beam for localizing a breathing crack. Smart Mater Struct 26(2):025009
12. Kashyap K, Sharma AK, Joglekar M (2020) Nonlinear dynamic analysis of aniso-visco-hyperelastic dielectric elastomer actuators. Smart Mater Struct 29(5):055014
13. Kim GW, Johnson DR, Semperlotti F, Wang K (2011) Localization of breathing cracks using combination tone nonlinear response. Smart Mater Struct 20(5):055014
14. Mitra M, Gopalakrishnan S (2006) Wavelet based spectral finite element modelling and detection of de-lamination in composite beams. Proc R Soc A: Math Phys Eng Sci 462(2070):1721–1740
15. Munian RK, Mahapatra DR, Gopalakrishnan S (2018) Lamb wave interaction with composite delamination. Compos Struct 206:484–498
16. Ramadas C, Balasubramaniam K, Joshi M, Krishnamurthy C (2009) Interaction of the primary anti-symmetric lamb mode (Ao) with symmetric delaminations: numerical and experimental studies. Smart Mater Struct 18(8):085011
17. Ramadas C, Padiyar J, Balasubramaniam K, Joshi M (2012) Propagation of the fundamental symmetric lamb mode in a symmetrically delaminated composite laminate. Int J Veh Struct Syst (IJVSS) 4(3)

18. Rauter N, Lammering R, Kühnrich T (2016) On the detection of fatigue damage in composites by use of second harmonic guided waves. Compos Struct 152:247–258
19. Rose JL (2014) Ultrasonic guided waves in solid media. Cambridge University Press
20. Schaal C, Samajder H, Baid H, Mal A (2015) Rayleigh to lamb wave conversion at a delamination-like crack. J Sound Vib 353:150–163
21. Shen Y (2017) Numerical investigation of nonlinear interactions between multimodal guided waves and delamination in composite structures. In: Health Monitoring of Structural and Biological Systems 2017, vol 10170. International Society for Optics and Photonics, p 101701Z.
22. Soleimanpour R, Ng CT (2017) Locating delaminations in laminated composite beams using nonlinear guided waves. Eng Struct 131:207–219
23. Watkins R, Jha R (2012) A modified time reversal method for lamb wave based diagnostics of composite structures. Mech Syst Signal Process 31:345–354
24. Yelve NP, Mitra M, Mujumdar P (2015) Detection of stiffener disbonding in a stiffened aluminium panel using nonlinear lamb wave. Appl Acoust 89:267–272
25. Yelve NP, Mitra M, Mujumdar P (2017) Detection of delamination in composite laminates using lamb wave based nonlinear method. Compos Struct 159:257–266
26. Zaitsev VY, Matveev LA, Matveyev A (2011) Elastic-wave modulation approach to crack detection: Comparison of conventional modulation and higher-order interactions. NDT & E Int 44(1):21–31

Parametric Study on Resolution Achieved Using FMC-TFM-Based Phased Array Ultrasonic Imaging

Sumana and Anish Kumar

1 Introduction

The linear array transducers are commonly used in phased array ultrasonic testing which has many advantages over conventional mono-element transducers such as versatility, ability to focus in contact mode, and steering/sweeping of the sound beam. These result in large inspection coverage on both sides of the transducer with a minimal mechanical movement of the transducer [1, 2]. Focusing leads to enhancement in both resolution and sensitivity [3]. Focusing in contact mode could also be achieved by applying Synthetic Aperture Focusing Technique (SAFT), which was developed in early 1950s to improve the lateral resolution in radar images by post-processing the B-scan data acquired using diverging beams [4]. Later on, many approaches, such as Virtual Source with SAFT (VS-SAFT) [5], multi-SAFT [6], etc., were developed based on SAFT. Despite the fact that SAFT has ability to focus at all the depths, the limitations of this method include poor Signal-to-Noise ratio (SNR), high grating lobes, reduced resolution at larger depths, and limited information due to data acquired in the pulse-echo mode only [7, 8]. Such limitations could be overcome by adopting a data acquisition scheme, called as Full Matrix Capture (FMC), comprising of collecting data of each transmit/receive pairs of N element linear phased array transducer resulting in N^2 signals [9–11]. The data is post processed by Total Focusing Method (TFM). The basic principle of TFM is delay and summation of N^2 signals synthetically. As a result, focusing at each point is achieved leading to improved SNR and sensitivity as compared to SAFT [12, 13]. FMC-TFM

Sumana · A. Kumar (✉)
Non-Destructive Evaluation Division, Metallurgy and Materials Group, Indira Gandhi Centre for Atomic Research, Homi Bhabha National Institute, Kalpakkam 603102, Tamil Nadu, India
e-mail: anish@igcar.gov.in

Sumana
e-mail: sumasonu10@gmail.com

C. K. Mukhopadhyay and R. Mulaveesala (eds.), *Advances in Non-destructive Evaluation*, Lecture Notes in Mechanical Engineering,
https://doi.org/10.1007/978-981-16-0186-6_32

is reported to exhibit better flaw characterization and resolution as compared to conventional phased array technique [14]. However, due to a large amount of data acquired and processed in case of FMC + TFM, it reduces the inspection speed [15]. Many variants of SAFT and FMC-TFM such as Virtual Source Aperture (VSA) [16], Technique for Imaging using Virtual Array of Sources (TIVAS) [17], Sparse Matrix Capture (SMC) [15], Synthetic Transmit Aperture (STA) [18], Autofocused Virtual Source Imaging (AVSI) [19], and Multiple Virtual Source Aperture (MVSA) [20] have been developed during the last two decades. The concept of the virtual source was first used by Karman et al. [21] in the medical field, where the data was acquired using a set of elements that produced a virtual source near to the surface. The sound beam created by the virtual source has greater energy enabling the inspection for large thickness with improved SNR [17]. The problem of side lobe and artifacts exists in Virtual Source Matrix (VSM) which may reduce the resolution at higher depths [22]. Recently, Carcreff et al. [23] utilized the Inverse Total Focusing Method (TFMp) to study the resolution in aluminum blocks containing SDHs of 1 mm diameter with the center to center spacing of 2 mm at 40 mm depth.

Resolution in ultrasonic imaging is generally expressed in terms of −6 dB width of a point reflector. For a mono-element immersion focused transducer, −6 dB width (Δd) is given by Eq. 1 [24].

$$\Delta d = \frac{2.44\lambda . L}{D} \tag{1}$$

where λ is wavelength, L is the focal depth, and D is the probe diameter.

In case of conventional phased array inspection where focusing is achieved using delay laws, resolution is given by Eq. 2 [25] as below:

$$\Delta d = \frac{0.11\lambda}{\sin\left[\frac{\theta}{2}\right]} \tag{2}$$

where β is the refraction angle, θ is the angle formed by the extreme ends of the active aperture as shown in Fig. 1, λ is wavelength, and f is frequency.

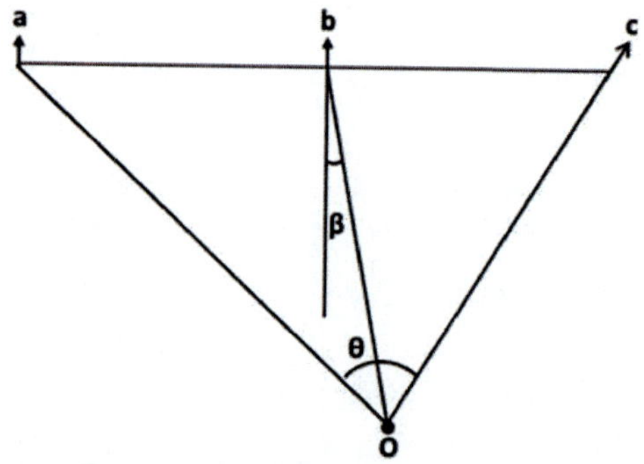

Fig. 1 Schematic of angle formed at any point by the extreme ends of an active aperture for phased array transducer

It can be seen in the above two equations that the resolution in ultrasonic imaging is essentially governed by the focusing depth, diameter of the transducer (aperture size), and wavelength. In case of SAFT using conventional mono-crystal ultrasonic transducer, the resolution is given by $D/2$, where D is the diameter of the transducer [26]. Even though resolution is systematically studied in conventional mono-element and phased array inspection, very few studies have been reported on the effect of experimental parameters on resolution achieved using FMC-TFM technique, particularly at larger depths. The present work aims to study the influence of various parameters such as frequency and aperture size on resolution achieved using FMC-TFM at different depths. The SAFT and FMC-TFM algorithms are explained briefly in the following sections.

1.1 Synthetic Aperture Focusing Technique (SAFT)

In 1950s, SAFT was developed primarily for improving the resolution of airborne radar mapping systems [4]. SAFT is a post-processing technique, which is based on the ray acoustic model. Figure 1 shows the principle of image reconstruction using SAFT. The basic principle of SAFT considers that the signal scattered from a defect is collected at each transmitter position. During the reconstruction process, the region of interest in the sample is divided into grids and two steps, i.e., delay and sum are applied at all the grid points as explained below.

Delay: Let us consider that the SAFT algorithm is being applied at point $S(i, j)$. For clarity, it is assumed that a point reflector is located at point $S(i, j)$. When a transducer with large beam divergence is scanned on the top surface of the specimen, reflection from point $S(i, j)$ is obtained in the A-scan signals acquired from all the different transducer positions indicated in Fig. 2a due to the beam divergence. The defect indication is obtained at the minimum time of flight when the transducer is kept at location X_i just above the point $S(i, j)$. As the transducer is moved away on either side, the time of flight corresponding to the defect signal increases. Hence, the

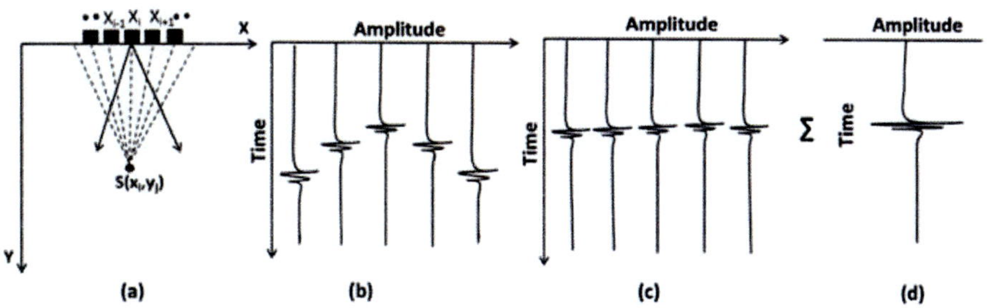

Fig. 2 Schematic of the image reconstruction with SAFT: **a** data acquisition, **b** received signals show the hyperbolic time of flight pattern, **c** time shifted A-scans to the reference point, and **d** correlated A-scan signal

time domain signals received from point $S(i, j)$ at different transducer positions make a hyperbolic time of flight A-scan signature as shown in Fig. 2b. The delay in time of flight (d_{TOF}^{SAFT}) for the signal obtained from transducer position x_T is calculated with reference to the signal obtained from X_ith position as given below:

$$d_{TOF}^{SAFT} = 2 \times \frac{\sqrt{\left[(x_T - x_i)^2 + y_j{}^2\right]} - y_j}{v} \tag{3}$$

The factor 2 indicates the pulse-echo mode, i.e., the same travel path during both transmission and reception. The time shifted individual A-scan signals appear as if they have arrived from the reflector at the center position X_i as shown in Fig. 2c.

Sum: Once the A-scan signals are shifted to the reference level, the averaging of signals lead to constructive interference of the defect signals obtained in all the A-scan signals as shown in Fig. 2d. When the same SAFT algorithm is applied for any other position, where the defect is not present, constructive interference does not happen or even destructive interference may take place. Hence, improvement in the sensitivity and lateral resolution is achieved in the SAFT reconstructed image.

1.2 Full Matrix Capture-Total Focusing Method (FMC-TFM)

FMC consists of capturing the time domain signals for each transmit-receive pairs in an array transducer from an N × N matrix. Hence, it yields N^2 A-scan signals. The post-processing of FMC is performed by Total Focusing Method (TFM). Similar to the SAFT processing, the area of interest is divided into grids, and time delays are calculated for each and every grid points. However, the delay (d_{TOF}^{TFM}) calculation is slightly different as given in Eq. 3:

$$d_{TOF}^{TFM} = \frac{\sqrt{\left[(x_T - x_i)^2 + y_j{}^2\right]} - y_j}{v} + \frac{\sqrt{\left[(x_R - x_i)^2 + y_j{}^2\right]} - y_j}{v} \tag{4}$$

where x_T and x_R indicate transmitter and receiver locations. It can be seen in Eqs. 3 and 4 that $d_{TOF}^{TFM} = d_{TOF}^{SAFT}$ when the receiver is at the same location as transmitter. In other words, SAFT is a small subset of FMC-TFM. Due to the large amount of data obtained and processed in both pulse-echo as well as pitch-catch mode, FMC-TFM provides imaging of misoriented planar indications also unlike SAFT imaging which provides improved resolution for only point reflectors.

2 Experimental Method

2.1 Equipment and Data Acquisition Details

Experiments were performed using a multi-2000 phased array ultrasonic equipment supplied by M/s. M2M, France having the maximum number of 64 simultaneous transmitting and receiving channels. Ultrasonic linear array transducers of 2.25, 5, and 10 MHz frequencies having 128 elements with pitches (p) of 0.75, 1, and 0.5 mm, respectively, were used. To study the influence of total aperture size on resolution without changing any other parameter, experiments were carried out by using 16 elements (N) with different numbers of missing elements ($n = 0, 1, 2, 3, 4$, and 6) between active elements leading to the different total aperture sizes equal to $[(N - 1)(n + 1)p]$. The same elements were also used in reception with null delay law for different transducers, similar to FMC-TFM. A total of 256 (16 transmission sequences × 16 receivers in each transmission) A-scans were acquired in each configuration. In the case of SFLA, a single element is utilized for transmission as well as the reception; hence, only 16 A-scans were acquired. Further, for comparison, more number of A-scan signals for the same aperture size was also obtained by using 64 elements without any missing element. This corresponded to the aperture size of $63p$, which is close to using 16 elements with 3 missing elements (aperture $= 60p$). The signals acquired were processed using software developed in-house in LabVIEW incorporating SAFT and TFM algorithms in time domain.

2.2 Specimen Details

An aluminum specimen of 200 mm thickness with side drill holes (SDHs) of 1 and 2 mm diameters at different locations was used for the study. The SDHs at different depths in the range of 25–175 mm were machined using the Electrical Discharge Machined (EDM) technique. Figure 3 shows the schematic of the specimen used for the study with SDHs at different depths. The data were acquired from top, left, and right in order to study the defects located at different depths as shown in Fig. 3.

3 Results and Discussion

Figure 4a shows the B-scan image obtained after SAFT processing of the 16 A-scan signals acquired using 16 continuous elements [total aperture = (16–1) × 0.75 = 11.25 mm] of 2.25 MHz transducer. The SDH at 25 mm depth could be detected clearly; however, the SDHs at other depths could not be detected. Figure 4b shows the SAFT processed B-scan image using the 16 A-scan signals acquired using the 16 elements of the same transducer, but with a gap of three elements between the

Fig. 3 Schematic of the specimen containing SDHs of 1 and 2 mm diameters

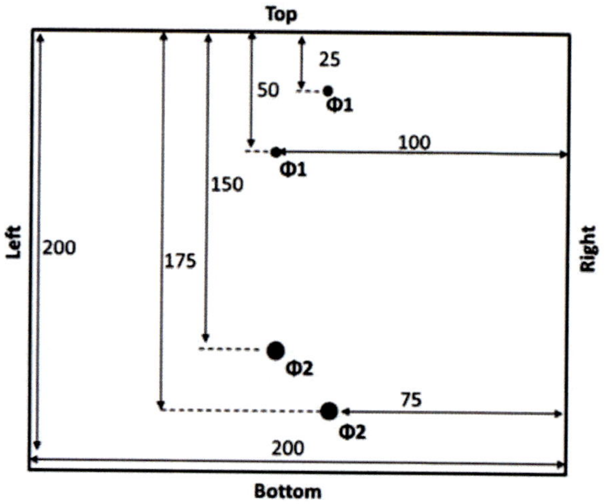

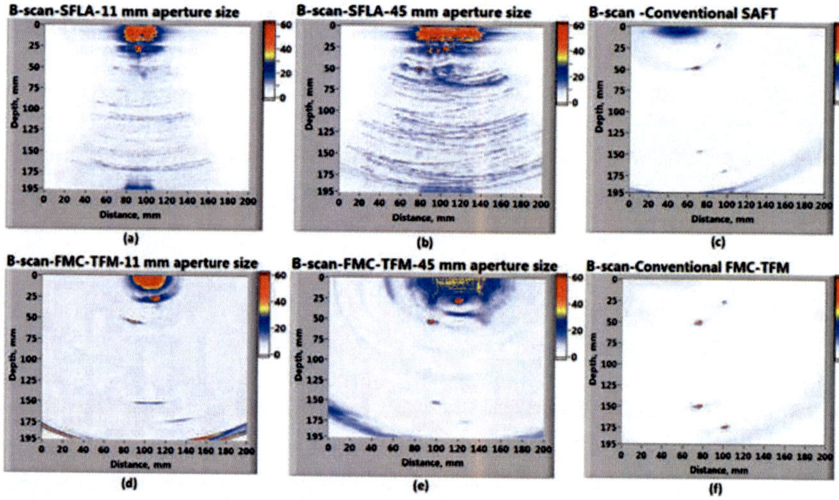

Fig. 4 B-scan images for SDHs obtained by **a–c** SFLA and **d–f** FMC-TFM processing for different aperture sizes of **a, d** 11 mm, **b, e** 45 mm, and **c, f** 47 mm. The pitch was 0.75 mm in **a, c, d, f**, whereas it was 3 mm in **b, e**

active elements. The resultant pitch is 3 mm (4 × 0.75 mm), and the total aperture size is equal to 45 mm [(16–1) × 3 = 45 mm]. It can be seen in Fig. 4b that the SDHs located at 25 and 50 mm depths could be clearly detected, indicating the advantage of increasing the total aperture size. However, with increasing gap between the active elements, constructive interference at undesired locations also take place causing spurious indications in the B-scan image. Spurious indications near the actual SDH locations at both 25 and 50 mm depths can be clearly seen in Fig. 4b. For similar total aperture size of ~47 mm, when 64 continuous elements were used to acquire 64

A-scan signals in pulse-echo mode, the SAFT processed B-scan image clearly shows all the four SDHs at 25, 50, 150, and 175 mm depths (Fig. 4c). The study indicated that increasing the total aperture size by using the same active elements, i.e., by increasing pitch is useful in increasing defect detectability at intermediate depths in SFLA; however, it also leads to spurious signals. On the other hand, increased aperture size at finer pitch leads to enhanced detectability and resolution at larger depths.

In case of FMC-TFM, all the four SDHs could be detected even with the small aperture size of 11 mm, i.e., when 16 consecutive elements were used to acquire 256 A-scans in FMC mode (Fig. 4c). However, with increasing the aperture size for the same 16 numbers of active elements (i.e., with a gap of 3 elements between active elements as discussed above in the case of SFLA), a drastic improvement in the resolution can be observed particularly for the SDHs at larger depths of 150 and 175 mm. Figure 4f shows the TFM processed B-scan image obtained by using 64 consecutive elements to acquire 64 × 64 numbers of A-scans. Even though the total effective aperture size is almost similar, improvement in the amplitude for SDHs at larger depths can be clearly seen in Fig. 4f as compared to Fig. 4e. The study clearly demonstrates that resolution improves with effective aperture size in both SFLA and FMC-TFM with better sensitivity in the later. For the same effective aperture size, FMC + TFM provide better sensitivity for defects at larger depths. Further, SFLA suffers with spurious indications for larger pitch. However, for the same pitch and aperture, the spurious indications are not observed in FMC + TFM. Due to the above mentioned advantages of FMC-TFM over SFLA, further study is focused only on FMC-TFM mode.

Figures 5 and 6 show the variations in the −6 dB size of indications observed in the FMC-TFM B-scan images with the depth of the SDHs and aperture size, respectively, for three different frequencies. It can be seen that the −6 dB spread increases with increasing defect depth and decreasing aperture size for all the three frequencies. Further, the rate of increase in −6 dB spread with increasing depth is observed to be higher for smaller aperture size and lower frequency. For smaller aperture sizes in the range of 11–34 mm at 2.25 MHz, the −6 dB spread is observed to increase continuously for all defect depths. However, for larger aperture sizes, the −6 dB spread is observed to be independent of the defect depth initially, followed by a sudden increase beyond a critical value of the defect depth. The critical value is observed to be about 75 mm for 45 and 56 mm apertures and 150 mm for 79 mm aperture at 2.25 MHz.

At 5 MHz, a sudden increase in the −6 dB spread is observed beyond 100 mm for 15 mm aperture. This indicates that the critical value increases with aperture size and frequency. Hence, this critical value may be correlated to the effective near field considering the total aperture.

The above results indicate that the total effective aperture size is the most important parameter in SAFT and FMC-TFM-based imaging which govern the resolution at a given depth. This is attributed to the fact that the total effective aperture governs the maximum range of angles at which any point in the specimen is insonified. When all A-scan signals acquired from different transmitter locations are used in processing

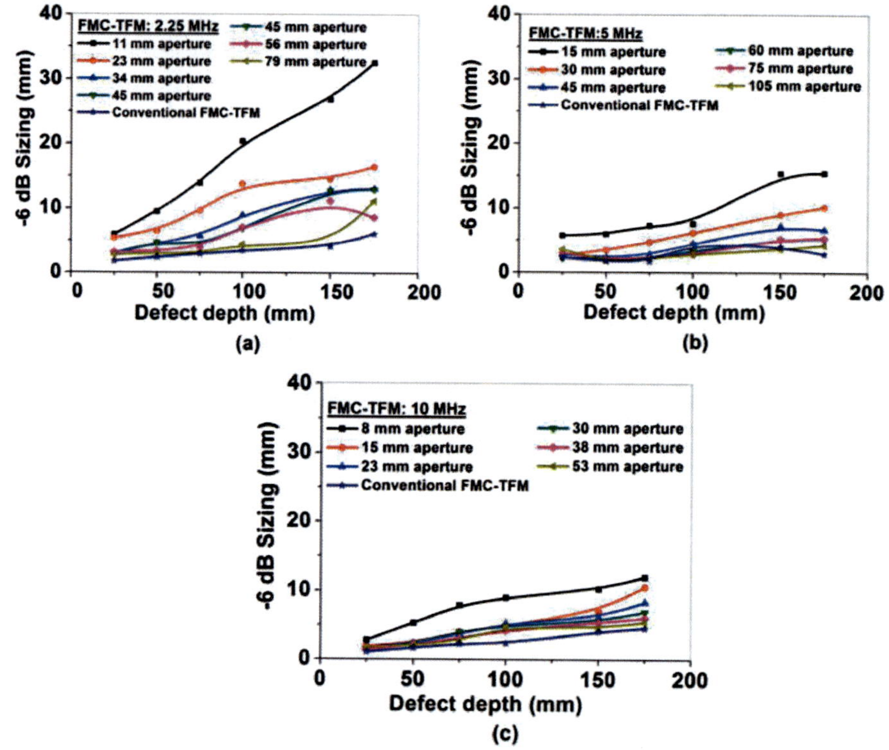

Fig. 5 Effect of defect depth on −6 dB spread in FMC-TFM at different frequencies

for SAFT/FMC-TFM, this angle acts similar to the angle formed at any point by the extreme ends of an active aperture in case of focusing in convention phased array ultrasonic imaging as shown in Fig. 1. This angle is inversely proportional to the resolution achieved. Even though the −6 dB spread primarily depends upon the aperture size, it can also be seen in Fig. 5 that it is less for 64 elements FMC-TFM as compared to the 16 elements FMC-TFM with equal or even larger total aperture size. This indicates that the resolution is also governed by the total number of A-scan signals utilized in images reconstruction.

4 Conclusions

In this paper, the effect of various parameters such as aperture size, frequency, and depth of the defects on resolution achieved in SFLA and FMC-TFM was studied. The study clearly demonstrated that resolution improves with effective aperture size in both SFLA and FMC-TFM with better sensitivity in the later. Further, SFLA suffers

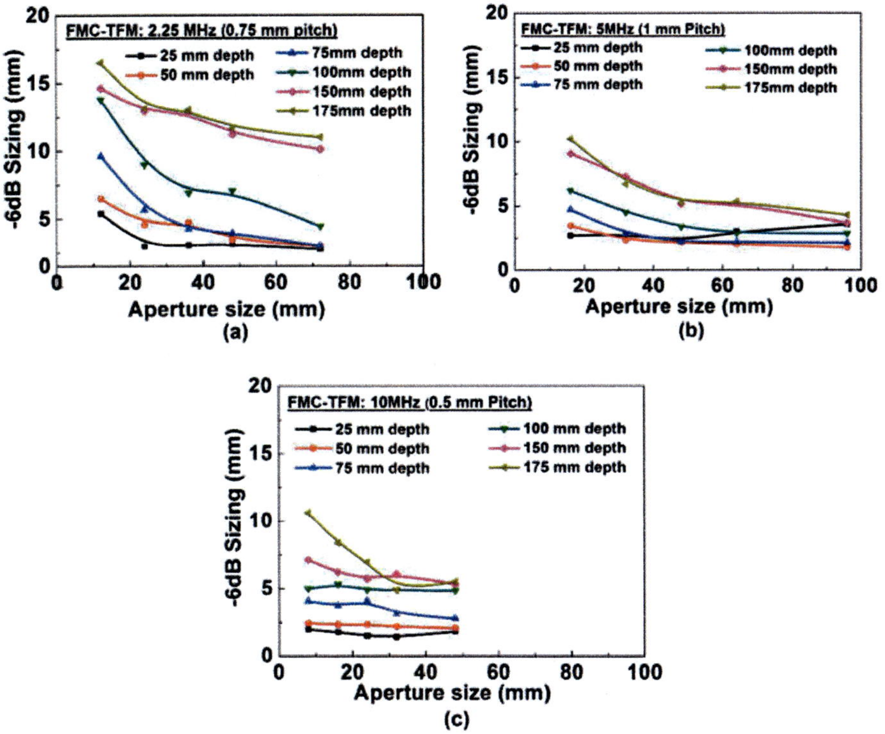

Fig. 6 Effect of aperture size on −6 dB spread in FMC-TFM at different frequencies

with spurious indications for larger pitch. However, for the same pitch and aperture size, the spurious indications are not observed in FMC-TFM. Hence, improved resolution in FMC-TFM can be effectively achieved by increasing effective aperture size by using probes with larger pitch or using active elements with gaps of a few elements. It is also observed that the resolution achieved is governed by the total number of A-scan signals utilized in image reconstruction. Further, the study also indicated that the best resolution in FMC-TFM-based imaging can be obtained only up to a critical depth which increases with aperture size and frequency.

References

1. Azar L, Shi Y, Wooh S-C (2000) Beam focusing behavior of linear phased arrays. NDT&E Int 33(3):189–198. https://doi.org/10.1016/S0963-8695(99)00043-2
2. Gebhardt W (1983) Improvement of ultrasonic testing by phased arrays. Nucl Eng Des 76(3):275–283. https://doi.org/10.1016/0029-5493(83)90111-5
3. Ginzel EA, Johnson D (2008) Phased-array resolution assessment techniques. NDT

4. Sherwin CW, Ruina JP, Rawcliffe RD (1962) Some early developments in synthetic aperture radar systems. IRE Trans Mil Electron 2:111–115. https://doi.org/10.1109/IRET-MIL.1962.5008415
5. Frazier CH, O'Brien WD (1998) Synthetic aperture techniques with a virtual source element. IEEE Trans Ultrason Ferroelectr Freq Control 45(1):196–207
6. Bae M-H, Jeong M-K (2000) A study of synthetic-aperture imaging with virtual source elements in B-mode ultrasound imaging systems. IEEE Trans Ultrason Ferroelectr Freq Control 47(6):1510–1519. https://doi.org/10.1109/58.883540
7. Nanekar P, Jothilakshmi N, Kumar A, Jayakumar T (2017) Synthetic focusing of sound beam using linear array. Mater Eval 75(6):813–821
8. Holmes C, Drinkwater BW, Wilcox PD (2004) The post-processing of ultrasonic array data using the total focusing method. Insight-Non-Destruct Test Condition Monit 46(11):677–680. https://doi.org/10.1784/insi.46.11.677.52285
9. Holmes C, Drinkwater BW, Wilcox PD (2005) Post-processing of the full matrix of ultrasonic transmit–receive array data for non-destructive evaluation. NDT&E Int 38(8):701–711. https://doi.org/10.1016/j.ndteint.2005.04.002
10. Peng J, Zhang Y, Gao X, Peng C, Wang Z (2013) Study on the railway wheel ultrasonic inspection method using the full matrix capture. In: 2013 far east forum on nondestructive evaluation/testing: new technology and application. IEEE (2013)
11. Holmes C, Drinkwater BW, Wilcox PD (2008) Advanced post-processing for scanned ultrasonic arrays: application to defect detection and classification in non-destructive evaluation. Ultrasonics 48(6–7):636–642. https://doi.org/10.1016/j.ultras.2008.07.019
12. Hunter AJ, Drinkwater BW, Wilcox PD (2008) The wavenumber algorithm for full-matrix imaging using an ultrasonic array. IEEE Trans Ultrason Ferroelectr Freq Control 55(11):2450–2462. https://doi.org/10.1109/TUFFC.952
13. Jobst M, Connolly GD (2010) Demonstration of the application of the total focusing method to the inspection of the steel welds. In: 10th European conference on non-destructive testing, pp 1–11
14. Dao G, Braconnier D, Gruber M (2015) Full matrix capture with customizable phased array instrument. AIP Conf Proc 1650:1001–1006. https://doi.org/10.1063/1.4914707
15. Bannouf S, Robert S, Casula O, Prada C (2012) Data set reduction for ultrasonic TFM imaging using the effective aperture approach and virtual sources. J. Phys. 457:1–14. https://doi.org/10.1088/1742-6596/457/1/012007
16. Sutcliffe M, Weston M, Charlton P, Dutton B, Donne K (2012) Virtual source aperture imaging for non-destructive testing. Insight 54(7):1–5. https://doi.org/10.1784/insi.2012.54.7.371
17. Alavudeen S, Krishnamurthy CV, Balasubramaniam K (2011) Technique for imaging using virtual array of sources (TIVAS). AIP Conf Proc 1335:1687–1694. https://doi.org/https://doi.org/10.1063/1.3592131
18. Lockwood GR, Li P-C, O'Donnell, Foster FS (1996) Optimizing the radiation pattern of sparse periodic linear arrays. IEEE Trans Ultrason Ferroelectr Freq Control 43(1):885–893. https://doi.org/https://doi.org/10.1109/58.484457
19. Camacho J, Cruza JF (2015) High resolution autofocused virtual source imaging (AVSI). In: IEEE international ultrasonics symposium proceedings, pp 1–4. https://doi.org/https://doi.org/10.1109/ULTSYM.2015.0048
20. Sutcliffe M, Charlton P, Weston M (2014) Multiple virtual source aperture imaging for non-destructive testing. Ultrasound Imaging 56:75–81
21. Karaman M, Li P-C, O'Donnell M (1995) Synthetic aperture imaging for small scale Systems. IEEE Trans Ultrason Ferroelectr Freq Control 42(3):429–442. https://doi.org/10.1109/58.384453
22. Brizuela J, Katchadjian P, Desimone C, Garcia A (2014) virtual ultrasound sources for inspecting nuclear components of coarse-grained structure, 40th annual review of progress in QNDE. AIP Conf Proc 1581:1053–1060

23. Carcreff E, Laroche N, Braconnier D, Bourguignon S, Duclos A (2017) Improvement of the total focusing method using an inverse problem approach. In: Ultrasonics symposium (IUS), 2017 IEEE international, pp 1–4. https://doi.org/https://doi.org/10.1109/ULTSYM.2017.8092258
24. Sharma GK, Kumar A, Rajkumar KV, Sahu M, Jayakumar T (2007) Ultrasonic spectral analysis methodology for thickness mapping of electromechanical hydrogen meter thimbles. Mater Eval 65(4):417–422
25. Olympus NDT (2007) Advances in phased array ultrasonic technology applications. Olympus NDT 48
26. Thomson RN (1984) Transverse and longitudinal resolution of the synthetic aperture focusing technique. Ultrasonics, pp 9–15

EMAT-Phased Array Inspection of Thick Austenitic Stainless Steel and Dissimilar Metal Welds

R. Dhayalan, Anish Kumar, and C. K. Mukhopadhyay

1 Introduction

Electromagnetic acoustic transducers (EMATs) are now being widely investigated for non-destructive testing (NDT) of solid materials. It can generate or detect ultrasound in electrically conductive or magnetic materials through the Lorentz force principle or magneto-elastic effects [1–4]. The main advantage of EMAT over conventional piezoelectric transducer (PZT) is that it does not need any couplant (non-contact) and can eliminate the inconsistency arising from the couplant use during the inspection. It permits making measurements at elevated temperatures, in corrosive and other hostile environments [5–7]. This type of transducer can be easily fabricated and is quite compatible as compared to the other transducers. The two primary components of an EMAT are a coil that is fed by a very large alternating current pulse and a magnet designed to induce a strong static magnetic flux within the skin depth of the test specimen directly below the EMAT. The pulsed alternating current fed to the coil induces eddy currents (j_e) within the skin depth of the test piece. In the presence of a large bias magnetic flux (B_S), these eddy currents lead to body forces (F_L) at the surface layer of the specimen,

$$F_L = B_S \times j_e \tag{1}$$

The Lorentz forces (F_L) on the eddy currents are transmitted to the solid by collisions with the lattice. These forces on the solid are alternating at the frequency of the driving current and act as a source of ultrasonic waves [8]. So, the generation of ultrasonic wave is provided by coupling between the electromagnetic field and

R. Dhayalan (✉) · A. Kumar · C. K. Mukhopadhyay
Non Destructive Evaluation Division, Indira Gandhi Centre for Atomic Research, Kalpakkam 603102, Tamil Nadu, India
e-mail: dhayalanr@igcar.gov.in

C. K. Mukhopadhyay and R. Mulaveesala (eds.), *Advances in Non-destructive Evaluation*, Lecture Notes in Mechanical Engineering,
https://doi.org/10.1007/978-981-16-0186-6_33

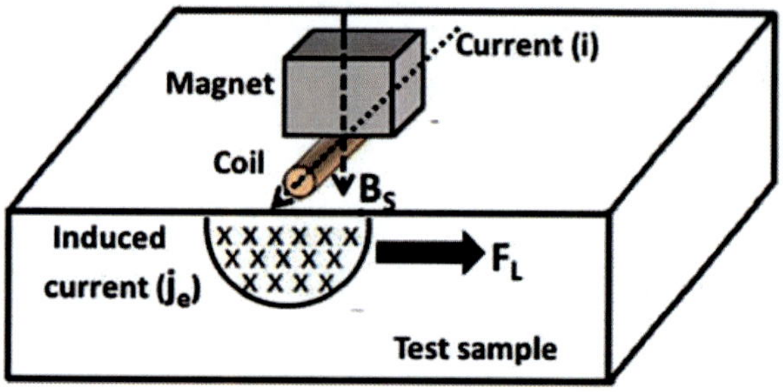

Fig. 1 Lorentz force mechanism

the elastic field in the surface skin. Figure 1 shows the schematic of a single coil and magnet leading to the generation of Lorentz force for EMAT. The type of wave mode generation depends upon the coil geometry, the operating frequency and the applied magnetic field. It can generate any specific ultrasonic mode including normal beam and angle-beam shear waves, Rayleigh waves and plate waves [9–11].

Austenitic stainless steel (AUSS) is widely being used in the nuclear industry due to its excellent corrosion resistance, superior mechanical properties at elevated temperature and good fabricability. There are many types of AUSS weld joints in the nuclear structures, and the most common are AUSS to AUSS, dissimilar metal weld between AUSS to low alloy/carbon steel and AUSS to a nickel base alloy. Conventional ultrasonic testing (UT) of AUSS welds is very difficult because of the metallurgy of the material; grains are elongated and large compared to those found in a ferritic steel, resulting a large degree of acoustic anisotropy, beam distortion and scattering [12, 13]. Moreover, these elongated grains are often organized in columnar structure in the welds, which can result in the elastic waves being skewed in an unexpected direction. Shear vertical (SV) wave commonly used in UT suffers most from the skew effect due to anisotropy of AUSS crystal structures. Longitudinal (L) waves skew significantly less than SV on AUSS weld, but still, experience strong mode conversion at structural and weld boundaries and require access to both sides of the weld. Early research in 1980s showed that the shear horizontal (SH) wave does not present mode conversion at structure boundaries and has a much smaller skew effect compared to L and SV waves. Figure 2a shows the amount of beam skew expected for L, SV and SH waves, and Fig. 2b shows an AUSS weld model with different ultrasonic beam patterns [14–16]. As a result, the SH wave has been recognized as potentially the best solution for the inspection of these welds.

Notwithstanding this, shear energy does not propagate through liquid couplants, and the horizontal polarization cannot be easily excited through mode conversion with a wedge, so it is very difficult to generate with PZT, and it is impractical in field use. EMAT on the other hand is an effective alternative to generate SH waves in ultrasonic testing. Although EMAT excitation of SH waves can be very efficient in many engineering materials such as ferromagnetic steels, aluminum and copper, it remains challenging in AUSS. Because, it has very low conductivity and low or no magnetism which affects the ability to generate eddy currents, hence sound, with

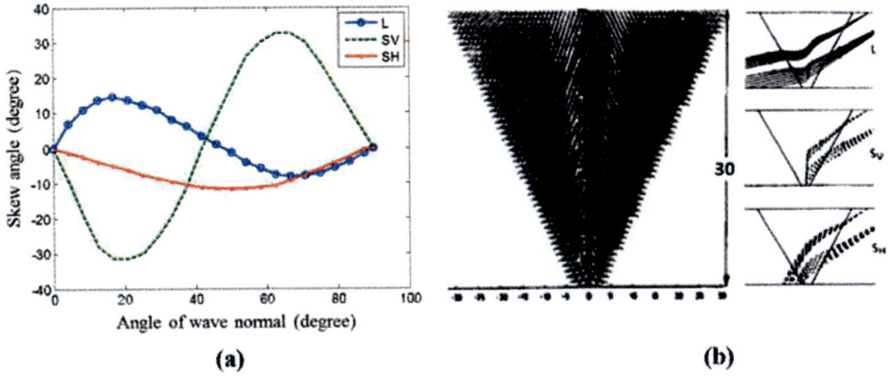

Fig. 2 **a** Relationship between skew angle and incident angle with respect to columnar direction and **b** modelling of different ultrasonic beam profiles in an austenitic weld [16]

EMAT. Compared to other non-ferromagnetic materials, AUSS is 10–15 times more resistive with proportional effects on signal-to-noise (SNR).

In the past, a number of different weld inspection schemes using SH wave EMATs have been proposed [17–19]. In the majority of cases, such schemes relied upon the periodic permanent magnet (PPM) EMAT configuration, and other transducer configurations have also been proposed [20, 21]. The main disadvantages of this single-channel SH wave EMAT are low transduction efficiencies, narrow beam angle for single frequency and poor SNR. In order to overcome these complexities, an EMAT-phased array (PA) sensor with a high-power tone-burst generator and signal amplifiers has been used to enhance the power level for improving the signal strength and SNR [21]. The EMAT PA sensor has been designed in tandem mode and radiates SH wave with almost equal amplitude from 0 to 90°. So, it is able to cover the entire volume of the weld including heat affected zone (HAZ) by scanning from one sensor position. In this paper, the EMAT PA sensor has been utilized for detection of defects in PFBR main vessel mock-up weldments. There were three different thick mock-up weldments made with artificial defects (notches) of varying depths and side-drilled holes (SDH) with different diameters. Of these two weldments were made of AUSS with two different thicknesses and weld configurations. The third weldment was made by two different metals of the same thickness (AUSS with ferritic steel) or dissimilar metal weld (DMW). The PA sensor provides enough sensitivity to detect 10% deep defects and is able to resolve closely spaced SDHs in both AUSS and DMW irrespective of weld configurations up to 30 mm thickness considered in the present study.

2 SH Wave EMAT PA Probe

In recent years, systems with PA technology have been widely used for inspection of welds to achieve better sensitivity and resolution. These systems typically employ PA probes in the frequency range from 1 to 5 MHz and utilize 16 or more elements (piezoelectric crystals) to steer beam within the base material for weld inspection. In most of the conventional PA probes, L or SV wave modes have been used for inspection of ferritic steel welds. For the reasons explained in the previous section, these types of high-frequency conventional PA have inherent limitations for AUSS weld inspection. To overcome the metallurgical issues, the operating frequency of the probe needs to be lowered (less than 1 MHz) to reduce scatter and attenuation, and by using SH waves, the beam skewing can be reduced significantly. Given the advantages associated with SH waves, the concept of PA functionality is replicated with EMATs for SH wave EMAT PA probe. An eight-channel SH wave EMAT PA probe (M/s. Innerspec Technologies, Spain) is used for this work which is made by using flexible meander RF coils and permanent magnet arrays [19]. Figure 3a, b shows the photograph and the pitch-catch tandem arrangement of the eight-channel SH wave EMAT PA probe.

In this EMAT PA probe, the eight transmitters are excited with independent time delays so the wavefronts constructively interfere with each other around the focal spot to achieve a wave field of strong intensity. The reflected signals from the focal region arrive at each element at a different time, and it is delayed according to focal laws so the signal from the focal region sums up in phase. As a result, the SH wave can be steered across a predefined range. Without the effect of mode conversion, SH waves can be focused at a large range of angles and can be used for both zero degree and angle-beam steering. Figure 4a shows the schematic of the principle of SH wave EMAT transmitter with meander coil and permanent magnet array. The flexible coils and magnet arrays permit complying from flat to any curved specimen surfaces. The magnet array determines the wavelength (channel pitch) which is about 3.2 mm. Figure 4b shows the frequency response of the EMAT PA probe for any given angle

(a)

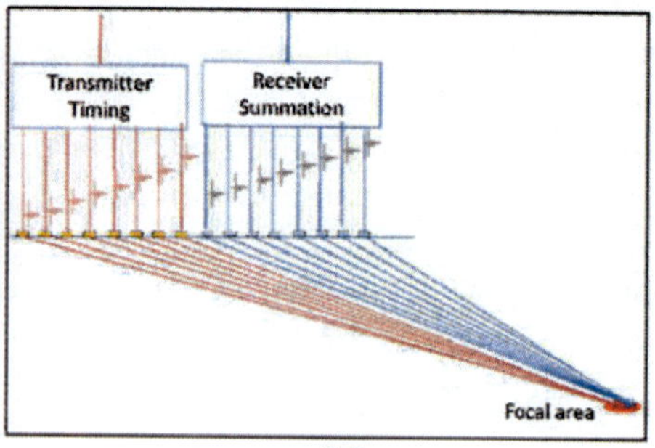

(b)

Fig. 3 **a** Eight-channel SH wave EMAT PA probe and **b** pitch-catch arrangement for beam focusing

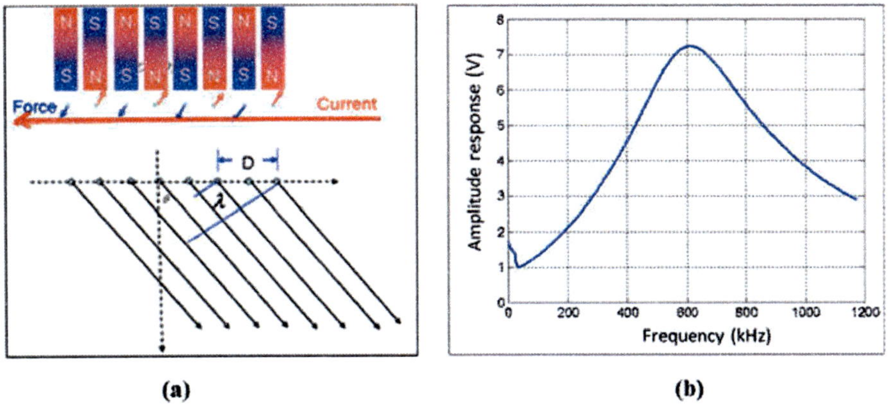

Fig. 4 **a** Schematic of SH wave EMAT with meander coil and magnet array and **b** frequency response of SH wave EMAT PA probe.

at which the peak or optimum frequency is 600 kHz, and it is recommended to operate within 500–700 kHz. The active area or footprint of the probe is about 58 mm length and 45 mm width. The larger aperture can also provide improved sensitivity and focused inspection of thick components.

3 Experimental Details

In order to excite the eight-channel SH wave EMAT PA probe, an eight-channel high-power tone-burst system temate® Power Box-8 (M/s. Innerspec Technologies, Spain) was used along with signal conditioning box to compensate the impedance mismatch between the system and PA probe. It can provide power level up to 20 kW or 2000 Vpp of peak power per channel at 1% duty cycle for frequency from 100 kHz to 7 MHz. Figure 5a, b shows the photograph and schematic of the experimental set-up used for inspection of thick AUSS welds and DMWs. This system was controlled through an external PC over an Ethernet connection. The input parameters were fed through the computer for the pulsing signal including tone-burst frequency of 600 kHz. After receiving the input, a low-voltage pulse train was generated and subsequently amplified to high voltage that was fed through the signal conditioning box to the transmitter. With the high voltage excitation, SH wave with a wide beam profile was generated into the material. The reflected waves were received and converted into electrical signals by the receiver. These signals were amplified and filtered by the signal conditioning box and sent to the system for further amplification and treatment. Finally, the signals were digitized and sent to the PC for further processing for display and storage.

There are different types of weld configurations and geometries in PFBR main vessel developed by various techniques, materials and welding parameters. Based

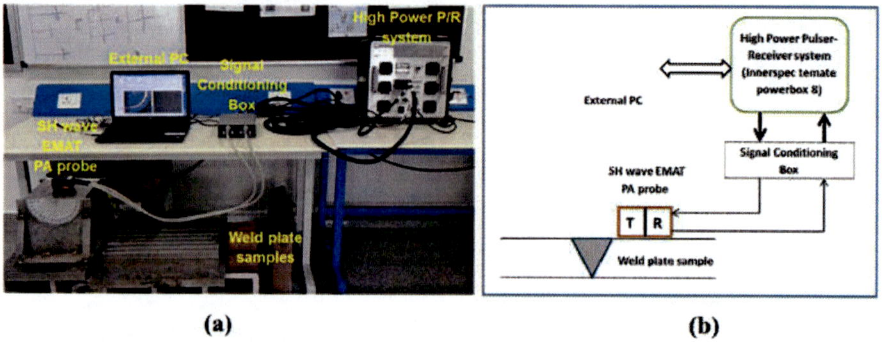

Fig. 5 **a** Photograph and **b** schematic of the experimental set-up

on the weld geometry, the scan plan for inspection was configured with the EMAT PA system. The input parameters like the selection of probe frequency, gain setting and scanning pattern were configured in the system. The PA probe was excited with three cycles square modulated sine wave tone-burst signals at 600 kHz. With proper selection of focal points and SH wave velocity, the probe delay calculation, sound path and surface distance were completed automatically. The sector scan representation of angle from 0 to 90° was selected with 1° angle increment. On the receiver side, a 32 dB gain was used to amplify the received signals. The generated SH waves were allowed to make three full skips (more than three legs/zoom) so that to cover the entire volume of the weld including the heat affected zones (HAZ) on both sides of the weld. The inspection procedure was validated by using two sets of AUSS and one DMW mock-up weld samples as shown in Fig. 6a–c. For the first mock-up sample shown in Fig. 6a, three plates were joined together by double V weld configuration, and the size of the plate sample was 500 × 500 × 28 mm^3. In this weld plate, 12 artificial defects (notches) of the same length (25 mm), width (2.5 mm) and varying depths were made parallel and perpendicular to the weld orientations. The depth of the notches as a percentage of thickness was 10, 20 and 30%, respectively. For the second mock-up sample shown in Fig. 6b, a 30 mm thick plate was welded with a 25 mm plate with K-type weld configuration in which six artificial notches of the same length (25 mm), width (3 mm) and varying depths were made parallel and

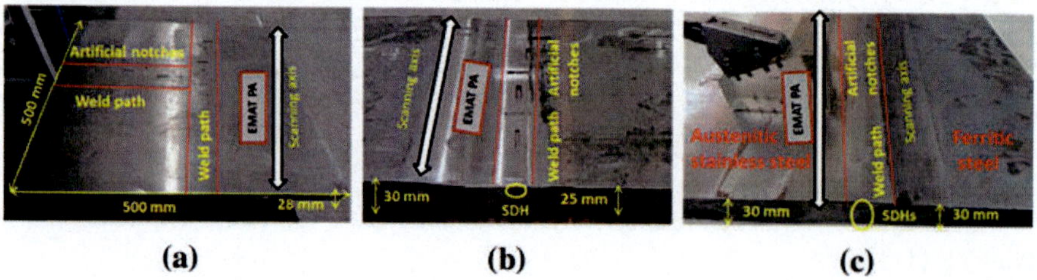

Fig. 6 PFBR main vessel mock-up weld samples with artificial defects. **a, b** AUSS weld plates and **c** DMW plate

perpendicular to the weld orientations. The depth of the notches as a percentage of thickness (25 mm) was 10, 20 and 30%, respectively. A side-drilled hole (SDH) of 2.5 mm diameter and 25 mm length was made on the side of the weld plate at 20 mm depth. The notches were made on both the weld and HAZ. For the DMW mock-up sample which is shown in Fig. 6c, a 30 mm thickness AUSS and a ferritic steel plate of the same thickness were joined together by double V weld configuration. In this DMW sample, 16 artificial defects (notches) of the same length (10 mm), width (1 mm) and varying depths were made on both AUSS and ferritic sides including HAZ. Though the PA sensor can able to detect less than 10% of wall thickness (WT), the maximum design allowable defect size cannot exceed 10% of WT for all the welds during inspection. In order to check the defect sensitivity and resolution capability, closely spaced SDHs with different diameters were made on both sides of the sample. For all the mock-up samples, the inspection procedure was to scan the PA probe parallel to the weld path keeping the probe at the same distance from the weld, pausing at regular intervals for data acquisition. Multiple trials were conducted to validate the repeatability and reproducibility of this inspection procedure with PA probe.

4 Results and Discussion

As mentioned earlier, the EMAT PA probe generates SH wave with almost uniform amplitude for a wide beam angle from 0 to 90°. In order to verify the radiation pattern of the SH waves, a semi-circular AUSS solid block of 120 mm length with 100 mm diameter was used.

The EMAT PA probe was kept exactly at the centre of the flat surface and by facing the curvature as shown in Fig. 7a. It was determined to generate the waves directly towards the curvature of the block and reflect back to the EMAT PA probe. Figure 7b shows the sector scan image of SH waves reflected from the curvature of the solid

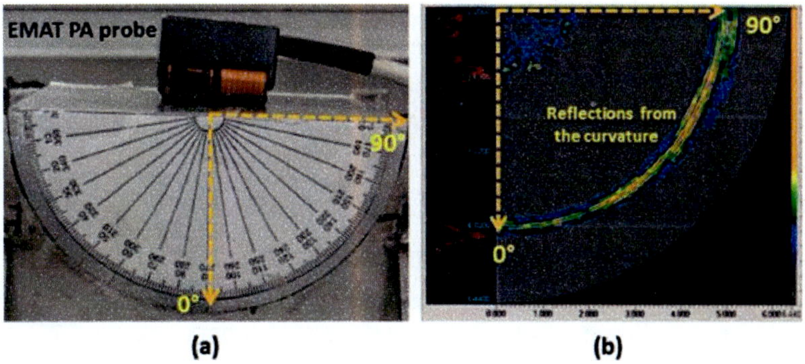

Fig. 7 **a** Photograph of the semi-circular block with EMAT PA probe arrangement and **b** sector scan image of SH waves obtained with the same set-up

block. From the sector scan, it was observed that the reflected energy or amplitude of the SH wave was almost similar at all the angles. The A-scan signal shown on the left side of the sector scan was corresponding to the wave at zero degree, and the initial random signals were the EM noise within the EMAT probe. From Fig. 7, it was confirmed that the EMAT PA probe generates SH waves at all angles, and it can cover the entire volume of the weld including HAZ. Also, it is possible to identify the location of the defects from the sector scan images.

For all the mock-up weld samples, the scanning was performed on both surfaces by using the EMAT PA probe. Figure 8 shows the schematic diagram of the test set-up and the typical sector scan images obtained from the weld sample shown in Fig. 6a. In the schematic diagram, the defects are shown on the same axis but in actual, all of them were well separated by about 50 mm laterally. The reflected energies for all the defects are highlighted as red colour dotted circles, and the corresponding angles represent the locations of the defects (notches and SDH). The reflected signals at 0° represent the multiple reflections from the base material (HAZ). The reflected signals at ~20°, ~60° and 90° are corresponding to the bottom notch (10% WT), SDH at the middle of the sample and the surface notch (20% WT), respectively. It was inferred that the reflected waves for the larger defects show higher amplitude as compared to the smaller defects. It has also been seen that the reflected signal from the SDH was slightly scattered compared to other defects.

Figure 9 shows the schematic diagram and the test results obtained with the second weld sample shown in Fig. 6b. Though mock-up sample was made with more number of notches with different sizes and orientations, the sector scan images of selective

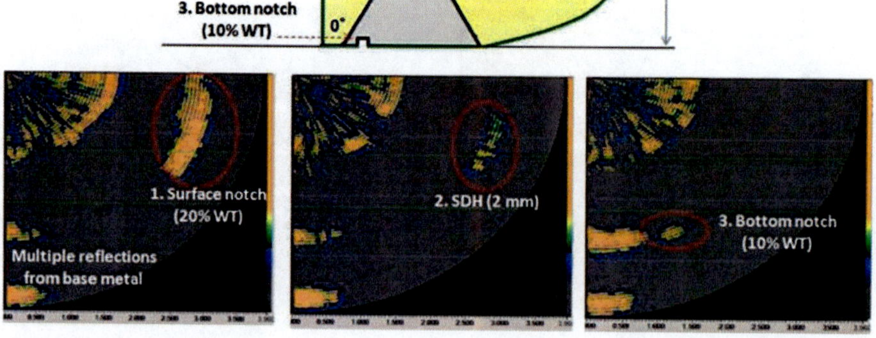

Fig. 8 Schematic diagram of the test set-up and sector scan images obtained from the first mock-up sample with defects on both sides of the weld

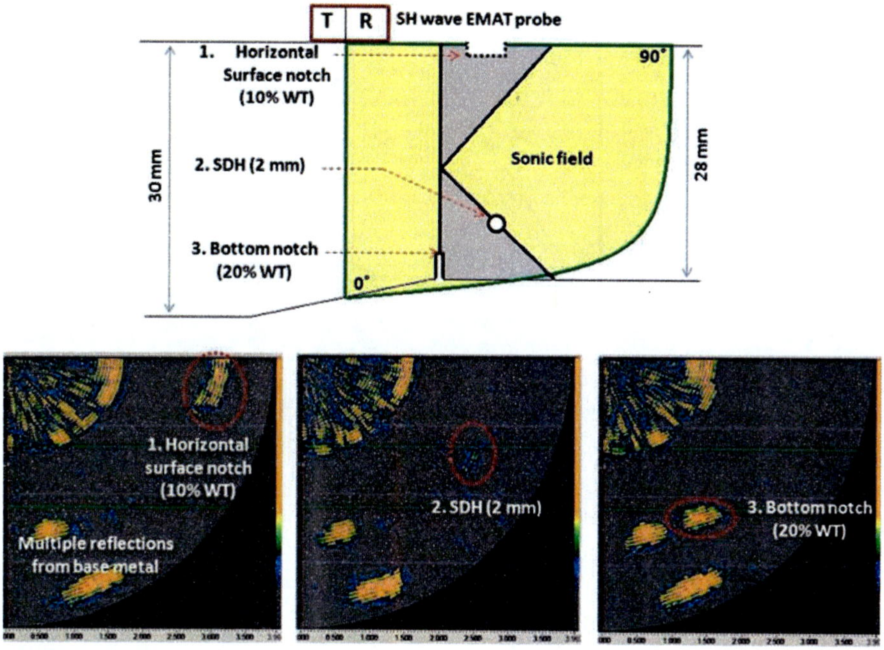

Fig. 9 Schematic diagram of the test set-up and sector scan images obtained from the second mock-up sample with defects on both sides of the weld

notches are shown in Fig. 9. Since the sample was made with two different thick plates and the probe was kept at the tapered portion, the multiple reflections from the base material were slightly inclined with respect to the normal. The reflected energies from all the defects were highlighted, and the corresponding locations were predicted from the beam angles. It has been observed that the intensity of the reflected wave for the largest defect shows the maximum intensity, similar to the first sample. The reflected energy from the surface notch was very less since it was made perpendicular to the weld path. The reflected signal for the SDH was highly scattered when compared to the previous sample because it was accessed about 20 mm height from the top surface. Therefore, the reflection amplitude was also affected by the location and orientation of the defect.

For the DMW mock-up sample, the EMAT PA probe was kept on the AUSS plate side since the permanent magnets get stuck if placed on the ferritic steel plate. Figure 10 shows the schematic picture and the sector scan images obtained with the DMW mock-up sample. In this sample, almost all the notches were made parallel to the groove faces in both ferritic and AUSS plates. Further, the notches were of 10 mm length as compared to 25 mm in case of the two AUSS weldments. Hence, the reflected energies from all the notches were less when compared to the other two AUSS mock-up samples, and the SNR was slightly low due to dissimilar weld joint. Though, the SDHs were in the same axis which could be detected and resolved clearly. Because of the tandem arrangement of the PA probe, the indications for the

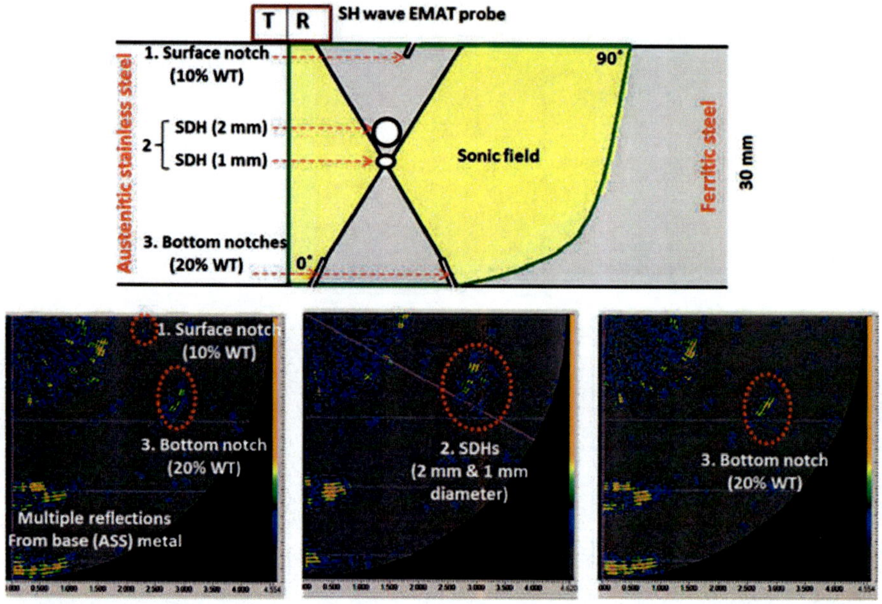

Fig. 10 Schematic diagram of the test set-up and sector scan images obtained from the DMW sample with defects on both sides of the weld

two SDHs were slightly inclined with each other. When two notches were on the same axis of both top and bottom surfaces of the sample, those could be detected with the same probe position and their locations could be identified from the beam angles.

From all the test results, it was inferred that the amplitude of the defect signal varies due to the defect geometry and orientation, but all the defects could be reproduced with measurable indications on both sides of the welds. Inspection from one side of the weld is highly desirable when access to the other side is difficult or impossible. In such cases, multiple angle inspection accompanied with probe scan perpendicular to the weld shall be sufficient for the entire volume of the weld including HAZ. Although the exciting frequency of the EMAT PA probe is very low, it provides enough sensitivity to detect defects down to 10% WT from both sides of the weld.

5 Conclusion

This paper has reported on test results of thick AUSS and DM weld inspection using an eight-channel SH wave EMAT PA probe. The pitch-catch tandem arrangement of the EMAT PA probe provides enhanced power levels with superior SNR compared to conventional EMAT transducers. It has been verified that the EMAT PA probe radiates SH wave with almost equal amplitude from 0 to 90°, and it provides the entire volume coverage of the weld including HAZ from one probe position. It has also been observed that the EMAT probe used in the present study generates SH wave with the maximum amplitude at 600 kHz. At this optimum frequency, it has been utilized for detection of artificial defects in both AUSS and DM weld mock-up samples and confirmed that the probe is capable of detecting defects as small as 10% WT of thick weld joints. The capability of detecting defects from one side of the weld confirms the possibility of using this probe in situations, where there is only one side accessibility.

References

1. Maxfield BW, Fortunko CM (1983) The design and use of electromagnetic acoustic wave transducers (EMATs). Mater Eval 41:1399–1408
2. Thompson RB (1990) Physical principles of measurements with EMAT transducers. In: Thurston RN, Pierce AD (eds) Physical acoustics, vol 19. Academic Press, San Diego, New York, pp 157–200
3. Ribichini R, Cegla F, Nagy PB, Cawley P (2012) Experimental and numerical evaluation of electromagnetic acoustic transducer performance on steel materials. NDT&E Int 45:32–38
4. Dhayalan R, Satyanarayanmurthy V, Krishnamurthy CV, Balasubramaniam K (2011) Improving the signal amplitude of meandering coil EMATs by using ribbon soft magnetic flux concentrators (MFC). Ultrasonics 51:675–682
5. Maxfield BW, Kuramoto A, Hulbert JK (1987) Evaluating EMAT designs for selected applications. Mater Eval 45(10):1166–1183
6. Jian X, Dixon S, Edwards SR (2005) Optimal ultrasonic wave generation of EMAT for NDE. Non-Destruct Evaluat 20: 42–62
7. Hirao M, Ogi H (2003) EMATs for science and industry—non-contacting ultrasonic measurements. Kluwer Academic Publishers, Boston
8. Dhayalan R, Balasubramaniam K (2011) A two-stage finite element model of a meander coil electromagnetic acoustic transducer (EMAT) transmitter. Non-Destruct Test Evaluat 26:101–118
9. Xie Y, Yin W, Liu Z, Peyton A (2015) Simulation of ultrasonic and EMAT arrays FEM and FDTD. Ultrasonics 66:154–165
10. Dhayalan R, Balasubramaniam K (2010) A hybrid finite element model for simulation of electromagnetic acoustic transducer (EMAT) based plate waves. NDT&E Int 43:519–526
11. Thring CB, Fan Y, Edwards RS (2016) Focused Rayleigh wave EMAT for characterisation of surface-breaking defects. NDT&E Int 81:20–27
12. Hudgell RJ, Gray BS (1985) The ultrasonic inspection of austenitic materials-state of the art report. OCED Nuclear Energy Agency, CSNI Report No. 94
13. Hubschen G, Salzburger HJ, Kroning M et al (1993) Results and experiences of ISI of Austenitic and dissimilar welds using SH-waves and EMUS-probes. Elsevier Science Publishers
14. Musgrave MJP (1954) On the propagation of elastic waves in aeolotropic media, II. Media of hexagonal symmetry. In: Proceedings of the Royal Society of London, Series A. vol 226, pp 356–366
15. Hudgell RJ, Seed H (1980) Ultrasonic longitudinal wave examination of austenitic welds. Br J NDT 25:78–85
16. Ogilvy JA (1987) The influence of austenitic weld geometry and manufacture on ultrasonic inspection of welded joints. Br J NDT 29:147–156
17. Ludwig VB, Werner R, Schmid R, Friedrich M, Kroning K (1994) Current in-service inspection of austenitic stainless steel and dissimilar metal welds in light water nuclear power plants. Nucl Eng Des 151:539–550

18. Sawaragai K, Salzburger HJ, Hubschen G, Enami K, Kirihigashi A, Tachibana N (2000) Improvement of SH-wave EMAT phased array inspection by new eight segment probes. Nucl Eng Des 198:153–163
19. Gao H, Lopez B (2010) Development of single-channel and phased array electromagnetic acoustic transducers for austenitic weld testing. Mater Eval 68(7):821–827
20. Hillis S, Dixon S (2014) Localisation of defects with time and frequency measurements using pulsed arrays. NDT and E Int 67:24–30
21. Isla J, Cegla F (2017) EMAT phased array: A feasibility study of surface crack detection. Ultrasonics 78:1–9
22. Arun K, Dhayalan R, Balasubramaniam K, Maxfield BW, Patrick P, Barnoncel D (2012) An EMAT based shear horizontal wave (SH) technique for adhesive bond inspection. In: Review of progress in quantitative nondestructive evaluation (QNDE 2012), AIP conference proceedings, vol 1430, pp 1268–1275
23. Petcher PA, Burrows SE, Dixon S (2014) Shear horizontal (SH) ultrasound wave propagation around smooth corners. Ultrasonics 54(4):997–1004

Computed Tomography (CT) Is an Asset to Ensure the Quality and Reliability of Parts in Aerospace Applications

Manu Joseph, M. Arumugam, Regi Varghese, and G. Narayanan

1 Introduction

ISRO is keen on ensuring high quality and reliability in all its missions. This trickles down even to the lowest levels of manufacturing and inspection thereby upholding the ideals of absolute quality. All parts, sub-assemblies, components, and modules used in launch vehicles and satellites need to be inspected, tested, certified, verified, and reliably ascertained that they are free of defects. NDE and metrology play vital roles in this regard.

CT is widely implemented in the field of medicine and health care for reliable diagnosis and pre-surgery planning of the human body as reported by Samuelson et al. and Dessel et al. [1, 2] With the advancements in electronics and computation, CT has become faster and more accurate. Its applications are now encroaching into the industry especially the aerospace sector viz a viz NDE [3], metrology [4–7], and failure analysis (FA) [8] where its superlative imaging capability is utilized to strengthen quality and increase reliability in routine inspection activities. This paper discusses in detail five unique cases of NDE where CT was used, namely;

M. Joseph (✉) · M. Arumugam (✉) · R. Varghese
Quality Control and Non-destructive Evaluation Group, Liquid Propulsion Systems Centre, Indian Space Research Organisation, Thiruvananthapuram 695547, India
e-mail: manu_joseph@lpsc.gov.in

M. Arumugam
e-mail: m_arumugam@lpsc.gov.in

R. Varghese
e-mail: regivarughese@lpsc.gov.in

G. Narayanan
Systems Reliability and Quality Assurance Entity, Liquid Propulsion Systems Centre, Indian Space Research Organisation, Thiruvananthapuram 695547, India
e-mail: g_narayanan@lpsc.gov.in

C. K. Mukhopadhyay and R. Mulaveesala (eds.), *Advances in Non-destructive Evaluation*, Lecture Notes in Mechanical Engineering,
https://doi.org/10.1007/978-981-16-0186-6_34

a. Electron beam (EB) weld of test ring hardware
b. Port welds of housing sub-assembly
c. Bucking of aluminum rivets
d. End fittings realized through additive manufacturing (AM)
e. Carbon–carbon (C–C) composite samples.

CT is also used for NDE in FA of assembled components. Two such cases are presented in detail;

a. Investigation of reduced stroke in isolation valve
b. Investigation of failed pressure regulator.

The scope of CT in dimensional metrology is huge but is not ventured into much. CT can be implemented on a wide range of parts for dimensional inspection. A couple of cases of dimensional inspection with a performance evaluation study are presented. The paper concludes by summarizing the merits of implementing CT in the aerospace sector.

2 CT for Non-destructive Evaluation

2.1 Inspection of Electron Beam (EB) Weld of Test Ring Hardware

Before any hardware goes into production, it is qualified by various levels of trials and tests. The test ring sub-assembly is utilized in the upcoming semi-cryogenic engine of ISRO. This has three rings that are joined by means of EB welding as per the schematic in Fig. 1. NDE by RT and UT of the test ring welds revealed linear indications in certain locations along the length which were, initially, wrongly interpreted as cracks. Upon CT examination of the samples extracted (Fig. 2) from the hardware, the true nature of the defect was revealed as shrinkage cavities (Fig. 3).

Fig. 1 Schematic of the cross section of the welded test ring assembly. The EB welds are highlighted. The three rings are denoted as 1, 2, and 3

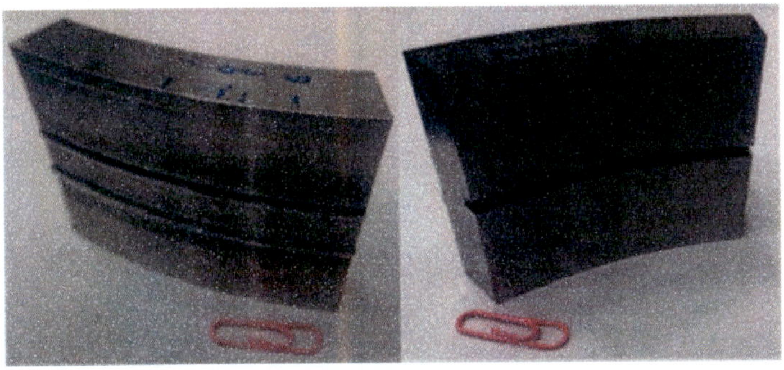

Fig. 2 Samples extracted from hardware for CT scan

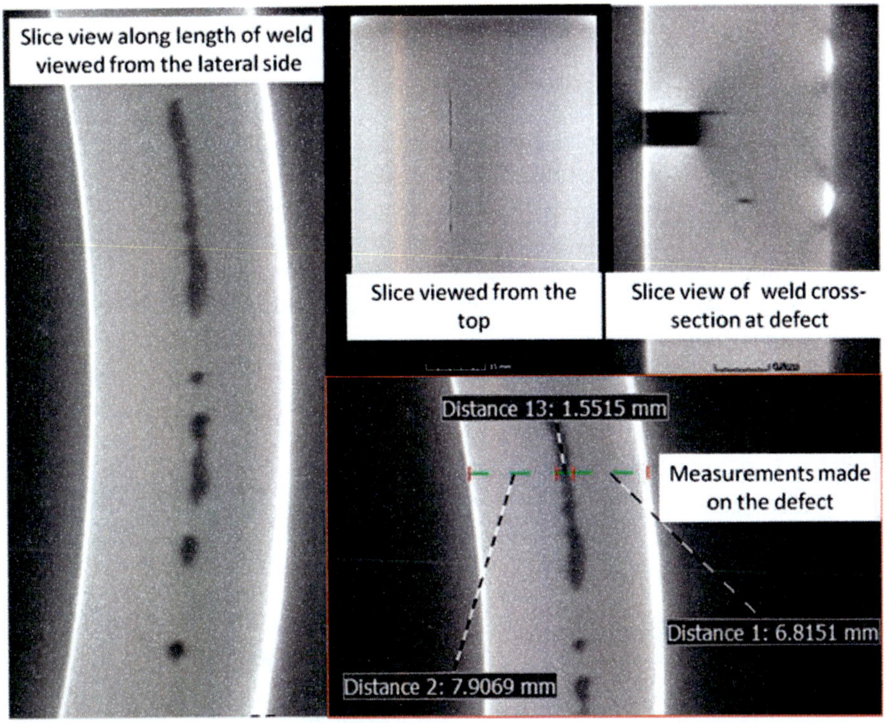

Fig. 3 Slice views along mutually perpendicular orientations. The minimum distance of the defect from surface and the depth of defect could be measured

2.2 *Inspection of Port Welds of Housing Sub-assembly*

The housing sub-assembly consists of a forged main body with three ports welded by GTAW to the body (Fig. 4). The ports are realized by welding three solid rods to the main body followed by drilling the nozzle cavity in a later machining stage.

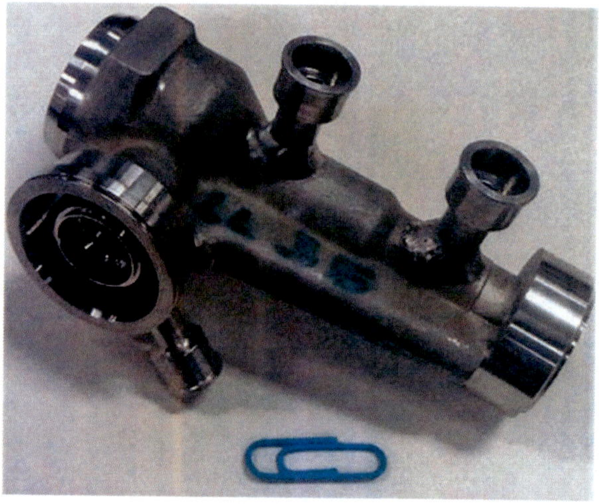

Fig. 4 Photograph of the housing sub-assembly

The process flow is illustrated in Fig. 5. Of the three ports, the outlet port withstands pressures as high as 20 bar during flight. Thus, it is essential to evaluate the integrity of the outlet port weld. The current practices involve visual inspection of the weld from the interior of the nozzle cavity, by endoscopy, and liquid penetrant testing on the outside to ensure absence of surface discontinuities. Radiography, though is essential to ensure absence of volumetric discontinuities, cannot be performed as the complex geometry of the part makes it impossible to produce interpretable radiographs. Endoscopy of the outlet port revealed surface irregularities and discoloration on the weld at the root area.

CT with its superlative imaging capability was utilized to image the weld area in 3D to ensure that the observed discontinuities were not volumetric in nature. On the contrary, it was revealed that the defects observed were shrinkage cavities that were formed during initial welding which opened into the nozzle cavity after machining. The CT screenshots are provided in Fig. 6.

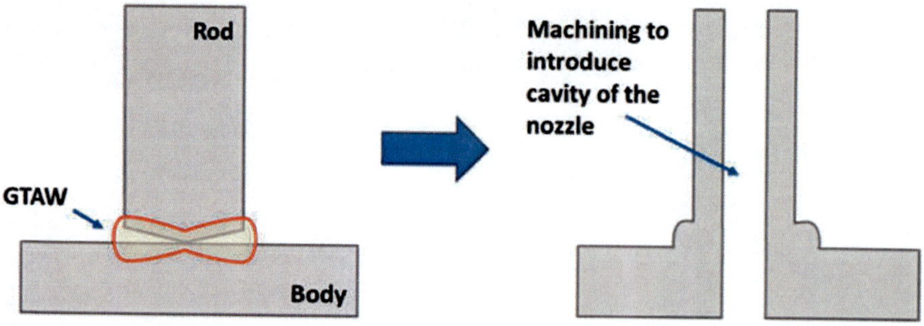

Fig. 5 Representation of process flow adopted to realize the nozzles of the sub-assembly

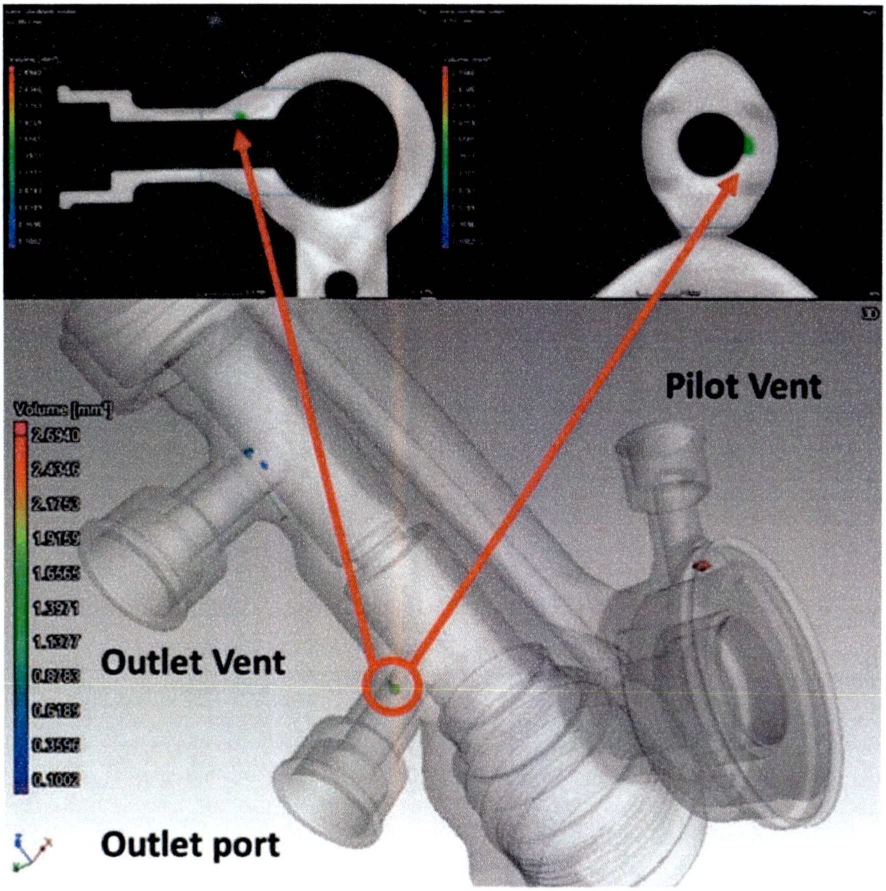

Fig. 6 3D view with voids in all ports especially the outlet port which holds highest pressure. The slice views of the defect are also shown

Another highlight is that the defect dimensions (projection length on x-, y-, and z-axes and volume) and orientation information served as direct input for numerical flight simulation of the defective hardware which aided in further decision making.

2.3 *Inspection and Measurement of Bucking of Aluminum Alloy V65 Rivets*

As part of qualification of rivet designs, a trial specimen with several rivets of V65 material of different configuration was made. The requirement was to ensure proper bucking of rivets with the substrate and sheet (both made of aluminum alloy AA7075). Radiographic testing was not possible due to limitations in facility. But CT was able

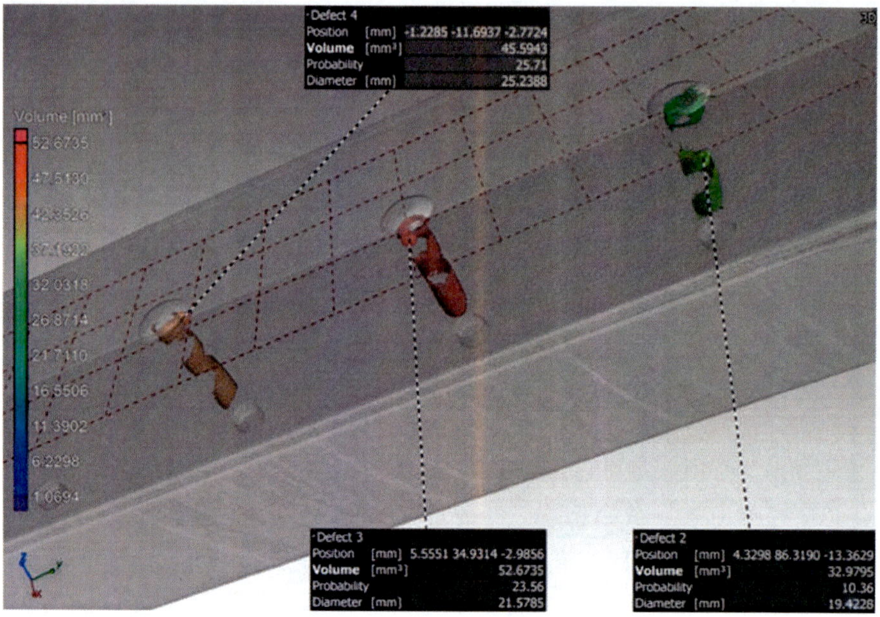

Fig. 7 Porosity analysis to find volume of the defects

to detect, measure, and quantify the improper bucking with the porosity algorithm. The void could be visualized as spiraling along the rivet hole as illustrated in Fig. 7. The volume of the gap was measured, and this information served as input for some critical decisions on future riveting processes.

2.4 *Inspection of End Fittings Realized Through Additive Manufacturing (AM)*

Additive manufacturing (also known as 3D printing) is revolutionizing the manufacturing industry. The relative ease with which parts and components can be manufactured in much less time makes it very viable for the space industry where costs and time for manufacturing are high. As the technology is relatively new, extensive study, analysis, and testing are required before 3D printed parts/systems can be inducted for space application. CT is effectively utilized as an aid for evaluation of 3D printed components in ISRO. The case of CT inspection of the end fitting (Fig. 8) is discussed to demonstrate this.

The reconstructed CT model of the end fitting was subjected to porosity and thickness analyses. It revealed voids in the ring area as depicted in Figs. 9 and 10 which are in the form of intermittent rings. This was observed in three different levels along the built direction. Initial speculation for cause of such a defect was attributed

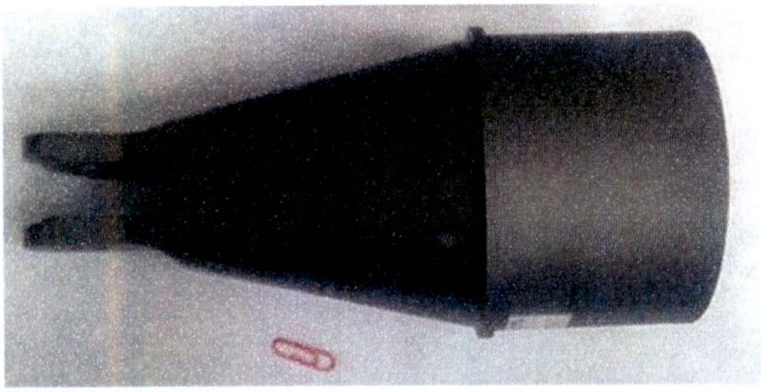

Fig. 8 Photograph of the end fitting

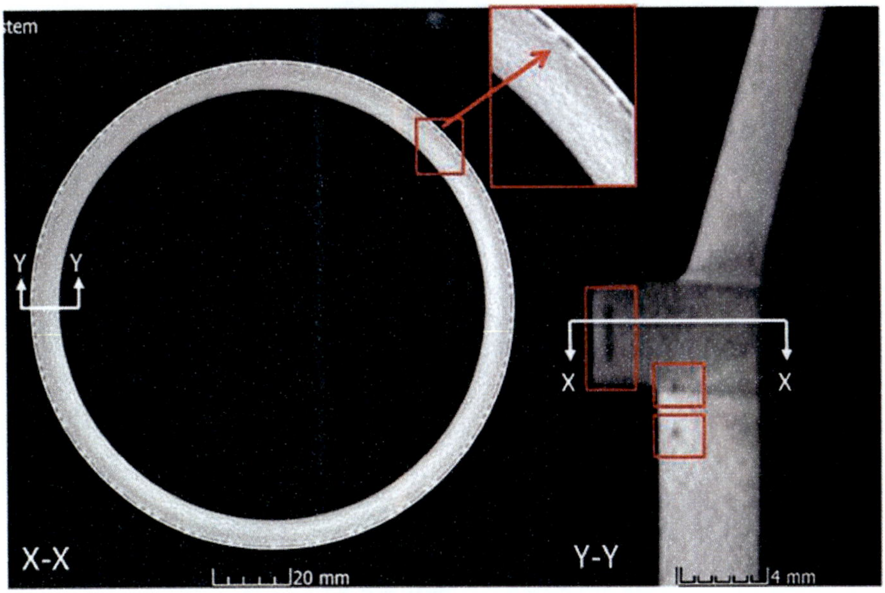

Fig. 9 Three intermittent rings of voids at different layers observed in the end fitting. The X–X cross section provides top view and Y–Y provides side view

to improper laser tracking. Nevertheless, the true cause can be arrived at only after extensive study.

Figure 11 shows results of thickness analysis that was done on the hardware. The wall was observed to be thinner at the bottom than at the top of the cylindrical portion. As the hardware was built from bottom to top, there is requirement for the ring to have enough support as it is protruding out like a cantilever. To ensure this, the wall thickness is increased at each layer of build to transmit the load due to weight of every subsequent layer, finally leading to the ring. This activity is not done on the conical portion as it is self-supporting by nature.

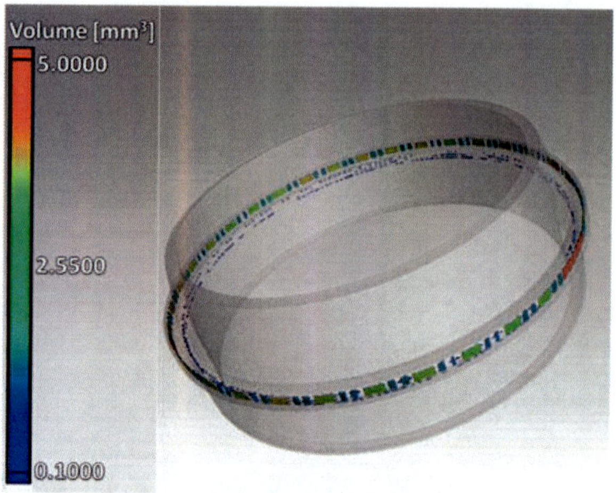

Fig. 10 3D distribution of voids. Color scheme based on varying volume

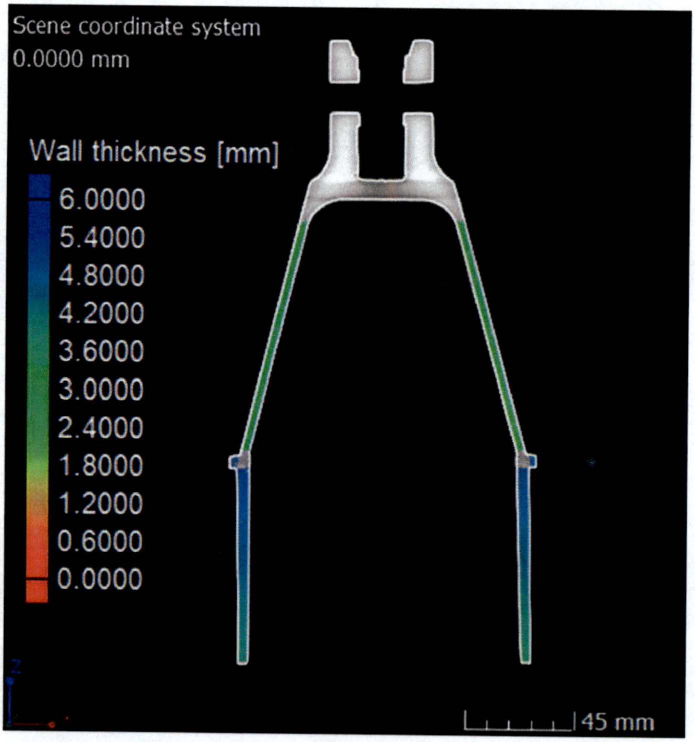

Fig. 11 Thickness analysis of the end fitting

2.5 *Inspection of Carbon–Carbon (C–C) Composites*

Inspection of pre-form Carbon-carbon composites are made of reinforcement and matrix which are both carbon based. The final product is obtained by impregnating the pre-form (reinforcement) with an epoxy (matrix) followed by curing. Hence, evaluating the lay of the individual fibers in the pre-form will provide an insight into the integrity of the reinforcement in the final product. The pre-form that underwent CT examination has a 3D tri-weave structure. It consists of a carbon-based 120° tri-weave stacked into several layers with carbon fiber-based rods reinforcing the layers perpendicular to the plane of weave. CT examination revealed distortion of the weave at internal layers and defects in the carbon fiber rods (Fig. 12).

Inspection of siliconized C–C composite Siliconization is the process of introducing silicon carbide (SiC) in the composite volume to improve its mechanical properties. This is done after the impregnation and curing stages. A siliconized 3D tri-weave C–C composite specimen was inspected using CT to evaluate the extent of SiC deposited on it.

It was observed that the siliconization was not achieved uniformly throughout the volume. This is illustrated in Fig. 13. To understand the distribution, an inclusion

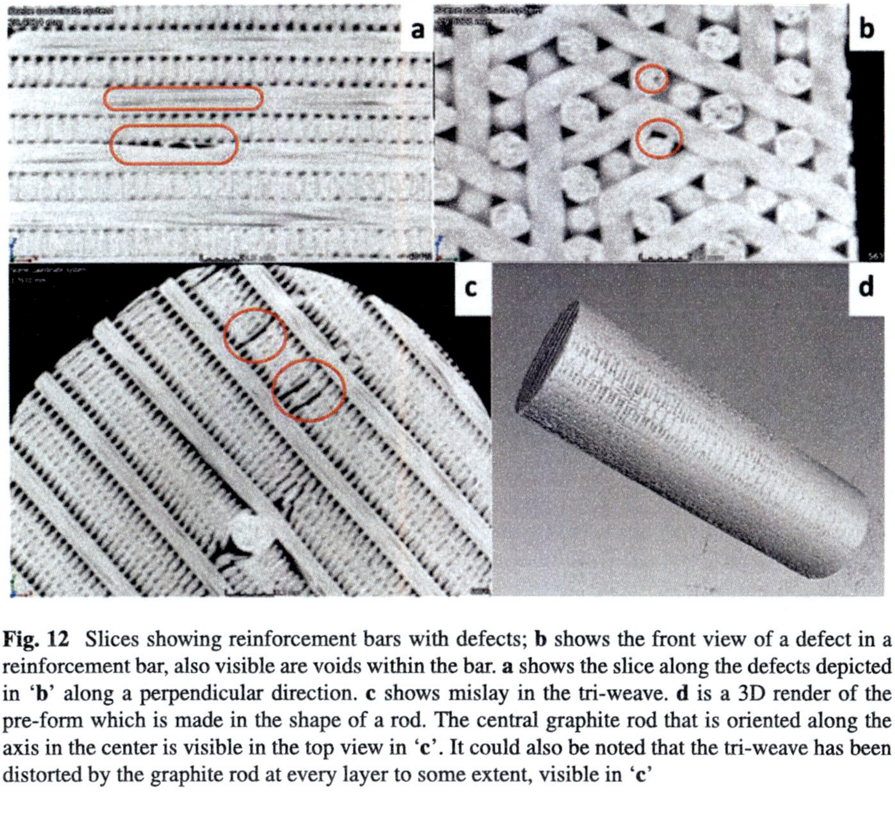

Fig. 12 Slices showing reinforcement bars with defects; **b** shows the front view of a defect in a reinforcement bar, also visible are voids within the bar. **a** shows the slice along the defects depicted in '**b**' along a perpendicular direction. **c** shows mislay in the tri-weave. **d** is a 3D render of the pre-form which is made in the shape of a rod. The central graphite rod that is oriented along the axis in the center is visible in the top view in '**c**'. It could also be noted that the tri-weave has been distorted by the graphite rod at every layer to some extent, visible in '**c**'

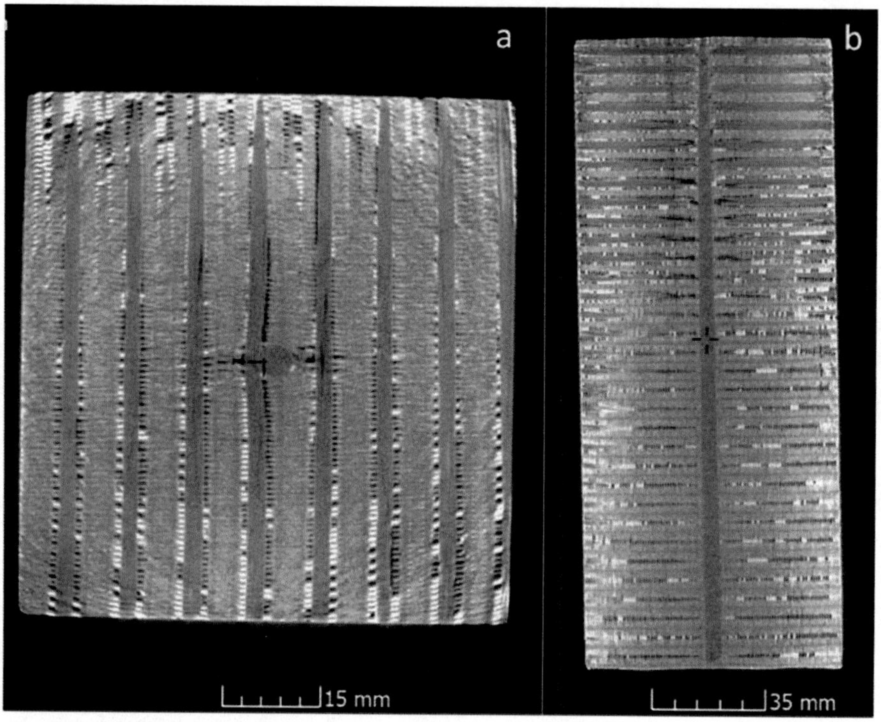

Fig. 13 Slice views of a siliconized C–C composite—**a** top view and **b** front view at the center of the volume. The higher gray values (higher density) correspond to SiC

analysis was performed subsequently (Fig. 14). The results suggest that the SiC deposited prominently toward the external faces of the composite sample.

3 CT as an Aid for Failure Analysis (FA)

3.1 Investigation of Reduced Stroke in Isolation Valve

Failure of a component used in the upper stages of launch vehicles could lead to failure of the whole mission. All mission critical components undergo rigorous and mandatory tests to ensure freedom from defects before they are integrated into the launch vehicle.

The isolation valve used in the upper stage of launch vehicle is one such component. During tests, this component exhibited reduction in stroke. Upon investigation using CT, the cause was pin pointed to the tilted ball connector in the assembly (Fig. 15) which limited the movement of the poppet thereby reducing the stroke of the isolation valve.

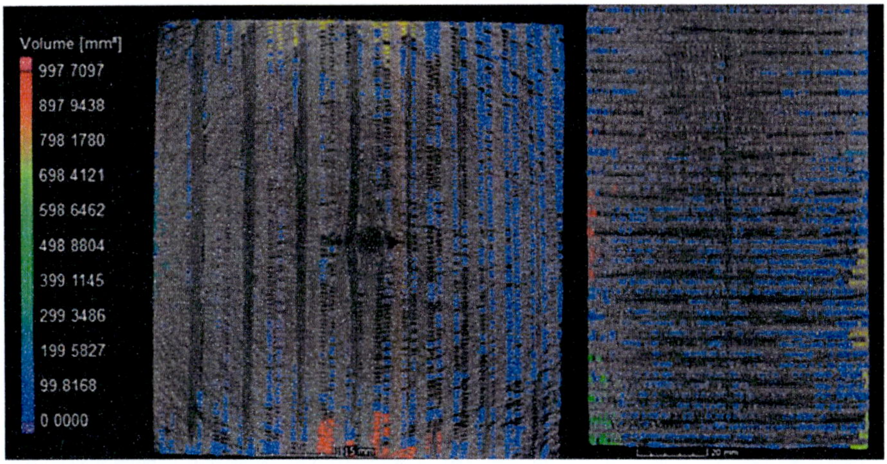

Fig. 14 Inclusion analysis performed on the siliconized C–C composite revealed that the SiC deposits are prominent on the external faces

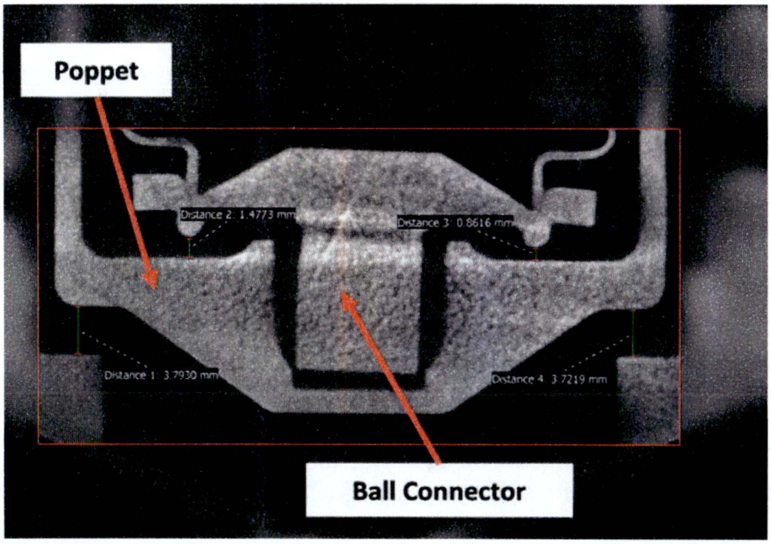

Fig. 15 Tilted ball connector has reduced the stroke of the isolation valve, which was revealed by a CT scan. The tilt is visible only along a certain direction

3.2 Investigation of Failed Pressure Regulator

The pressure regulator used in the upper stage of launch vehicle is a critical component. A failed pressure regulator was subjected to CT scan to evaluate the cause of failure.

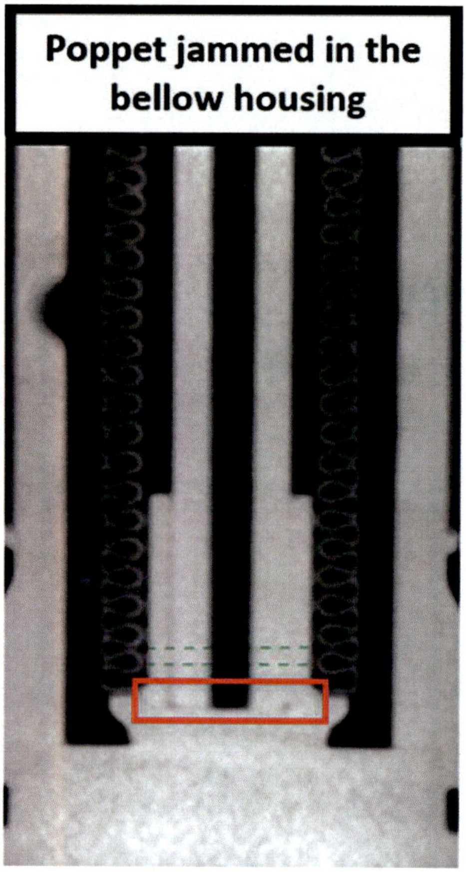

Fig. 16 Expanded view of the cause of failure which is the lack of clearance between poppet and bellow housing. It also depicts undesirable contact between bellow and housing

The cause was narrowed down to lack of clearance between poppet and its guide, as illustrated in Fig. 16, which blocked the communication hole and led to failure. Also, the number of contact points between mating parts could be quantified, and gaps between mating surfaces could be measured accurately. Such a quantitative approach is impossible using 2D X-ray radiography.

4 CT for Dimensional Metrology

CT is a versatile tool for dimensional inspection of parts used in the aerospace industry. The reconstructed 3D model is a precise and accurate rendering of the scanned part. Hence, measurements can be made directly on the model. As the CT uses X-ray, all inaccessible internal features can be measured [7] at any virtual cross section. With the use of micro-focal X-ray, granite-based manipulators, and precise numerical control, length measurement accuracy of sub-micron levels can be achieved.

4.1 Dimensional Measurement of Parts

CT is used for reliable dimensional inspection at ISRO. Complete dimensional inspection performed on poppet seat of reference pressure regulator (earth storable stage) and poppet of fuel injection valve (semi-cryogenic stage) is discussed. The measurements can be made either on the slices or directly on the 3D rendered model after reconstruction (as in Figs. 17 and 18). The colored circles in front of the measured values, in Fig. 17, represent whether the measured features are within tolerance. A comprehensive report with all the measured values can also be generated.

Comparison of Measurements—CT Against Conventional For comparison, some features from both parts were selected and subjected to measurements using conventional instruments like Vernier calipers, digital height gauges, and co-ordinate measuring machines. The results of comparison are given in Table 1. Figure 19

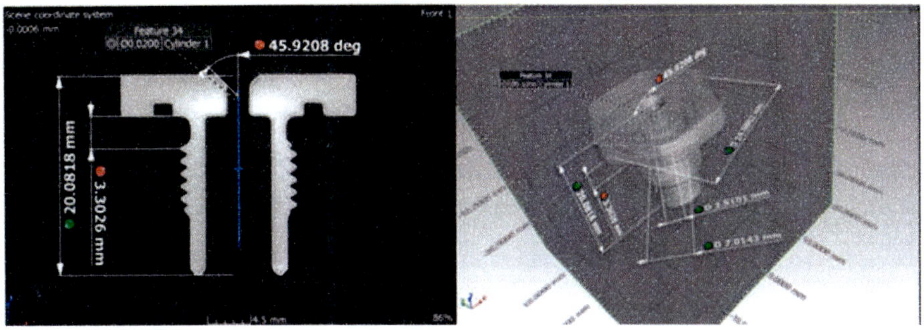

Fig. 17 Slice views with some dimensional measurements made on poppet seat of reference pressure regulator

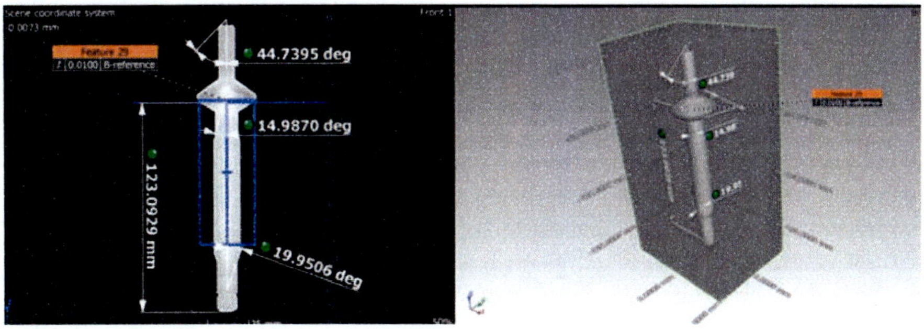

Fig. 18 Slice views with some dimensional measurements made on poppet of fuel injection valve

Table 1 CT measurements against conventionally measured values

Part	No.	Feature dimension (mm)	CT route (mm)	Conventional route (mm)	Absolute deviations (μm)
Poppet seat	1	Ø10 (−0.1/0)	9.9543	9.97	**15.7 μm**
	2	Ø7 (0.013/0.028)	7.0143	7.013/7.016	**1.7 μm**
	3	22 (−0.2/0)	21.8881	21.90	**11.9 μm**
	4	Ø24 (± 0.2)	24.0002	24.01	**9.8 μm**
	5	Ø2.5 (0/0.05)	2.5101	2.507/2.508	**3.1 μm**
Poppet	1	2 (± 0.2)	1.8637	1.88	**16.3 μm**
	2	32 (−0.25/0.0)	31.8582	31.861/32.000	**2.8 μm**
	3	35°	35°1′	35°	**0°1′**
	4	28°(−30′/0)	34°	34°	**0°**
	5	60°	59°59′	60°	**0°1′**

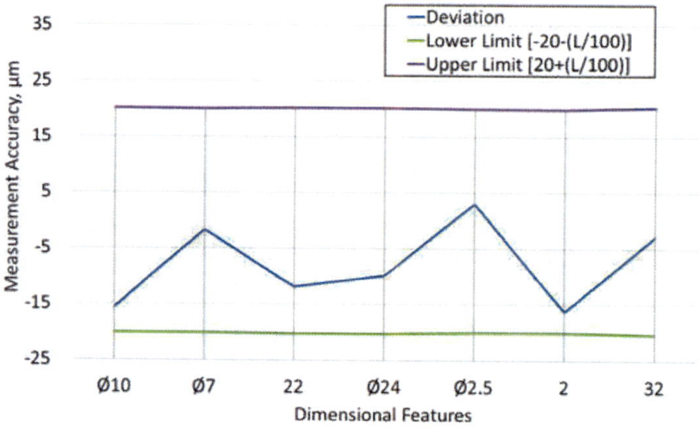

Fig. 19 Performance evaluation of CT by comparison with conventional metrology

shows the results of the performance evaluation performed with data received from comparison. It may be observed that the deviations in measurement are well within the specification of the German VDI 2630-1.3 standard which mandates a requirement of $\pm 20 + (L/100)\mu m$ in length measurement accuracy for mini focus X-ray CT systems (like the one installed at ISRO).

It must also be mentioned that the relatively higher deviations observed in measurements made with CT compared to those made with manual instruments like Vernier calipers can be attributed to operator skill. This is evident from the fact that CMM and digital height gauge measurements are closer to that provided by CT (Table 1) as operator intervention in both cases is minimum. Table 2 gives details of the features that were considered for comparison. Figure 20 illustrates drawing

Table 2 Type of features selected for performance evaluation and the conventional instruments that were used for measurement

Part	No.	Feature dimension (mm)	Feature type	Device
Poppet seat	1	Ø10 (−0.1/0)	Outer diameter	Vernier calipers
	2	Ø7 (0.013/0.028)	Inner diameter	CMM
	3	22 (−0.2/0)	Distance	Vernier calipers
	4	Ø24 (±0.2)	Outer diameter	Vernier calipers
	5	Ø2.5 (0/0.05)	Inner diameter	CMM
Poppet	6	2 (±0.2)	Distance	Profile projector
	7	32 (−0.25/0.0)	Distance	Digital height gauge
	8	35°	Angle	Profile projector
	9	28°(−30′/0)	Angle	Form tester
	10	60°	Angle	Profile projector

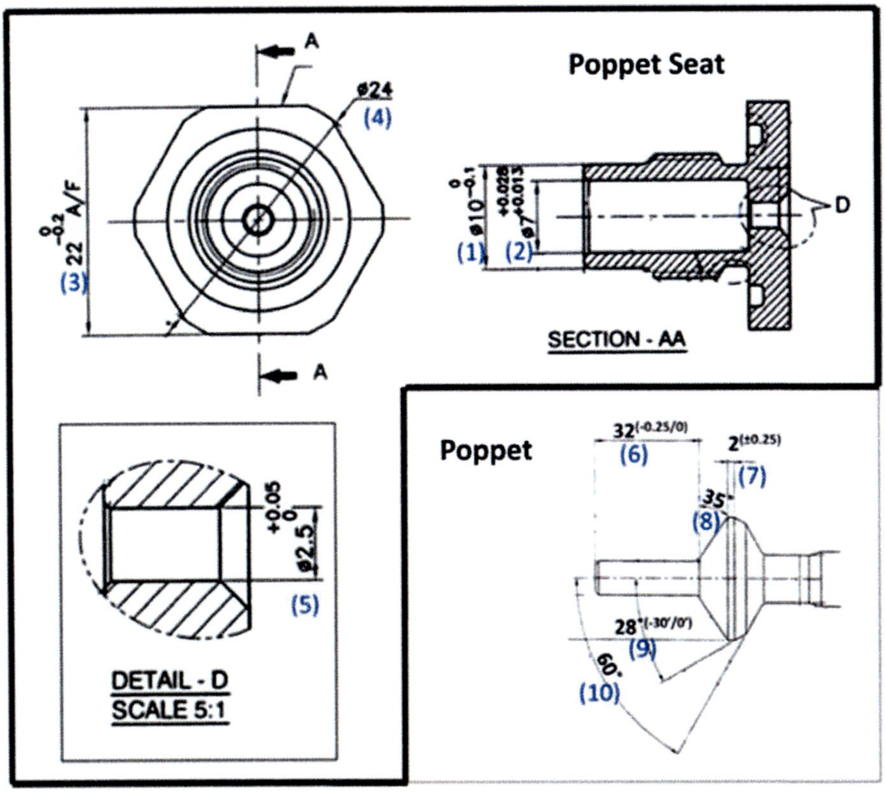

Fig. 20 Drawing snippets showing the features of the hardware that were considered for performance analysis. Numbering is as per Table 1

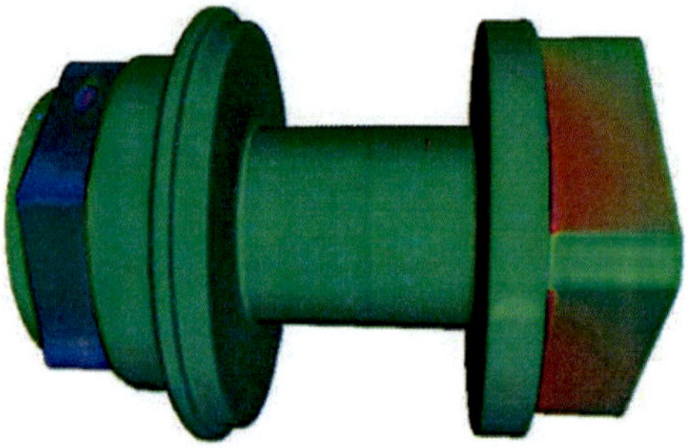

Fig. 21 Actual to nominal comparison of solenoid bobbin—open tolerance

Fig. 22 Actual to nominal comparison of solenoid bobbin—close tolerance

snippets of the features on the hardware that were considered for comparison and subsequent performance analysis.

4.2 3D CAD Comparison

As CT systems provide 3D data, it is possible to perform an actual to nominal comparison using the CAD model of the part. The actual data is received from the CT scan and the nominal data from the CAD model of the part. This can be compared with close or open tolerances, and a general qualitative inspection of the part can be performed. A solenoid bobbin of solenoid valves used in launch vehicles was subjected to an actual–nominal comparison, and results are provided in Figs. 21 and 22. 3D comparison is used as a screening tool for parts made in large scale.

5 Merits of CT

CT is a proven inspection tool with applications ranging widely from NDE to reverse engineering. Merits of using CT in the aerospace industry are summarized below:

- Whole volumes can be scanned for non-destructive evaluation—3D information of defects is made readily available.
- Various types of analyses can be utilized to extract volume, orientation, and severity (size) of all discontinuities.
- Since the object is rotated, defects not favorably oriented for 2D radiography will be aligned for better detection in CT. This qualitatively increases the probability of detection (POD) of planar defects.
- CT makes failure analysis (FA) less complex as more information is available for decision making. There is lesser need for guess work and speculation on the cause of failure as CT generates a one-to-one model which can be sliced along any desirable plane and analyzed.
- Accurate non-contact dimensional metrology is possible with granite-based manipulators and precision-controlled CNC.
- Internal and external features can be measured without compromise on accuracy and precision.
- Quick 3D CAD comparison of manufactured parts is possible as a preliminary screening tool.
- CT is made use of in AM to verify if the part can be manufactured as per design and to aid in design modification to cater to the manufacturing process.
- As the reconstructed 3D model has all dimensional and orientation information of the defects and other related features, it is possible to provide real numerical data for simulation and analysis.

6 Conclusions

Computed tomography is a versatile tool, as was demonstrated, for comprehensive NDE, accurate and precise dimensional metrology and extensive failure analysis. Hence, it goes without saying that CT is an asset to ensure quality and reliability of parts in the aerospace industry.

References

1. Samelson EJ, Christiansen BA, Demissie S, Broe KE, Zhou Y, Meng CA, Yu W, Cheng X, O'Donnell CJ, Hoffmann U, Genant HK, Kiel DP, Bouxsein ML (2011) Reliability of vertebral fracture assessment using multidetector CT lateral scout views: the Framingham osteoporosis study. Osteoporos Int 22:1123–1131. https://doi.org/10.1007/s00198-010-1290-6
2. Dessel JV, Nicolielo LFP, Huang Y, Coudyzer W, Salmon B, Lambrichts I, Jacobs R (2017) Accuracy and reliability of different cone beam computed tomography (CBCT) devices for structural analysis of alveolar bone in comparison with multislice CT and micro-CT. Eur J Oral Implantol 10:95–105
3. Huang R, Ma K-L, McCormick P, Ward W (2003) Visualizing industrial CT volume data for nondestructive testing applications. IEEE Vis 547–554. https://doi.org/10.1109/visual.2003.1250418

4. Hiller J, Maisl M, Reindl LM (2012) Physical characterization and performance evaluation of an x-ray micro-computed tomography system for dimensional metrology applications. Measur Sci Technol 23:18. https://doi.org/10.1088/0957-0233/23/8/085404
5. Kasperl S, Hiller J, Krumm M (2009) Computed tomography metrology in industrial research & development. Mater Test. https://doi.org/10.3139/120.110053
6. Lifton JJ, Malcolm AA, McBride JW, Cross KJ (2013) The application of voxel size correction in x-ray computed tomography for dimensional metrology. In: Singapore International NDT Conference & Exhibition SINCE2013
7. Villarraga H, Morse E, Hocken E, Smith S (2014) Dimensional metrology of internal features with x-ray computed tomography. Am Soc Precis Eng Ann Meet 59:684–689
8. Bossi RH, Shepherd W (1993) X-ray computed tomography for failure analysis investigations, Interim report. Wright Laboratories, Boeing Defense and Space Group, Seattle

Improvements in Quality of Neutron Radiography Images of Pyro Components Used in Aerospace Applications Using Image Processing Tools

Girish N. Namboodiri, V. Shaheer Ali, M. C. Santhosh Kumar, K. K. Moideenkutty, M. Nallaperumal, S. Umasankar, and G. Levin

1 Introduction

Neutron radiography (NR) can be used to detect a low density material present inside a sealed metallic device. NR set-up using a low flux neutron source (D–T-based neutron generator) which can be used in a radiographic enclosure is preferred in organizations where it is not practical to have a nuclear reactor as neutron source. Safe handling and operation of such neutron generators are much easier in comparison to nuclear reactors. Similar to X-ray generators, the D–T neutron generators can be used for generation of neutrons as and when required, ensuring all safety parameters.

A 12-bit cooled CCD-based neutron imaging camera only can be used for imaging using the above set-up due to the low neutron flux and hence low collimator (L/D) ratio, where L is the distance between aperture of collimator and image plane and D is the diameter of aperture. The noise levels in the NR images are much higher whereas sharpness and contrast are much lower in comparison to images obtained from similar neutron imaging cameras used in NR facilities with nuclear reactor as neutron source. Hence suitable image processing has to be used to obtain better quality images, wherever required.

G. N. Namboodiri (✉) · K. K. Moideenkutty · M. Nallaperumal · S. Umasankar · G. Levin
Rocket Propellant Plant, Vikram Sarabhai Space Centre, Thiruvananthapuram 695022, India
e-mail: girish_nn@vssc.gov.in

G. N. Namboodiri · V. Shaheer Ali · M. C. Santhosh Kumar
National Institute of Technology, Tiruchirappalli 620015, Tamil Nadu, India
e-mail: santhoshmc@nitt.edu

C. K. Mukhopadhyay and R. Mulaveesala (eds.), *Advances in Non-destructive Evaluation*, Lecture Notes in Mechanical Engineering,
https://doi.org/10.1007/978-981-16-0186-6_35

1.1 Image Processing for Grayscale Images

The short exposure times used in neutron radiography causes severe noise due to the statistical variation of the neutron flux. Time averaging can be applied to reduce such statistical noise from the original images. Image degradation in neutron radiography can be influenced by a number of system components such as object scattering degradation, geometric unsharpness due to divergence of neutron beam, statistical noise due to short exposure time, etc. [1].

The value of the filtered image at a given location is a function of the values of the input image in a small neighbourhood of the same location. In Gaussian low-pass filtering, the weights decrease with distance from the neighbourhood centre and a weighted average of pixel values in the neighbourhood is computed. Spatial and temporal filters which are mainly used for image processing fail at edges of the items imaged. This is due to the assumption of slow spatial variations used in such filters and are consequently blurred by low-pass filtering [1, 2].

Huang et al. [3] proposed the use of bilateral filter, a nonlinear filter which takes both the spatial distance and pixel similarity into consideration. Unlike Gaussian filter, median filter, etc., which may blur the edges of image, bilateral filter is able to reduce the Gaussian noise while preserving the edges of image since it takes pixel similarity into consideration. The pixels which are dissimilar to the centre pixel will have small weightings and thus the edges will not be blurred.

In most of the cases, image enhancement is done by the removal of various noises present in the image. Identification of the best-suited denoising filter is the most difficult task. Proper analysis of the image and identification of type of noises present in the image are the key to this. Additive noises or impulsive noises like salt and pepper noises and gamma spots in NR CCD images can be partially removed by the median filtering [3, 4]. Noise generated due to the non-uniformity in the neutron flux as well as in the image detecting system can be reduced with the help of averaging of multiple images taken by the repeated shots [5]. Images with Gaussian noise are efficiently handled by the filtering methods like Gaussian filter, anisotropic diffusion, bilateral filtering, etc. [2, 3, 6–9]. Once the efficient filtering method is identified, proper parameters should be chosen for carrying out the filtering process to obtain better quality images. In the case of median filter, window size or neighbourhood size plays a major role in the image quality. As the window size increases, more noise gets removed, but the sharpness of the image gets reduced [4, 10]. While in the case of bilateral filtering, variance values of geometric function and similarity function are the parameters, which determine the quality of images [1, 3, 8–10]. The images are processed using MATLAB and ImageJ software.

2 Experimental Details

The NR set up used for imaging dummy pyro devices is shown in Fig. 1.

Fig. 1 Neutron radiography facility

The exposure parameters used for carrying out NR of the samples are:

Exposure time: 25 s, fast neutron flux: 1×10^{10} n/cm^2 s at 160 kV and 900 μA neutron output.

The neutron imaging camera software has image processing options in the form of background reduction, flatfield corrections, bright spot removal, averaging, etc., which are utilized to obtain an image of reasonable quality so as to perform the purpose of detecting the presence of charge at the required location. All the images are taken with background subtraction and flatfield correction so as to remove the gamma noise and unevenness due to flux variation, respectively. Figure 2a shows the initial image and Fig. 2b shows the software averaged image of the sample after carrying out averaging of ten such frames.

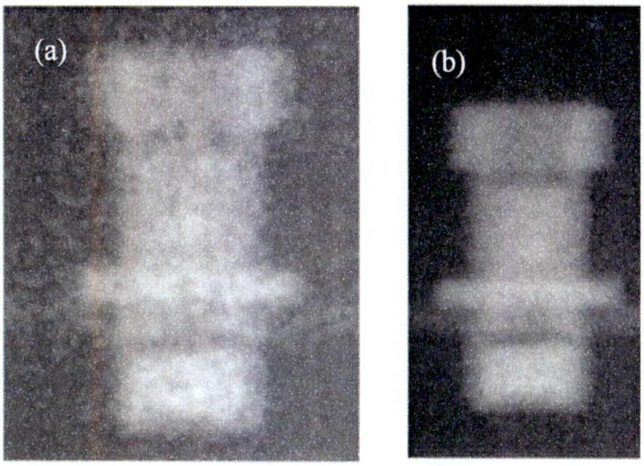

Fig. 2 **a** Initial NR image of dummy samples. **b** NR image with 10 frames averaged

From Fig. 2a, it can be noticed that the initial image is having enough noise which reduces the quality of the image and as a result, interpretation becomes difficult. On averaging of ten such frames, the noise due to the non-uniformity of the neutron flux and due to imaging system will get reduced and hence the image becomes better (Fig. 2b).

Averaging equation:

$$\text{Averaged image} = \frac{i/p(1)}{n} + \frac{i/p(2)}{n} + \cdots + \frac{i/p(n)}{n} \tag{1}$$

where *i/p*(1), *i/p*(2)…*i/p*(*n*) are the input images and n is the number of images for averaging. To improve the image quality to better levels and to obtain maximum information from the system, further image processing is carried out.

Various image processing tools are identified from referred literatures that can help in improving the image quality. Median filtering and Gaussian filtering are applied to the images in the beginning since they are the most common noise filters. It is noticed that along with the filtering process which is reducing the noise, the edge also is getting distorted. This reduces the image quality further which is not desirable. Bilateral filtering which had the advantage of doing the Gaussian filtering without distorting the edges is identified to be a better option to use.

3 Results and Discussion

3.1 Median Filtering

Median filtering is a spatial filtering technique in which image pixel values are modified by considering the neighbourhood pixel values. Each pixel value in the image is modified by the median of all the neighbourhood pixel values. The window size or the neighbourhood size plays a vital role in the filtering process and thereby in the image quality. More amount of noise can be removed from the image by choosing a higher window size, but it negatively affects the sharpness of image. Whereas the smaller window preserve the sharpness of image, but it fails to remove large-sized noise from the image. Square windows of size 3×3 and 5×5 are normally used for carrying out the filtering process. The filters removes noise but at the cost of losing the edge sharpness as seen in Fig. 3a and b.

3.2 Gaussian Filtering

Gaussian filtering is also a linear spatial filtering method, which is used for the removal of Gaussian noise from the image. Each pixel value in the image is modified

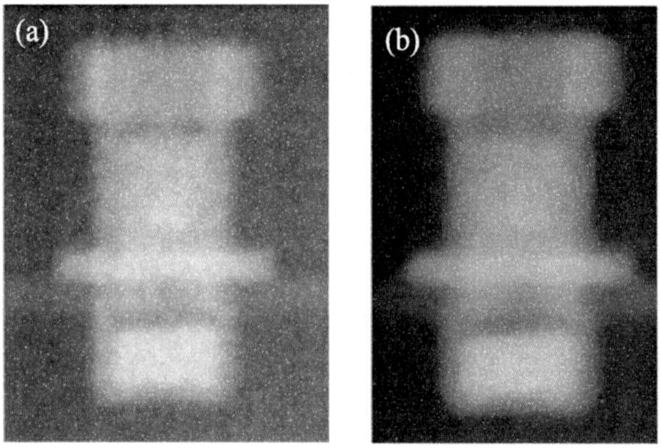

Fig. 3 **a** Median-filtered image with window size 3 × 3. **b** Median-filtered image with window size 14 × 14

by computing the weighted average of the corresponding neighbourhood pixels. In this case, neighbourhood pixel weight decreases with increase in the distance from the pixel at the centre of the window.

Gaussian filter can be represented by the following equation,

$$g(s) = Ae^{-\left(\frac{x^2+y^2}{2*\sigma^2}\right)} \tag{2}$$

where $g(s)$ represents the weight of the neighbourhood pixel and (x, y) represents the relative position of the neighbourhood pixel from the centre pixel of the window. In the equation, A is a constant and σ stands for the standard deviation. The filtering process can be controlled by varying the parameter sigma (σ). Increase in the values of σ is almost similar to the increase in the neighbourhood size.

Gaussian filter is applied on the averaged image with size 3, kernel with weighting factors used for filtering are

$$w = \begin{matrix} 0.0113 & 0.0838 & 0.0113 \\ 0.0838 & 0.6193 & 0.0838 \\ 0.0113 & 0.0838 & 0.0113 \end{matrix} \tag{3}$$

On applying Gaussian filter with size 5, kernel with weighting factors used for filtering are

$$w = \begin{matrix} 0.0318 & 0.0375 & 0.0397 & 0.0375 & 0.0318 \\ 0.0375 & 0.0443 & 0.0469 & 0.0443 & 0.0375 \\ 0.0397 & 0.0469 & 0.0495 & 0.0469 & 0.0397 \\ 0.0375 & 0.0443 & 0.0469 & 0.0443 & 0.0375 \\ 0.0318 & 0.0375 & 0.0397 & 0.0375 & 0.0318 \end{matrix} \tag{4}$$

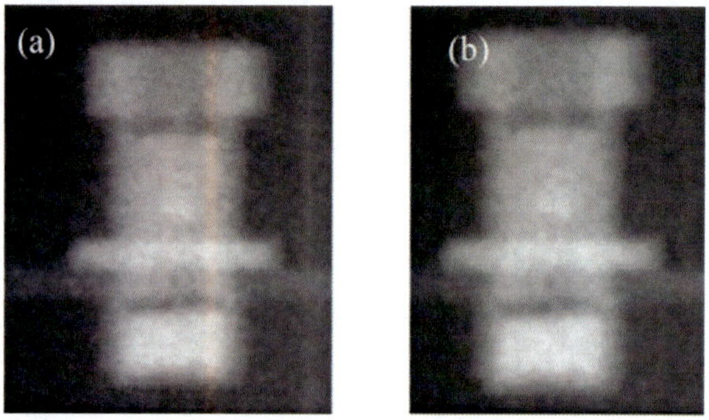

Fig. 4 **a** Gaussian-filtered image (with window size 3 and σ = 0.5). **b** Gaussian-filtered image (with window size 5 and σ = 3)

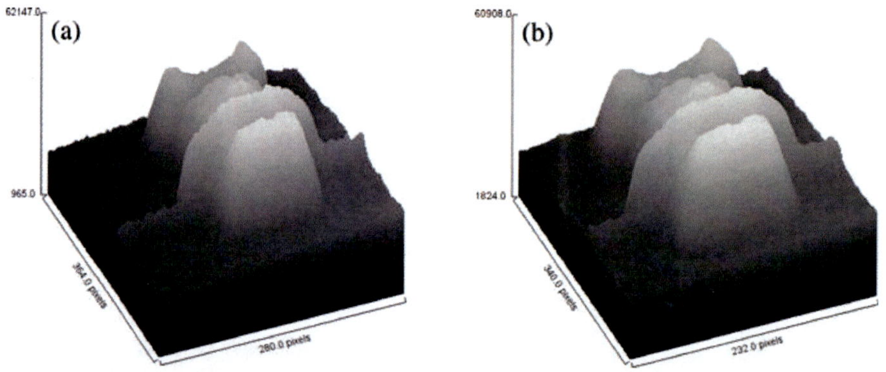

Fig. 5 **a** Surface plot of Gaussian-filtered image (with window size 3 and σ = 0.5). **b** Surface plot of Gaussian-filtered image (with window size 5 and σ = 3)

The results obtained by applying the above filters are shown in Fig. 4a, b. The Gaussian filter also reduces the noise to little extend only and that too by reducing the edge sharpness. Surface plot of the filtered images are shown in Fig. 5a, b.

3.3 *Bilateral Filtering*

Bilateral filtering is an image enhancement method used for removing Gaussian noise by preserving the edges in the image. Unlike the other spatial filtering processes such as median filtering and Gaussian filtering, the edges in the image are not affected by the filtering process. Bilateral filtering is a combination of domain filtering and range filtering. Domain filtering is done on the basis of geometric similarity and the range filtering is done on the basis of pixel similarity. Consideration of the pixel similarity in the filtering process enables the filter to preserve the edges in the image. Here, filtering of the edges, present inside the window is not carried out by the filter, because of the large variation in the pixel values across the edges.

Modified image pixel values (I_c') by the bilateral filtering can be represented as,

$$I_c' = \frac{\sum_{q\in w} f(|c-q|)g(|I_c - I_q|)I_q}{\sum_{q\in w} f(|c-q|)g(|I_c - I_q|)} \tag{5}$$

where I_c is the image pixel which is to be processed by the bilateral filtering, and w is the window centred at position c. $f(-)$ and $g(-)$ indicates the spatial weighting function and similarity weighting function, respectively.

$$f(|c-q|) = Ae^{\frac{|c-q|^2}{2\sigma_f^2}} \tag{6}$$

$$g(I_c - I_q) = Be^{\frac{|I_c - I_q|^2}{2\sigma_g^2}} \tag{7}$$

where A and B are normalization constants, σ_f^2, σ_g^2 are the variances of Gaussian kernels, σ_f and σ_g indicates the variances of geometric function and similarity function, respectively. From the above equations of spatial weighting function and similarity weighting function, it is clear that the neighbourhood pixels are assigned a low weight by the filter, if the pixel shows the large variation from the pixel to be modified. Hence, the distant neighbourhood pixels and dissimilar pixels do not influence much on the modified value of the centre pixel.

In bilateral filtering, $\sigma_g = 50$ and $\sigma_f = 3$ are used at first to study the changes in the image quality. Various trials are carried out by varying the σ_f values. It is noted that at $\sigma_f = 10$, the image quality is becoming very good and the noise factor reduces by a good extend.

Bilateral-filtered images with varying $\sigma_f = 3$ and $\sigma_f = 10$ are shown in Fig. 6a, b. Surface plots of the filtered images are shown in Fig. 7a, b.

It is noticed that bilateral filtering is suitable for improving the image quality without affecting the edges. As a result, the edges of the items where there is a

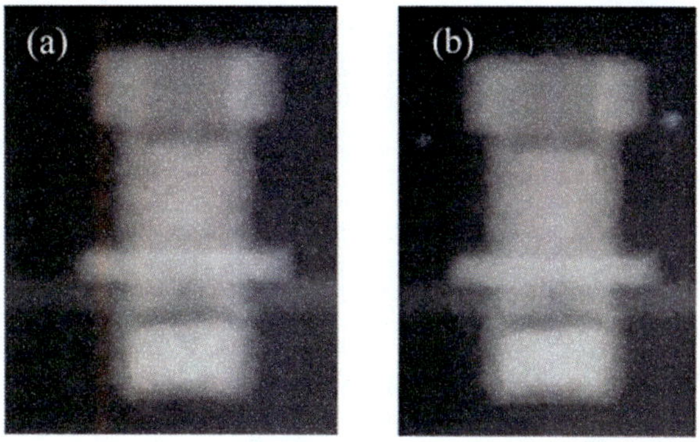

Fig. 6 **a** Bilateral-filtered image with $\sigma_g = 50$ and $\sigma_f = 3$. **b** Bilateral-filtered image with $\sigma_g = 50$ and $\sigma_f = 10$

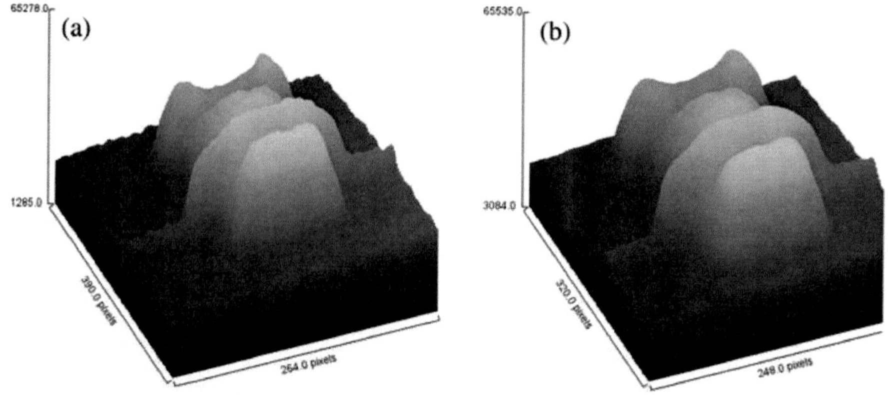

Fig. 7 **a** Surface plot of bilateral-filtered image with $\sigma_g = 50$ and $\sigma_f = 3$. **b** Surface plot of bilateral-filtered image with $\sigma_g = 50$ and $\sigma_f = 10$

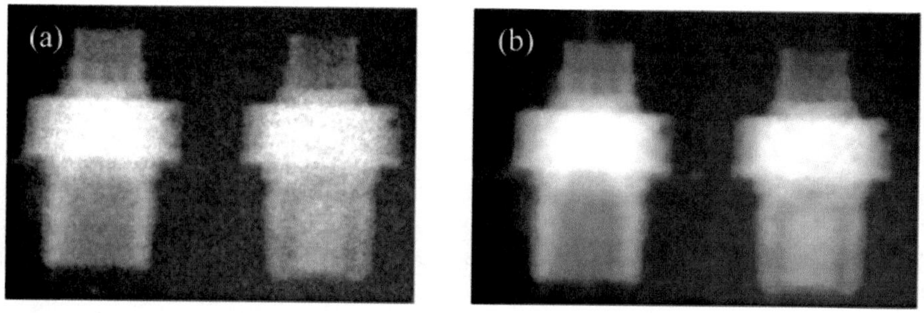

Fig. 8 **a** NR image of a pyro cartridge with ten frames averaged. **b** Bilateral filtering done on averaged image

sharp variation in grayscale value (Intensity) is preserved by the filter while the other regions, noise will be completely removed.

Further, the filter is applied to a few other pyro devices and found to be capable of improving the image quality to better levels.

Figure 8a, b shows the averaged image (ten frames) of a pyro cartridge and a bilateral filtered image averaged over ten frames, respectively. The charged region is having much reduced noise after bilateral filtering as is evident from the images.

4 Conclusion

The improvements in quality of image brought about by the usage of bilateral filter in the individual images followed by averaging are quite significant when compared to that brought about by median or Gaussian filtering. It becomes very important to select the optimum parameters for the image processing tool that is applied, so as to obtain the required image quality. Bilateral filter with parameters $\sigma_g = 50$ and $\sigma_f = 10$ was applied to many of the pyro cartridge images and it yielded good reduction in noise without losing the edge sharpness which helps in viewing the Region Of Interest(charged region) in a better way. Further studies are in progress to improve the image quality to better levels.

Acknowledgements Authors acknowledge ASOE, VSSC for their support to carry out the present study. We also acknowledge Dr. Amar Sinha, Dr. Tushar Roy and their team from BARC, Mumbai, who had helped in developing the NR facility at RPP.

References

1. Saito Y, Ito D (2015) Image enhancement for high frame-rate neutron radiography. Phys Proced 69:265–270. https://doi.org/10.1016/j.phpro.2015.07.037
2. Lee JS (1980) Digital image enhancement and noise filtering by use of local statistics. IEEE Trans ASSP 27(1):13–18
3. Huang R-Y, Dung L-R, Chu C-F, Wu Y-Y, Noise removal and contrast enhancement for X ray images. https://doi.org/10.1109/nssmic.2009.5401610
4. Li H, Schillinger B, Calzada E, Yinong L, Muehlbauer M (2006) An adaptive algorithm for gamma spots removal in CCD-based neutron radiography and tomography. Nucl Instrum Methods Phys Res A 564:405–413. https://doi.org/10.1016/j.nima.2006.04.063
5. Wang L, Jin X, Image YF (2010) Noise reduction based on superposition algorithm used in X-ray imager. Proced Eng 7:286–289. https://doi.org/10.1016/j.proeng.2010.11.046
6. Ito K, Xiong K (2000) Gaussian filters for nonlinear filtering problems control. IEEE Trans Autom Control 45(5)
7. Perona P, Malik J (1990) Scale-space and edge detection using anisotropic diffusion. IEEE Trans Pattern Anal Mach Intell 12(7)
8. Tomasi C, Manduchi R (1998) Bilateral filtering for gray and color images. In: Proceedings of the 1998 IEEE international conference on computer vision, Bombay, India
9. Paris S, Kornprobst P, Tumblin J, Durand F (2007) A gentle introduction to bilateral filtering and its applications. In: ACM SIGGRAPH 2007 Courses
10. Hwang H, Haddad RA (1995) Adaptive median filters: new algorithms and results. IEEE Trans Image Process 4(4)
11. The Math Works, Inc., Matlab 2018a online help documents and software
12. Imagej 1.52n
13. Digital image processing using MATLAB. McGraw Hill Educations (2010)

The Promise of Metamaterials for Ultrasonic Nondestructive Evaluation

Mohamed Subair Syed Akbar Ali and Prabhu Rajagopal

1 Introduction

Among nondestructive evaluation (NDE) methods, ultrasonic testing (UT) techniques are widely used in industries and medical diagnostics for their affordability and lack of radiation. However, conventional UT systems suffer by the constraints of the laws of physics such as diffraction limits and so on. Thus, there is much interest in materials that can overcome these limitations. In this context, metamaterials (MMs) have been a research attraction in the last two decades as they exhibit exotic properties such as negative refraction, negative compressibility, negative bulk modulus, etc. Unlike natural materials, the properties of MM are not just defined by the material, but by their structure and topology. The first demonstration of MM using a wire structure was proposed for achieving negative permeability in the electromagnetic regime [1]. Further efforts led to the realization of MM with negative refractive index resulting in perfect lenses overcoming diffraction limits [2]. Since then, several novel MM concepts have been proposed for resolution beyond the diffraction limit [3–11].

Although the concept of MM originated in electromagnetic waves, a translation to acoustics began only in the past decade. The acoustic MM concepts are centered towards controlling the propagation of acoustic waves by achieving properties such as negative density, negative compressibility, etc. For example, if either of the density or compressibility is negative, the wave speed becomes imaginary which leads to exponential decay resulting in sound absorption. If both the properties are negative, the wave speed is still positive but the energy travels in the opposite direction which results in negative refraction [12, 13]. The first acoustic MM demonstration was shown for attenuation application using locally resonant structures [14]. Following

M. S. Syed Akbar Ali (✉) · P. Rajagopal
Center for Nondestructive Evaluation, Department of Mechanical Engineering, Indian Institute of Technology Madras, Chennai 600036, Tamil Nadu, India
e-mail: zubair.musam@gmail.com

C. K. Mukhopadhyay and R. Mulaveesala (eds.), *Advances in Non-destructive Evaluation*, Lecture Notes in Mechanical Engineering,
https://doi.org/10.1007/978-981-16-0186-6_36

this work, various acoustic MM concepts were introduced for applications which include super resolution imaging [15–18], wave steering [19–21], cloaking [22–24] and bandpass filters [25, 26].

More recently, the concept of MM has been introduced in the field of ultrasonics. Ultrasonic MMs were reported for various applications such as wave focusing and manipulation [27–29], cloaking [30, 31], band gaps [32, 33] and superlensing below diffraction limit [34–38]. The ultrasonic MM creates different avenues in industrial NDE as well as non-invasive medical diagnosis due to their dynamic properties. This article highlights the fascinating applications of MM in ultrasonic NDE such as superlensing, control of wave propagation and topological lensing, based on the work at our research group at the Center for Nondestructive Evaluation (CNDE), Indian Institute of Technology Madras, Chennai, India.

2 Super Resolution Imaging

In classical wave applications, higher resolution imaging beyond the diffraction limit is enabled only when the evanescent waves carrying the finer information of the object are captured successfully which otherwise decay within the nearfield [39]. Several methods have been proposed to overcome the diffraction limit in both the electromagnetic and acoustic wave regimes, such as the time reversal technique [40, 41], Bragg Scattering [42, 43], super lens [5–9, 17], and hyperlens [10, 11, 15, 36]. Holey-structured metamaterials are super lenses that use the Fabry-Perot resonances in the holes to successfully transmit the decaying evanescent waves to the image plane [16]. At Fabry-Perot conditions, the transmission coefficient tends to one and thus enables complete transmission of all the scattered waves including the evanescent waves to the image plane. The first subwavelength ($\lambda/5$) imaging in the ultrasonic regime was demonstrated using such a holey-structured metamaterial (see Fig. 1) [35].

However, the Fabry-Perot resonant conditions greatly influenced by the design of the MM structure. Hence, the optimal geometrical parameters of the MM lens were determined based on the transmission characteristics obtained through finite element (FE) simulations and the use of such an optimized MM lens resulted in much deep subwavelength imaging ($\lambda/25$) (see Fig. 2, [36]).

Although the periodic metamaterials can effectively transmit high-frequency components of the scattered wavefield, they suffer by "Wood anomaly" which refers to the presence of amplitude drops at certain frequencies in transmission spectrum [44, 45]. Recent studies, particularly in the field of acoustics, demonstrated the use of quasi-periodic or aperiodic structures as a promising alternative to minimize or eliminate Wood anomaly [46, 47]. The use of such aperiodic or porous metamaterials (see Fig. 3) for deep subwavelength imaging in the ultrasonic regime has been demonstrated more recently [34]. Imaging with a resolution of $\lambda/36$ was achieved which perhaps is the highest resolution of deep subwavelength imaging reported in the ultrasonic regime worldwide (see Fig. 4).

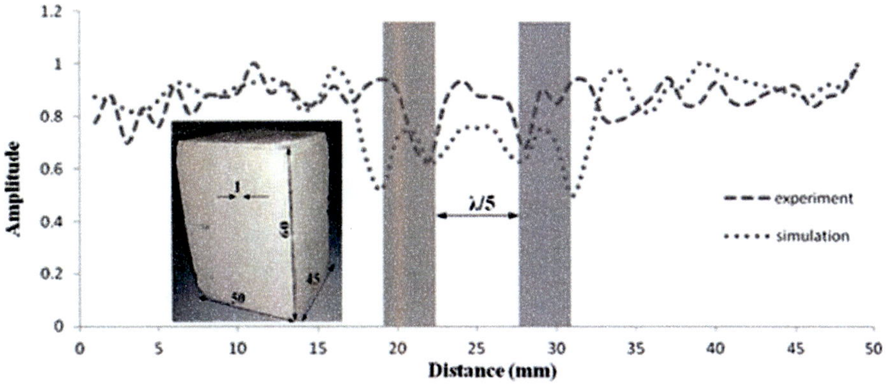

Fig. 1 Measured and simulated results for the normalized amplitude variation across the defects in the sample. The shaded boxes indicate the actual positions of the defects in the sample. Inset shows the photograph of the holey-structured MM lens used for experiments with details of the geometric dimensions (in mm) [35]

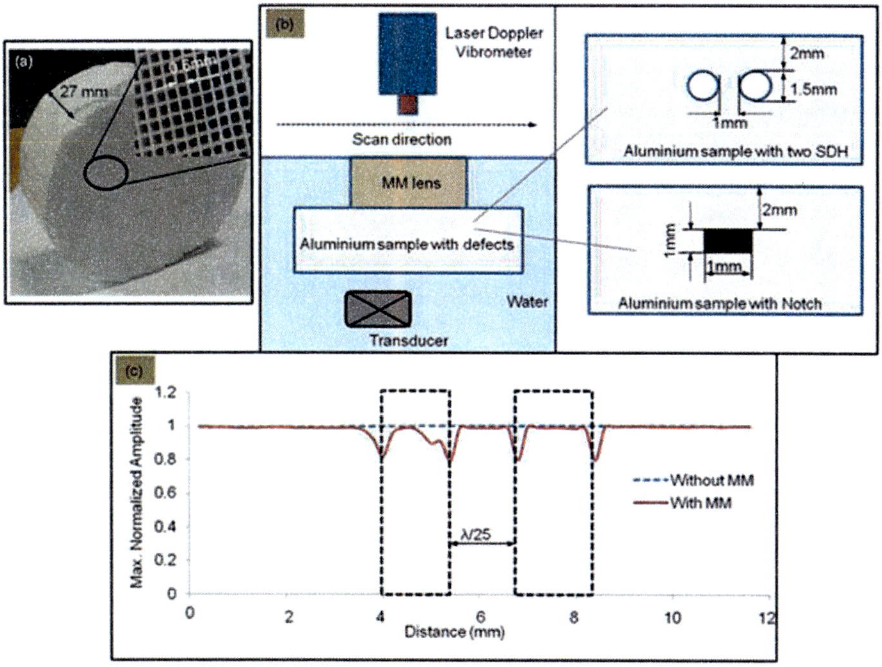

Fig. 2 **a** Photograph of the optimized holey-structured MM lens; **b** Illustration of experimental set-up with details of defects considered in a sample; **c** Experimental results of the line scan performed across the defects: the dotted rectangular boxes indicate actual positions of the defects in the sample [36]

Fig. 3 Photograph of the porous MM lens used for the experiments [34]

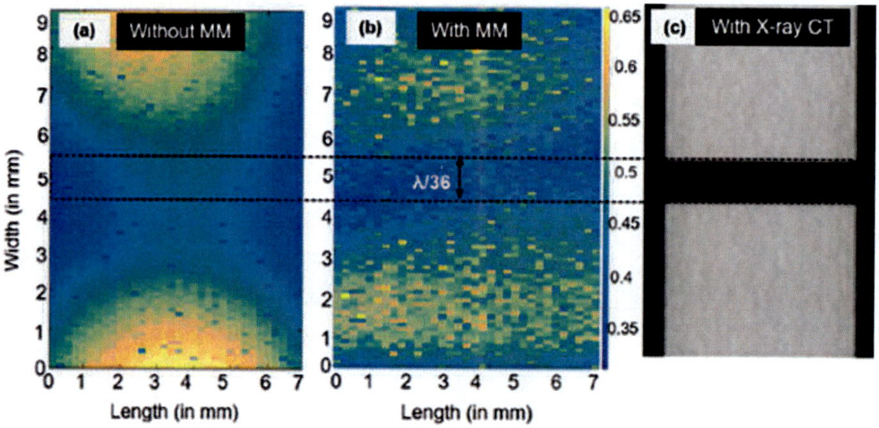

Fig. 4 C-scan image of a sample having through-notch of size λ/36: **a** without and **b** with the porous MM lens: the dotted rectangular box indicates the actual positions of the notch in the sample; **c** X-ray computed tomography image obtained for the same defect [34]

The extraordinary properties of these holey-structured metamaterials can lead to potential applications in NDE techniques such as ultrasonic imaging of metallic materials and non-invasive diagnostics in the medical field. Miniaturization of the metamaterial parameters to micro- or nanoscales can eventually produce an image quality comparable to micro X-ray CT scan.

3 Wave Manipulation

An interesting application of MM is the control of propagating waves inside the media. Among the various methods developed, transformation methods [48–51] have been found to be very effective where the properties of specially engineered buffers are used to steer waves around a target. These are also attractive for seismic vibration isolation, as this approach can also yield shielding/cloaking of the target in addition to stable bandgaps [31]. As compared to acoustics or optics where much of the work on cloaking has been demonstrated, transformation elastodynamics is more challenging as, for elastic waves, the wave equation is not invariant under a transformation [52]. This restricts the use of transformation-based methods for manipulation of elastic waves.

Nonetheless, under certain types of mappings and for high frequencies as those in the ultrasonic regime, certain approximations can be made for using the transformation-based methods. Recently, a metamaterial composed of customized hole pattern in a plate has been proposed for manipulating the wavefront of propagating elastic waves [53]. The medium (in this case a plate with holes) is designed by using a conformal map for the transformation and is implemented through gradient refractive index phononic crystals (GRIN PCs) [54]. Figure 5 shows the transformation of a finite discrete grid under the mapping.

The experimental configuration and the results obtained in FE simulations and experiments are shown in Figs. 6, 7 and 8, respectively. These results demonstrated the use of such metamaterials to improve the wave incidence on curved surfaces, for example, in pipe inspections. In general, the transformation methods through metamaterials open the possibility of manipulation of the waves to achieve arbitrary wavefronts as well as engineering the wave path as desired.

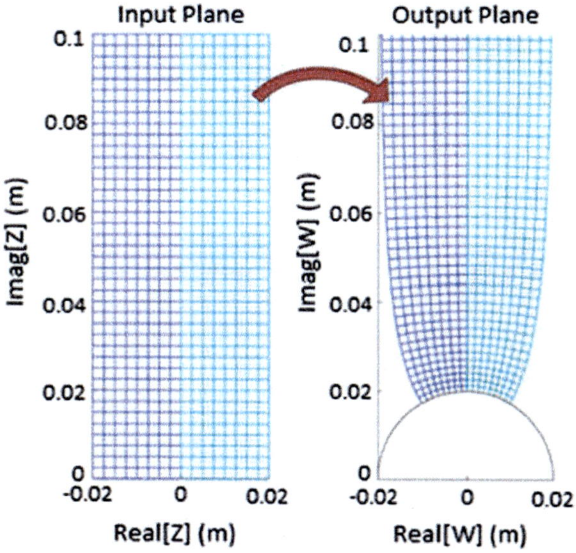

Fig. 5 Illustration of the transformation of grid lines under the conformal mapping [53]

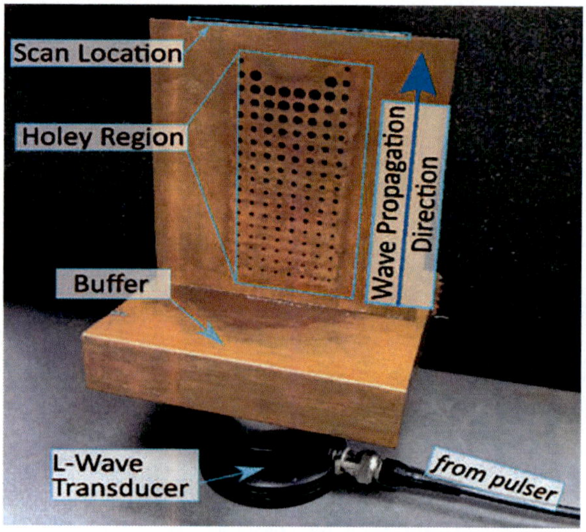

Fig. 6 Photograph of the holey MM plate specimen in experimental set-up [53]

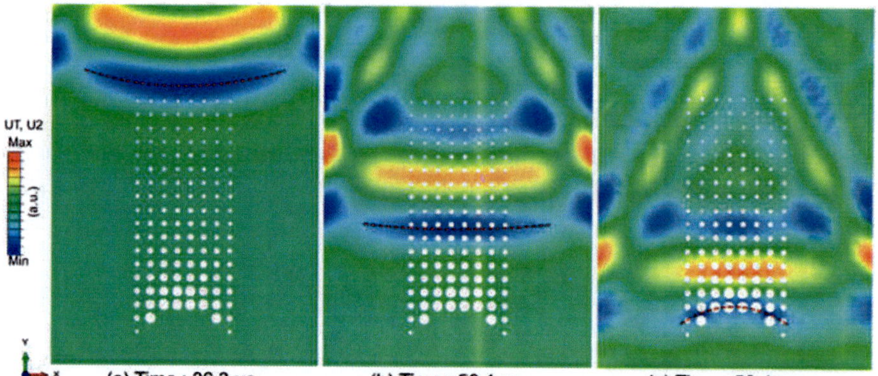

Fig. 7 Snapshots showing displacement contour in FE simulation: **a** Entering; **b** While passing through; **c** Exiting the holey region in MM plate. The dashed lines illustrate the wavefront profile [53]

4 Topological Lensing

Superlenses which utilize the negative refraction effect to achieve higher resolution are, typically, designed using phononic crystals and resonant structures [2, 17, 18, 55, 56]. However, such structures have drawbacks associated with complex design and manufacturing. Waveguides possess the property of negative refraction, which overcomes these drawbacks. Lamb waves, exhibit backward propagation [57–59], through which negative refraction can be achieved, and hence such waveguides seem to be a feasible alternative for improving the resolution of ultrasonic inspection.

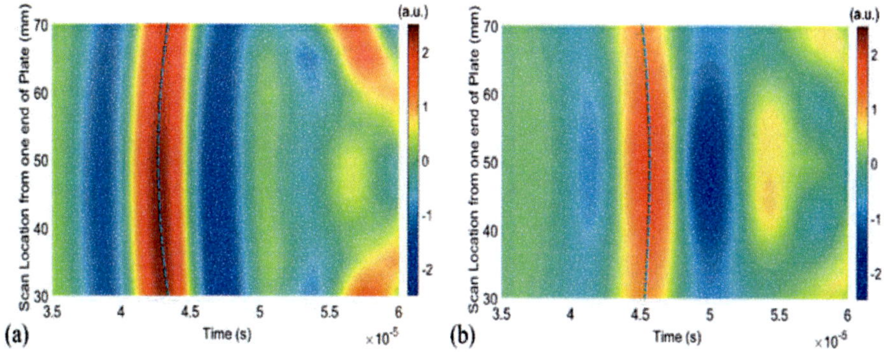

Fig. 8 Experimentally obtained B-scans for: **a** Plate without holes; **b** Holey MM plate [53]

Recent work on this concept demonstrated the occurrence of negative refraction in Lamb waves due to the conversion of backward-propagating wave into a forward-propagating wave in a plate at an interface with thickness change [60]. This work utilized the mode conversion between the backward-propagating S2b branch in the thin side of the plate and the forward-propagating S2 mode in the thick side (where S2 denotes second order Symmetric mode of Lamb waves). The two different thicknesses were selected such that dispersion curves of these two modes meet at a certain frequency. The dispersion curves for the selected thickness of the plate are shown in Fig. 9 and the schematic of the lens made of aluminium is shown in Fig. 10. FE simulation results obtained for the step plate and the constant thickness plate (without step) are shown in Fig. 11. The results show the focusing effect achieved through the occurrence of negative refraction due to the step change.

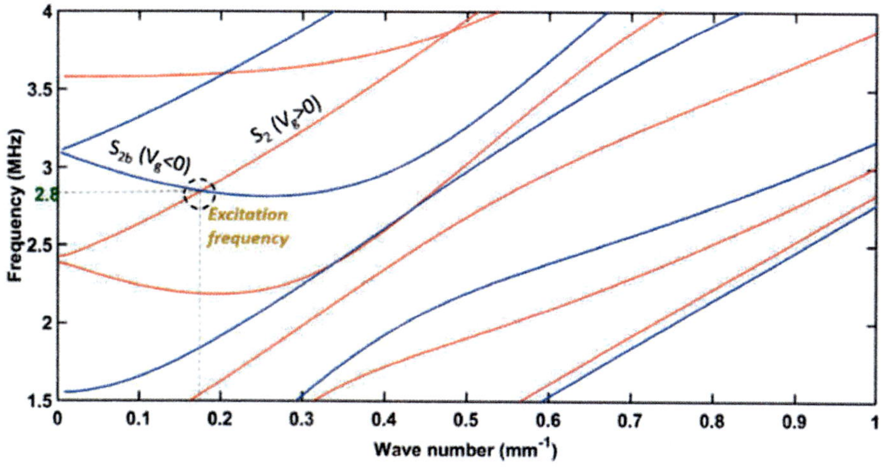

Fig. 9 Dispersion curves for aluminium: frequency versus wavenumber [60]

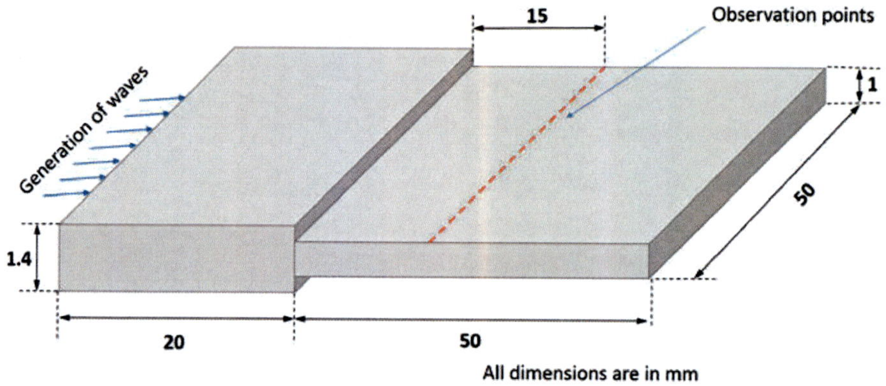

Fig. 10 Schematic representation of the aluminium plate with a change in thickness [60]

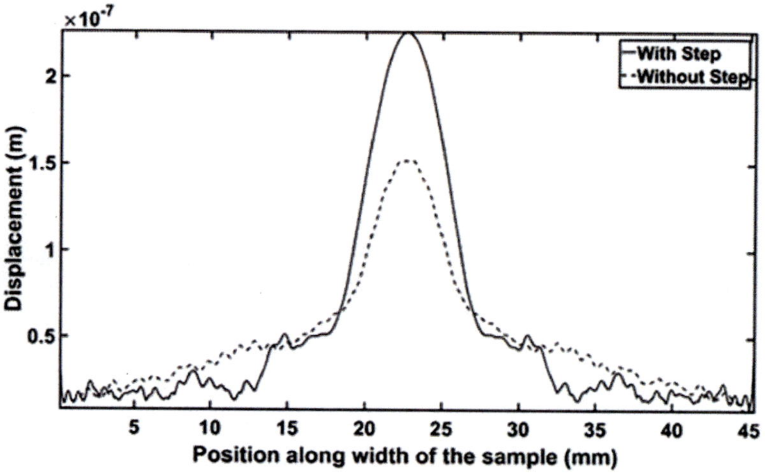

Fig. 11 Comparison of the FE simulation results obtained for the amplitude variations along the width of the sample [60]

Since in this waveguide approach, the impedance mismatch plays the key role to achieve the focusing, any mechanism which generates an acoustic impedance mismatch at an interface can also be used. This was successfully demonstrated by a material contrast lens composed of two different materials, namely aluminium and molybdenum [61]. As similar to the plate with step change, the intersection of the forward and backward S2 modes of aluminium and molybdenum can be observed from the dispersion curves shown in Fig. 12.

The experimental result of the focusing achieved through the proposed materials contrast lens is shown in Fig. 13a while the numerical result highlighting the focal spot is presented in Fig. 13b. The application of the proposed lens was also demonstrated by numerical simulation of imaging two defects separated by a subwavelength

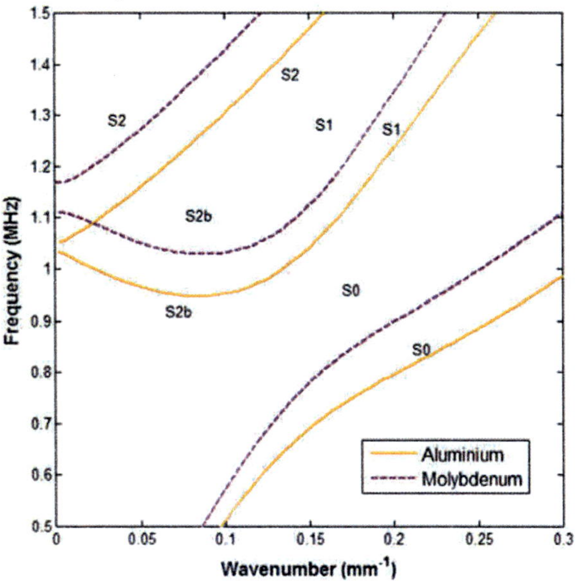

Fig. 12 Dispersion curves highlighting the intersection of S2 mode of aluminium and S2b mode of molybdenum for a plate thickness of 3 mm [61]

distance (λ/3) and the obtained result is shown in Fig. 14. The results open up the possibilities of developing new lensing devices which are easy to fabricate and find application in subwavelength imaging and resolution.

5 Conclusion

This paper sheds light on the potential application of metamaterials in the field of nondestructive evaluation (NDE). Metamaterials can be specially engineered to have desired properties, which can lead to NDE applications such as super resolution imaging, wave steering, wavefront control, mode filtering and trapping, and lensing. Thus, the use of metamaterials can improve the inspection capabilities of NDE techniques in industries as well as in non-invasive medical diagnostics. There is much scope to miniaturize metamaterials that have a great promise in day-to-day NDE applications eventually reducing the costs associated with inspection and avoiding radiation hazards.

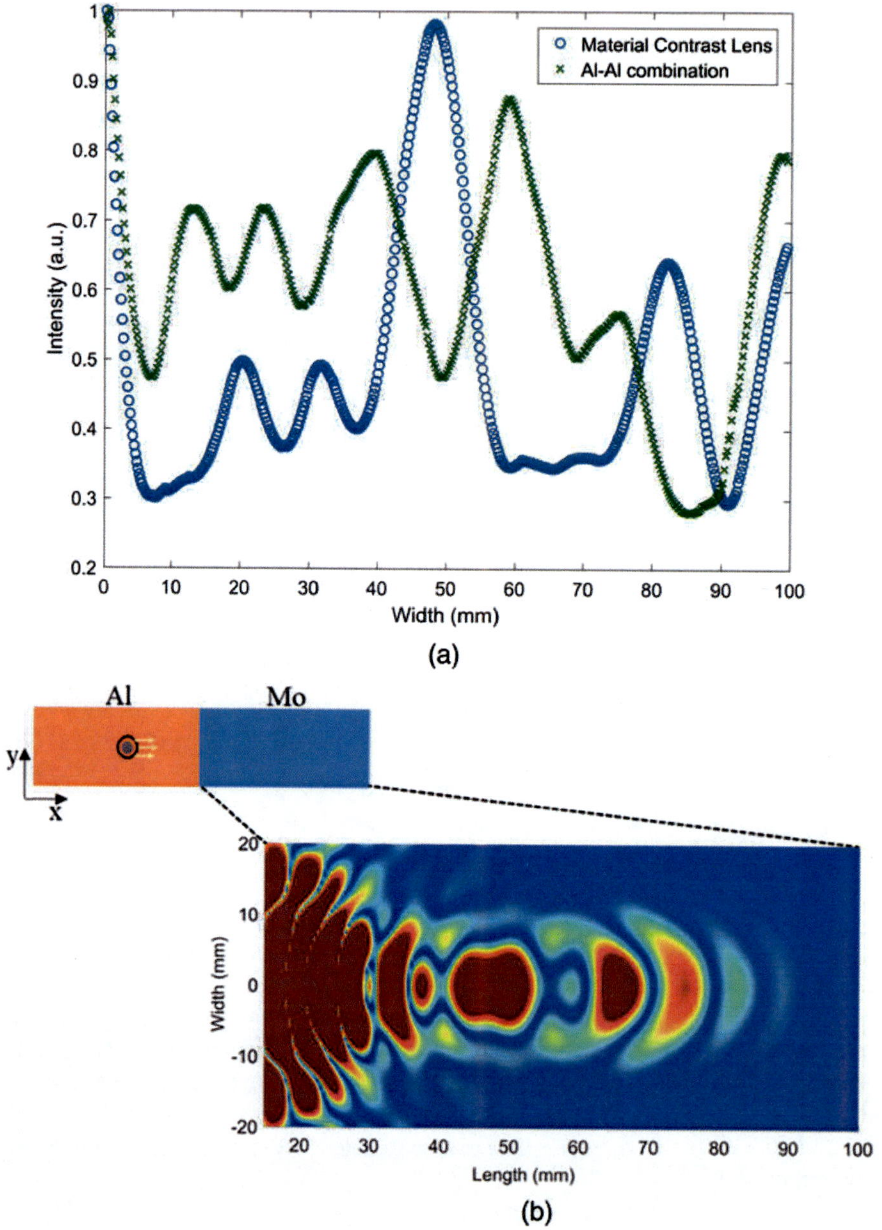

Fig. 13 **a** Measured amplitude variations of the out-of-plane displacements along the width for material contrast lens compared with aluminium–aluminium combination. **b** Numerical result highlighting the observed focal spot in the molybdenum section of Al–Mo material contrast lens [61]

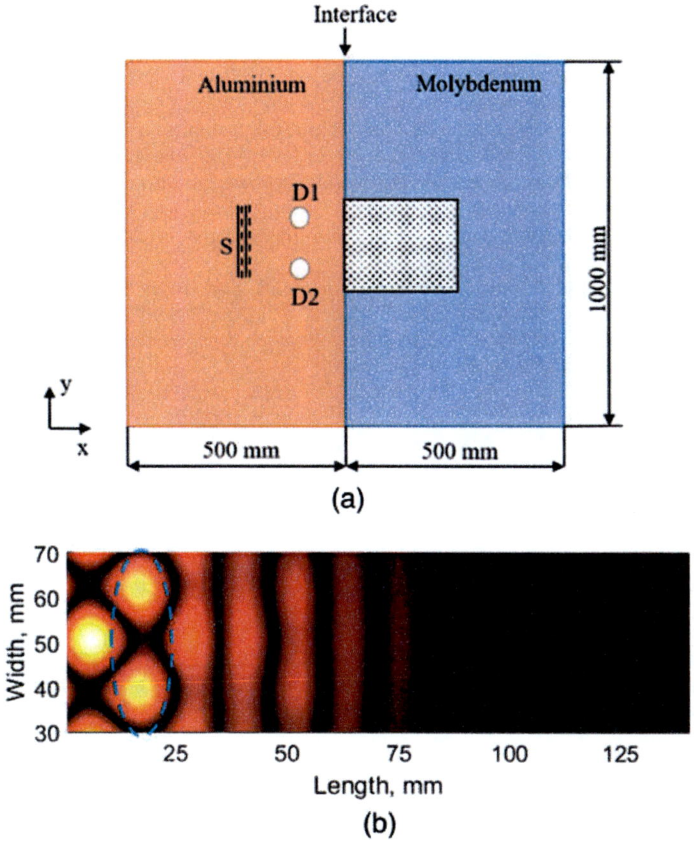

Fig. 14 **a** Schematic illustration of the configuration used to study the subwavelength imaging of defects using numerical simulations. S is the plane wave source, D1 & D2 are the defects. The dotted area shown on the molybdenum side is the region selected for imaging of defects. **b** Simulation results showing the successful imaging of the defects on Molybdenum side of the material contrast lens [61]

References

1. Pendry JB, Holden AJ, Stewart WJ, Youngs I (1996) Extremely low frequency plasmons in metallic mesostructures. Phys Rev Lett 76(25):4773–4776
2. Pendry JB (2000) Negative refraction makes a perfect lens. Phys Rev Lett 85(18):3966–3969
3. Smith DR, Pendry JB, Wiltshire MCK (2004) Metamaterials and negative refractive index. Science 305:88–792
4. Schurig D, Smith DR (2004) Negative index lens aberrations. Phys Rev E 70(6):065601
5. Fang N, Lee H, Sun C, Zhang X (2005) Sub-diffraction-limited optical imaging with a silver superlens. Science 308:534–537
6. Taubner T, Korobkin D, Urzhumov Y, Shvets G, Hillenbrand R (2006) Near-field microscopy through a SiC superlens. Science 313:1595
7. Liu Z et al (2007) Far-field optical superlens. Nano Lett 7:403–408

8. Ma CB, Liu ZW (2010) A super resolution metalens with phase compensation mechanism. Appl Phys Lett 96:183103
9. Ma CB, Liu ZW (2010) Focusing light into deep subwavelength using metamaterial immersion lenses. Opt Expr 18:4838–4844
10. Jacob Z, Alekseyev LV, Narimanov E (2006) Optical hyperlens: far-field imaging beyond the diffraction limit. Opt Expr 14:8247–8256
11. Rho J et al (2010) Spherical hyperlens for two-dimensional sub-diffractional imaging at visible frequencies. Nat Commun 1:143
12. Cummer SA, Christensen J, Alù A (2016) Controlling sound with acoustic metamaterials. Nat Rev Mater 1:16001
13. Haberman M, Guild M (2016) Acoustic metamaterials. Phys Today 3(12):31–39
14. Liu Z et al (2000) Locally resonant sonic materials. Science 289:1734–1736
15. Ao XY, Chan CT (2008) Far-field image magnification for acoustic waves using anisotropic acoustic metamaterials. Phys Rev E 77:025601
16. Li JS, Fok L, Yin XB, Bartal G, Zhang X (2009) Experimental demonstration of an acoustic magnifying hyperlens. Nat Mater 8:931–934
17. Zhu J, Christensen J, Jung J, Martin-Moreno L, Yin X, Fok L, Zhang X, Garcia-Vidal FJ (2011) A Holey-structured metamaterial for acoustic deep-subwavelength imaging. Nat Phys 7(1):52–55
18. Yang X, Yin J, Yu G, Peng L, Wang N (2015) Acoustic superlens using Helmholtz-resonator-based metamaterials. Appl Phys Lett 107(19):193505
19. Zhao J, Li B, Chen ZN, Qiu CW (2013) Redirection of sound waves using acoustic metasurface. Appl Phys Lett 103(15):151604
20. Cai L, Wen J, Yu D, Lu Z, Chen X, Zhao X (2017) Beam steering of the acoustic metasurface under a subwavelength periodic modulation. Appl Phys Lett 111(20):201902
21. Lee D, Cho C, Mun J, Park N, Rho J (2018) Demonstration of steering acoustic waves by generalized Eaton lens. Appl Phys Lett 113(16):161904
22. Cummer SA, Schurig D (2007) One path to acoustic cloaking. New J Phys 9:45
23. Chen H, Chan CT (2007) Acoustic cloaking in three dimensions using acoustic metamaterials. Appl Phys Lett 91:183518
24. Kan W et al (2016) Three-dimensional broadband acoustic illusion cloak for sound-hard boundaries of curved geometry. Sci Rep 6:36936
25. Zhu H, Semperlotti F (2013) Metamaterial based embedded acoustic filters for structural applications. AIP Adv 3(9):92121
26. Ma G, Yang M, Yang Z, Sheng P (2013) Low-frequency narrow-band acoustic filter with large orifice. Appl Phys Lett 103(1):11903
27. Guenneau S, Movchan A, Pétursson G, Ramakrishna SA (2007) Acoustic meta-materials for sound focusing and confinement. New J Phys 9:1367–2630
28. Zhang S, Yin L, Fang N (2009) Focusing ultrasound with an acoustic metamaterial network. Phys Rev Lett 102:194301
29. Colombi A et al (2017) Enhanced sensing and conversion of ultrasonic Rayleigh waves by elastic metasurfaces. Sci Rep 7(1):6750
30. Zhang S, Xia C, Fang N (2011) Broadband acoustic cloak for ultrasound waves. Phys Rev Lett 106(2):24301
31. Lee MK, Kim YY (2016) Add-on unidirectional elastic metamaterial plate cloak. Sci Rep 6:20731
32. Lee HJ, Lee JK, Kim YY (2015) Elastic metamaterial-based impedance-varying phononic bandgap structures for bandpass filters. J Sound Vib 353:58–74
33. Kim MS, Lee WR, Kim YY, Oh JH (2018) Transmodal elastic metasurface for broad angle total mode conversion. Appl Phys Lett 112(24):241905
34. Amireddy KK, Balasubramaniam K, Rajagopal P (2018) Porous metamaterials for deep subwavelength ultrasonic imaging. Appl Phys Lett 113:124102-1–124102-5
35. Amireddy KK, Balasubramaniam K, Rajagopal P (2016) Holey structured metamaterial lens for subwavelength resolution in ultrasonic characterization of metallic components. Appl Phys Lett 108:224101-1–224101-5

36. Amireddy KK, Balasubramaniam K, Rajagopal P (2017) Deep subwavelength ultrasonic imaging using optimized holey structured metamaterials. Sci Rep 7:1–8. https://doi.org/10.1038/s41598-017-08036-4
37. Syed Akbar Ali MS, Amireddy KK, Balasubramaniam K, Rajagopal P (2019) Characterization of deep sub-wavelength sized horizontal cracks using holey-structured metamaterials. Trans Indian Inst Met 72:2917–2921. https://doi.org/10.1007/s12666-019-01684-2
38. Lee HJ, Kim HW, Kim YY (2011) Far-field subwavelength imaging for ultrasonic elastic waves in a plate using an elastic hyperlens. Appl Phys Lett 98:241912
39. Zhang X, Liu Z (2008) Superlenses to overcome the diffraction limit. Nat Mater 7:435
40. de Rosny J, Fink M (2002) Overcoming the diffraction limit in wave physics using a time-reversal mirror and a novel acoustic sink. Phys Rev Lett 89:124301
41. Lerosey G, deRosny J, Tourin A, Fink M (2007) Focusing beyond the diffraction limit with far-field time reversal. Science 315(19):1120–1122
42. Sukhovich A et al (2009) Experimental and theoretical evidence for subwavelength imaging in phononic crystals. Phys Rev Lett 102:154301
43. He Z, Cai F, Ding Y, Liu Z (2008) Subwavelength imaging of acoustic waves by a canalization mechanism in a two-dimensional phononic crystal. Appl Phys Lett 93:233503
44. Wood RW (1935) Anomalous diffraction gratings. Phys Rev 48:928–936
45. Estrada H, Gómez-Lozano V, Uris A, Candelas P, Belmar F, Meseguer F (2012) Ultrasonic transmission through multiple-sublattice subwavelength holes arrays. Ultrasonics 52(3):412–416
46. Hao R, Qiu C, Hu Y, Tang K, Liu Z (2011) Resonant transmission of acoustic waves through an elastic plate quasiperiodically corrugated on surfaces. Phys Lett A 375(45):4081–4084
47. Hao R et al (2010) Exotic acoustic transmission through hard plates perforated with quasiperiodic subwavelength apertures. EPL 92(2):24006
48. Chen H, Chan CT (2010) Acoustic cloaking and transformation acoustics. J Phys D Appl Phys 43. https://doi.org/10.1088/0022-3727/43/11/113001
49. García-Meca C, Carloni S, Barceló C, Jannes G, Sánchez-Dehesa J, Martínez A (2014) Space-time transformation acoustics. Wave Motion 51:785–97
50. Garcia-Meca C, Carloni S, Barcelo C, Jannes G, Sanchez-Dehesa J, Martinez A (2013) Analogue transformation acoustics: Generalizing transformation techniques to non form-invariant equations. In: 7th International congress on advanced electromagnetic materials in microwaves and optics. IEEE Press, Talence, pp 235–237
51. Craster RV, Guenneau S (2012) Acoustic metamaterials: negative refraction, imaging, lensing and cloaking. Springer Science & Business Media, Springer
52. Milton GW, Briane M, Willis JR (2006) On cloaking for elasticity and physical equations with a transformation invariant form. New J Phys 8(248):1–20. https://doi.org/10.1088/13672630/8/10/248
53. Kuchibhatla SAR, Rajagopal P (2019) A transformation elasticity based device for wavefront manipulation. NDT E Int 102:304–310
54. Zhu Q, Jin L, Fu Y (2015) Graded index photonic crystals: a review. Ann Phys 527:205–218. https://doi.org/10.1002/andp.201400195
55. Shelby RA, Smith DR, Schultz S (2001) Experimental verification of a negative index of refraction. Science 292:77–79
56. Zhang XD, Liu ZY (2004) Negative refraction of acoustic waves in two-dimensional phononic crystals. Appl Phys Lett 85:341–343
57. Meitzler AH (1965) Backward-wave transmission of stress pulses in elastic cylinders and plates. J Acoust Soc Am 38:835
58. Bramhavar S et al (2011) Negative refraction and focusing of elastic Lamb waves at an interface. Phys Rev B 83:014106
59. Philippe FD, Murray TW, Prada C (2015) Focusing on plates: controlling guided waves using negative refraction. Sci Rep 5:11112

60. Manjunath CT, Rajagopal P (2017) Topographic metamaterials for ultrasonic non-destructive evaluation. In: 38th Prog. Electromagn. Res. IEEE Press, St. Petersburg, Russia, pp 2688–2693
61. Manjunath CT, Rajagopal P (2019) Lensing in the ultrasonic domain using negative refraction induced by material contrast. Sci Rep 9:6368. https://doi.org/10.1038/s41598-019-42655-3